PRAISE FOR PSYCHE UNBOUND

"Few scholars have as thoroughly explored and theorized about the depths of the human psyche as Stan Grof has in his lustrous career. *Psyche Unbound* is a gift not only to Grof but to all of us who are standing on the shoulders of giants. "

—Kile Ortigo, PhD
Author of *Beyond the Narrow Life: A Guide for Psychedelic Integration and Existential Exploration*

"The Festschrift is literally a tour de force. It weaves together insights into the transformative nature of consciousness throughout—and beyond—our lifespan. Wisdom is shared by leaders in this fascinating field and they open the shudders to an expanded horizon of human potentials."

—Marilyn Schlitz, PhD, MA
Professor, Sofia University; Author of *Death Makes Life Possible* and *Living Deeply: The Art and Science of Transformation*

"The contributions of this remarkable collection of theorists and commentators recognize Stan's vortexical brilliance, his boundless curiosity, and his buoyant spirit. The stature and expertise of the contributors and the robust intelligence of their commentaries are a fitting tribute, but can represent only a fraction of the immense impact of Stan's work."

—Mariavittoria Mangini, PhD, FNP
Nurse, Author, Investigator of Psychedelic-Assisted Therapies

"Brilliant and deep, this volume celebrates and illuminates Stan Grof's genius. His profound vision has created a powerful new paradigm for psychology, medicine, human and social development, and for understanding the nature of consciousness. A truly important worthy tribute."

—Jack Kornfield, PhD
Author of *A Path With Heart*

"*Psyche Unbound* is the long-awaited book that surveys the paradigm-shifting work of Stanislav Grof. His decades-long research projects and his voluminous theoretical contributions are indispensable resources for anyone wanting to keep informed regarding the directions being taken by psychology, psychiatry, and psychotherapy in the 21st century."

—Stanley Krippner, PhD
Affiliated Distinguished Faculty, California Institute of Integral Studies; Co-editor, *Varieties of Anomalous Experience: Examining Scientific Evidence*

"The Psyche, Birth, Death, Cosmic Infinity, are the most mysterious themes humans have always been and are still grappling with. This book is a fascinating compilation of essays by the deepest thinkers of our time. In the eternal search for an experience of Transpersonal Consciousness as the Holy Grail of humanity, Stan Grof has shone the light of our most divine essence and our most touching existence. He deserves this gathering of voices as a respectful and loving honoring."

— Françoise Bourzat, MA
Co-author of *Consciousness Medicine: Indigenous Wisdom, Entheogens, and Expanded States of Consciousness for Healing and Growth*

"This landmark volume celebrates the extraordinary depth and breadth of Stan Grof's contribution to contemporary culture—one which will resonate down through the 21st century as we forge deeper understandings of the nature of reality and the evolutionary imperative embodied in the psychospiritual process of birth, death and transformation. The galaxy of articles written by major figures opens up new vistas for a creative future that we must dare to embrace—a real treasure trove!"

— David Lorimer
Program Director at The Scientific and Medical Network; Editor Paradigm Explorer

"An inspiring, highly personal look at Stanislav Grof's groundbreaking work, and how it has created the modern foundation for what we now know as transpersonal psychology. This inspiring Festschrift compiles twenty-two anecdotal and analytical essays from leaders in their fields whose work has been influenced by Grof, and relays the breadth and significance of Grof's insights and theories throughout the years."

— Zaya and Maurizio Benazzo
Science and Non-Duality (SAND) Founders and Organizers

"Few people have contributed to our understanding of the psyche as much as Stan Grof. His pioneering research on multiple topics ranging from psychedelics to shamanism to death and dying, addiction, and more, coupled with his wide-ranging cross-disciplinary knowledge and formidable synthesizing capacities allowed him to make multiple breakthrough discoveries, and even to found new fields of discovery such as holotropic breathwork and spiritual emergencies. How wonderful that his life's work is being celebrated in this outstanding Festschrift."

— Roger Walsh MD, PhD
Professor of Psychiatry, Philosophy, and Anthropology, University of California

PSYCHE UNBOUND

Essays in Honor of Stanislav Grof

EDITED BY
Richard Tarnas
Sean Kelly

MULTIDISCIPLINARY ASSOCIATION
FOR PSYCHEDELIC STUDIES

Co-published by Synergetic Press & the Multidisciplinary Association for Psychedelic Studies (MAPS)

Library of Congress Cataloging-in-Publication Data is available.

ISBN 9780998276526 (hardcover)
ISBN 9780998276533 (ebook)

Cover design by Amanda Müller & Howie Severson
Cover illustration by Stanislav Grof
Book design by Howie Severson
Managing Editor: Amanda Müller
Printed in the USA

Table of Contents

SECTION II: PSYCHOTHERAPY AND CONSCIOUSNESS RESEARCH

SECTION III: COMPARATIVE AND THEORETICAL STUDIES

Foreword

Rick Doblin

Stanislav Grof is turning ninety in the midst of a renaissance in psychedelic research for which he was the primary catalyst and holder of the flame during dark decades. It is an ideal time for the publication of this Festschrift with its edited collection of essays from an illustrious group of friends and colleagues, all of whom Stan has profoundly influenced.

Collectively, these essays honor Stan's many contributions to psychiatry, transpersonal psychology, psychotherapy, spirituality, and the understanding of art. They honor his creation with the late Christina Grof of a new therapeutic approach, holotropic breathwork, and their creation of the International Transpersonal Association and the Spiritual Emergence Network. And they honor Stan's work with leading edge thinkers in many disciplines to craft a new worldview that synthesizes science, mysticism, and healing technologies. As the essays in this Festschrift illuminate, Stan's lifetime body of work provides key insights and techniques that can help twenty-first century humanity as a whole evolve and survive the birth pangs of a more globalized, interconnected, and environmentally challenged world.

Stan has been a leading advocate for psychedelic research for well over half a century, beginning in the early days of LSD research in the 1950s in Prague, through joining the flowering of research in the late 1960s and early 1970s at Johns Hopkins and the Maryland Psychiatric Research Center, then through several decades of the backlash and into the current renaissance. Stan consistently presented promising data, brilliant theories, and dramatic individual stories of the therapeutic potential of LSD, other psychedelics, and non-ordinary states of consciousness generally, despite substantial cultural and political pressure to disavow their value.

In the early stages of the global suppression of psychedelic research, Stan helped sustain scientific discussions and created opportunities for community. In 1972, he began to organize a series of ever larger international transpersonal conferences at which he and other psychedelic researchers presented their data and hypotheses as part of a fascinating program of scientists, scholars, spiritual leaders, artists, students, and activists. The most recent international transpersonal conference was in 2017 in Prague with new organizers, at which Stan was a keynote speaker.

In the mid-1970s, Stan and his late wife Christina found a courageous way forward and developed holotropic breathwork as a legal way for people to generate non-ordinary states of consciousness. Holotropic breathwork wasn't the same as LSD but similar enough and remarkably valuable therapeutically. They introduced holotropic breathwork into their workshops at Esalen Institute in Big Sur, CA and at events globally, starting a new therapeutic modality and training holotropic breathwork practitioners around the world.

In the early 1980s, Stan was part of the group of psychedelic researchers and therapists who worked to keep the therapeutic use of MDMA legal in the face of efforts, ultimately successful, by the US Drug Enforcement Administration to criminalize both recreational and therapeutic uses of MDMA. In the 1990s and onward as psychedelic research began to be approved again, Stan helped train new psychedelic researchers.

During the past five years, Stan has been joined by his wife Brigitte Grof in his several most recent contributions. Traveling to several continents around the world, they have run large holotropic breathwork workshops, worked on *The Way of the Psychonaut* documentary film directed by Susan Hess Logeais, and shepherded into being his major summation of his lifework, the two-volume *Way of the Psychonaut*. Finally, they launched Grof Legacy Training, a global training program for facilitators of holotropic states of consciousness, including psychedelic sessions.

Personally, in a very real sense, my entire adult life has been a Festschrift honoring Stan's work, beginning in 1972 when I was an eighteen-year-old college freshman at New College in Sarasota, Florida. I was struggling with integrating a series of challenging LSD and mescaline experiences and went to speak to a guidance counselor. Fortunately, he took my struggles with psychedelics seriously and handed me a manuscript copy of Stan's first book, *Realms of the Human Unconscious: Observations from LSD Research,* which wasn't even published until 1975.

My life's mission crystallized as I read Stan's book in which he looked at consciousness and mysticism through a scientific lens, with his approach grounded and reality-tested by being at its core about therapy and healing. Stan provided scientific evidence of the therapeutic potential of LSD-catalyzed transpersonal, spiritual experiences. He also suggested that profound experiences of unity and interconnectedness could serve as an antidote to separateness, prejudice, dehumanization, genocide, and environmental destruction.

My guidance counselor provided me with Stan's address so I could write to him directly. Amazingly, Stan took the time to write back to a confused eighteen-year-old. He invited me to attend a week-long workshop he was offering with Joan Halifax in California during the summer of 1972, which I was fortunate to be able to attend.

The compelling combination of therapeutic efficacy and political, evolutionary implications of psychedelic therapy and research that Stan presented in his book and workshop inspired me to devote my life to engaging in my own psychedelic therapy, to learn to become a psychedelic therapist, and to bring psychedelic therapy back above ground, legal by prescription and covered by insurance.

Stan's impact on my life and work is but one of many such stories. Even all the essays in this Festschrift in honor of Stan's ninetieth birthday will present only a partial view of Stan's contributions to the emergence of a transpersonal worldview that feels necessary for humanity and life on earth as we know it to survive, and thrive.

Rick Doblin

Introduction

RICHARD TARNAS & SEAN KELLY

The chapters in this volume represent a Festschrift of essays by prominent scholars inspired by the pioneering work of Stanislav Grof, one of the founders of transpersonal psychology and the world's leading researcher and theoretician in the fields of psychedelic therapy and the transformative potential of non-ordinary states of consciousness. Published here in honor of Grof's 90th birthday, the contributions to this volume demonstrate the magnitude and continuing fertility of his work and vision. The essays range over the past half century, from Joseph Campbell's and Huston Smith's remarkable assessments as they first encountered Grof's discoveries and recognized their significance in the early 1970s, to personal reflections in 2021 by psychiatrists and researchers participating in the current renaissance of psychedelic therapy.

Grof is recognized by many as having both inherited and extended the great revolution in psychology begun by Freud and Jung. His investigations led him to propose a radically expanded model of the psyche that honors the full range of human experience. Unconstrained by the dogmatic prejudices of mainstream psychology and of the dominant—reductive, mechanistic, and materialistic—scientific paradigm, he has brought forth a liberating vision of the human psyche. This volume both honors and explores the nature of that vision, its implications for various scholarly fields, and its legacy for the healing professions.

Stanislav Grof was born in Prague, Czechoslovakia, in 1931, the son of Stanislav, a chemist and business executive, and Maria Petnik Grof, an accomplished classical pianist and painter. His younger brother, Paul, was born four years later. Both parents were free-thinking in their religious views, and his mother in particular had strong interests in esoteric philosophy and spirituality. Grof's youth and early adult years were enriched by that traditional Central European cultural milieu that has so often brought forth extraordinarily cultured polymaths who are as at home in Wagner and Mozart as in the Upanishads and the Tao Te Ching, in Greek myth as in contemporary physics, and who know ten languages ancient and modern. Growing up in the 1930s and 1940s, however, came at a high price, coinciding with the Great Depression, World War II, the Nazi invasion of Czechoslovakia, and the postwar Communist takeover and Soviet domination of the country. In 1949 at the age of seventeen, Grof was arrested for possessing a leaflet calling for democratic elections, interrogated, and imprisoned for several months. But after his release, following a trial at which he was conditionally acquitted for lack of evidence, he was able to return to school. Soon after, a chance encounter with the writings of Sigmund Freud inspired him to enter medical school with the intention of becoming a psychiatrist.

During the 1950s, Grof completed his medical and psychiatric education while pursuing clandestine training in psychoanalysis, which was at that time banned by the Communist government. One day in 1954 a package of ampules of LSD-25 from Sandoz Pharmaceuticals arrived at the Psychiatric Clinic of the Medical Faculty where he worked, along with a letter describing the undetermined potential value of the new drug. Prevented at first from experiencing an LSD session himself because he had not yet completed his medical training, he nevertheless volunteered to supervise such sessions over the next two years for many subjects, including psychiatrists and psychologists, artists, and scientists. Finally, in 1956, he had the opportunity to undergo his own LSD experience, which was dramatically transformative and set him on a new career direction and, indeed, a new life path.

For most of the next two decades, Grof supervised thousands of LSD sessions and conducted psychotherapy with psychedelic substances with hundreds of patients. As principal investigator first at the Psychiatric Research

Institute in Prague and then, from 1967 to 1973, at the Maryland Psychiatric Research Center in Baltimore, at that time the last surviving psychedelic research program in the United States, he was able to investigate the varieties of human experience and the depths of the human psyche with what proved to be an unprecedentedly powerful catalyst of psychic processes. As time passed, Grof came to recognize LSD as serving a role in consciousness research comparable to that of the microscope in biology and medicine or the telescope in astronomy.

But just as with those earlier revolutionary transformations of scientific fields, the new evidence, and the new categories of evidence, disclosed through the new approach produced a profound paradigm crisis, beginning with the researcher himself. Rigorously trained within classical psychoanalysis and a reductively materialistic science, Grof began to encounter experiences in his subjects and in his own sessions that could not be explained within the existing conceptual frameworks he had fully internalized. In the course of this crisis he was impelled to develop a radically expanded cartography of the human psyche, at the same time evolving radically new approaches to psychotherapy and new understandings of the healing process.

One can highlight two sets of factors making possible such an outcome to the conceptual crisis he faced. The first arose from the early protocol developed in Europe of using a long series of relatively low doses of LSD as an adjunct to psychoanalytic psychotherapy. For Grof, working systematically and daily with a wide range of patients suffering from diverse neurotic, psychosomatic, and psychotic symptoms, this method brought about a gradual, highly detailed archaeological uncovering of deeper and deeper layers of the psyche. Moving from initial experiences of a geometric and aesthetic nature into more affect-laden memories and fantasies representing recent memories, the sessions gradually brought forth earlier and earlier psychologically significant memories from childhood and infancy. Had the sessions stopped there, the evidence would have suggested confirmation of many key insights and principles from the psychoanalytic tradition as framed by Freud.

In time, however, with repeated sessions the subject's experiential field typically deepened into profound encounters with the existential realities of finitude and mortality, with the visceral intensity of experiences of dying, accompanied by the challenging emotional and philosophical aspects of facing death, inextricably intertwined with what was experienced as a dramatic reliving of biological birth. Moreover, these engagements with the extremes of individual life, which Grof termed "perinatal," were pervaded by or opened up to an enormously wide range of experiences that could not be reduced to the usual limits of an individual's memories of the present life. Mystical, archetypal, and collective experiences flooded in with a phenomenological vividness that far transcended any Freudian frame of reference and pointed to the existence of what Grof called a transpersonal domain of the psyche, which bore many important resemblances to (as well as some differences from) the major discoveries and conceptual innovations of C. G. Jung.

Yet the slower, systematic manner in which this gradual archaeological uncovering took place over the course of many serial sessions with the same subject in well-contained clinical conditions allowed Grof to trace with nuanced clinical precision the extraordinary structural connections linking different emotional and psychosomatic disorders with particular stages of the birth process, and these in turn with a vast range of phenomena of a transpersonal nature. Striking new insights emerged into the causal factors at work in different categories of depression, suicidal impulses, addiction, various phobias, sadomasochism and sexual disorders, mania and bipolar disorder, obsessive-compulsive syndrome, as well as a range of psychosomatic symptoms. Whereas high-dose LSD sessions often rapidly catapulted the subject's experiential field into the transpersonal domain, or brought forth a multidimensional complexity of experience that could not be retrospectively parsed with respect to the subject's particular psychological conditions, the years of work using the psycholytic method allowed underlying psychodynamic processes and etiologies to come more sharply into focus. Grof was thus able to connect the multiple levels of the psyche and articulate an architecture of emotional and psychosomatic disorders in a manner that offered the fields of depth psychology and psychiatry deeper and more compelling diagnostic explanations than were previously available, while providing often dramatic therapeutic relief and psychospiritually transformative experiences for subjects.

The other set of factors that undoubtedly played a critical role in making possible Grof's discoveries and expanded map of the psyche was the particular attitude and approach Grof himself adopted in his therapeutic practice. That attitude was one of deep respect for and radical openness to the mystery of the psyche, combined with a trust in the healing intelligence within each individual. This approach recognized that both the mystery of the psyche and its spontaneous healing powers greatly transcended the application of any pre-established psychological theory that the therapist might deploy as if universally applicable to every patient. Such a therapeutic approach resembled more that of a psychospiritual-somatic midwife skillfully supporting a unique and naturally unfolding process rather than that of a dominating obstetrician who confidently imposes controls and technologies that manage every birth process in a predictable way. Grof's openness and flexibility encouraged the emergence of experiences that did not fit his psychoanalytic and psychiatric expectations, experiences whose range and complexity were in turn reflected in his expanded cartography.

Grof combined this openness with a courageous willingness to explore, and help others explore, an astonishing range of extraordinary states of consciousness that could take excruciatingly painful, frightening forms as well as bring forth ecstatic spiritual awakenings. While other contemporary psychedelic researchers had recognized the revelatory power of LSD, Grof observed and embraced the transformative potential of both positive and negative kinds of psychedelic experiences. He witnessed how the fully embodied cellular experience and conscious integration of states of extreme suffering allowed a release and liberation from imprisoning and distorting unconscious complexes. As a consequence of LSD's singular potency, memories and events that other forms of therapy could only intimate and approach, often just verbally and without somatic and affective intensity, now emerged in their fullness and, as a result, with greater therapeutic consequence. The catalyzing of the psyche's energetic level helped overcome the blocks and resistances of the ego that normally kept the energies frozen within the armored body and psyche out of range of conscious engagement. The overcoming of psychological resistances in the patient in turn led to an overcoming of attachment to particular theoretical assumptions, opening both the therapist and the patient to the unpredictable spontaneity of the psyche's natural unfolding, including its eruption from the depths, since theoretical assumptions can so often serve as egoic defenses.

Many of the details of Grof's discoveries and conceptual innovations will be explored in the chapters that follow. These include the recognition of what Grof called COEX systems: multilayered, dynamic systems of condensed experience comprising emotionally and somatically connected memories, fantasies, and attitudes, either positive or negative in nature, that are organically connected in the unconscious by similar thematic patterns such as experiences of abandonment, positive merging experiences, oppression and constraint, and so forth. As the sessions progressed, these constellations of charged memories and feelings operating at the personal biographical level were discovered to be rooted more deeply at the perinatal level in one of four Basic Perinatal Matrices, each of which comprised a richly multidimensional but coherently related phenomenology associated with a specific stage of birth. This intensive encounter with birth, complexly intertwined with the encounter with death, emerged as both an experiential and a therapeutic threshold and pivot, a transduction point where the personal and the transpersonal, the biographical and the archetypal, the biological and the spiritual, intertwine and implicate one another. Finally, beyond the perinatal level, the psyche opened up to a vast universe of transpersonal experiences, which could take a seemingly unlimited variety of forms — ancestral, historical, collective, phylogenetic, cosmic, mystical, mythological, archetypal, as well as experiences that had a convincing quality of past life memories.

Grof's expanded cartography of the psyche offered the field of depth psychology an extraordinary resolution of its foundational schism between Freud and Jung, the traumatic divorce that occurred in the field's first decades, and that has influenced it ever since. Behind that divorce loomed modern culture's great polarity between the Enlightenment and Romanticism, and beyond that the ongoing modern tension between science and religion, all of which could now appear in a new light. Grof was also able to recognize how deeply his work paralleled, and was in some sense an unexpected contemporary continuation of, the ancient practices of shamanism, the indigenous use of vision plants, traditional rites of passage, and the initiations characteristic of mystery religions and esoteric traditions.

Perhaps one of the reasons that Grof's work in psychology has been so fertile is its "essential tension" (to use Thomas Kuhn's phrase for what is necessary to achieve scientific progress and a successful paradigm revolution) that drove his own intellectual journey and brought forth his breakthroughs—the tension between a conservative tradition, in this case psychoanalytic theory wedded to mainstream materialistic science, and the profoundly multidimensional and transpersonal phenomena that he encountered in his research and therapeutic practice. Grof held that tension of opposites, and brought forth a profound and revolutionary synthesis.

With the closing down of legal psychedelic research, Grof moved in the fall of 1973 to Esalen Institute in Big Sur, California, where he was offered the position of Scholar-in-Residence. There he found a supportive community of kindred spirits, of teachers, practitioners, and students, as well as a place of stunning natural beauty on the coastal cliffs overlooking the Pacific Ocean. During the next fourteen years he was able to write and publish his books while hosting month-long workshops to which he invited brilliant thinkers and pioneers from many disciplines to both teach and enter into fruitful dialogue with each other and with his own explorations. It was also there that Grof developed, with his late wife Christina, both the powerful non-drug method of self-exploration he called holotropic breathwork, and the important concept of spiritual emergencies and their transformational potential. With Christina and others he went on to train a generation of breathwork facilitators in many countries, while conducting large group breathwork sessions with thousands of subjects in the subsequent decades.

Grof also founded and led the International Transpersonal Association for the next thirty years, organizing large international conferences in different parts of the world where a pluralistic community of seekers, scholars, and spiritual leaders convened, each such gathering a combination of psychology conference, multicultural festival, and something resembling a world parliament of religions. In the years after leaving Esalen to live in Mill Valley and the Bay Area, he went on to teach annual seminars in the Philosophy, Cosmology, and Consciousness program at the California Institute of Integral Studies, which introduced his work to hundreds of graduate students working in many disciplines. Finally, in his most recent years, he and his wife Brigitte have established the Grof Legacy Training, which will carry his work into the future worldwide through the systematic training of skilled facilitators of holotropic and psychedelic states of consciousness.

Grof gazed long and hard through his powerful telescope into the inner universe of the human psyche. He engaged the darkest shadows of human experience as well as the noblest spiritual impulses and high mystical realms, without reductionism or repression of either, showing how interconnected are the heights and the depths. His lifework has carried a quality of genuine heroism, expressed in the courage and the generosity with which he helped so many enter into strange realms and cross perilous thresholds they would otherwise have never attempted. Grof somehow learned and carried within himself a profound trust in the mysteries of the soul, both human and cosmic, which helped make possible not only the personal transformation of countless individuals on their life journeys, but also the creation of brilliant maps and navigating methods that can orient many others in their quests to come.

Overview of the Chapters

Part One of this volume brings together essays from some early witnesses and allies of Stanislav Grof, people who felt strongly enough about the significance of Grof's research findings to state their enthusiastic support, and this despite the challenges these findings presented to the dominant scientific and cultural paradigms. The main theme of mythologist Joseph Campbell's opening essay, "No More Horizons," speaks directly to a central implication of the title of this volume, *Psyche Unbound*—namely, the revolutionary and liberatory potential of awakening to the depth dimension of the psyche as "Mind at Large." Campbell celebrates Grof's model of the psyche as providing empirical evidence for a kind of archetypally grounded perennial philosophy, a philosophy which might serve as an antidote to the forces of fragmentation in a time of planetary crisis.

The next essay by Huston Smith, one of the world's most distinguished authorities in religious studies whose views on mysticism were influenced by his own psychedelic experience as a participant in the famous Good Friday experiment in 1962, includes both a comprehensive summary of Grof's foundational research and reflections on its implications for understanding the true nature of human being. Echoing Campbell's invocation of the idea of the perennial philosophy, Smith concludes that Grof's findings confirm the "primordial anthropology" as found in the writings of the Vedanta, whereby the individual is seen to consist of the four progressively more subtle layers of (the physical) body, (personal) mind, (archetypal) soul, and (infinite) Spirit.

As with Smith, the life and career of author and pioneer transpersonal psychologist Frances Vaughan was profoundly transformed when she became a subject in one of the early legal studies of psychedelics. As she describes in her testimonial, her first experience with LSD granted her direct and lasting insight into what she came to see as the perennial philosophy or "esoteric teachings of all times": the nondual nature of absolute reality; the relativity of space, time, and all opposites; the boundless creative potential of consciousness; and the supreme power of love.

A pioneer and distinguished scholar in psychedelic studies and consciousness research, Ralph Metzner begins his essay with a consideration of the implications of Grof's theory of the Basic Perinatal Matrices (BPM) for understanding the nature of certain intense psychedelic states as well as for findings from the emerging field of pre- and perinatal psychology. He then shares some of his own conclusions concerning the prenatal epoch, offering an integrative model of formative influences on the incarnation of individual personality, including insights gained from comparisons with Jewish Wisdom Literature and the Tibetan Buddhist Bardo teachings.

The title of physicist Fritjof Capra's essay, "Journeys Beyond Space and Time," is taken from that of the weeklong seminars that he and Grof used to offer together at Esalen Institute beginning shortly after their first meeting in 1977. Recounting the early and abiding influence of Grof's work on his own "systems view of life," Capra highlights the main themes of their shared vision—a critique of the dominant Cartesian-Newtonian paradigm, reality as constituted by dynamic patterns of relationship, and the mystery of creative emergence.

This first section concludes with an essay by Richard Tarnas summarizing over four decades of collaborative research with Grof on the pragmatic value and revolutionary implications of archetypal astrology for conducting skillful and effective work with psychedelic therapy and holotropic states of consciousness, and more generally for a deeper understanding of the nature of the psyche. Honoring the overwhelming weight of empirical evidence, and despite considerable cultural prejudice and their own initial skepticism, Grof and Tarnas came to regard astrology, in Grof's words, as a kind of "Rosetta Stone" for deciphering the complex relations between the symbolic language of the psyche and the symbolic language of the cosmos. Tarnas and Grof were stunned to discover robust correlations between the phenomenology of the four Basic Perinatal Matrices and the symbolic meanings of the four outer planets held by contemporary astrologers. Hardly less astonishing was their finding of consistent synchronistic correlations of personal and world transits of these outer planets with the timing of when those matrices and associated systems of condensed experience (COEX systems) were activated in deep experiential work. These correlations constitute an unexpected bridge between ancient wisdom and modern science and provide compelling evidence for the fundamental unity of psyche and cosmos.

Part Two includes a series of essays that present results from research conducted in various settings—therapeutic, clinical, personal, social, and academic—which both confirm and extend the value of Grof's foundational contributions to the practice of psychotherapy and consciousness research. The first essay by transpersonal psychologist Jenny Wade is her first exploratory report on a major qualitative study of spontaneous transcendent experiences during sex. Confirming many of the phenomenological categories which Grof had proposed as part of his cartography of transpersonal states (transcendence of space, time, and the ordinary sense of identity), Wade emphasizes the liberatory potential of numinously charged, sexually-mediated non-ordinary states of consciousness as naturally occurring forms of embodied spirituality.

With the second essay we turn from sex to birth. Bringing together perspectives from social anthropology and his experience as a holotropic breathwork facilitator, Gregg Lahood focuses on the nature of birth and birth-giving as a ritual event, and more specifically as an initiatory rite of passage. He contrasts the trauma-inducing "bad ritual" of modern patriarchal and technocentric birth practices with the "good ritual" of holotropic breathwork therapy, the transpersonal dimensions of which in particular hold great potential for the healing of birth-related trauma.

Next, Dr. Paul Grof and his assistant, Arlene Fox, report on their results over a ten-year period using holotropic breathwork with sixty patients, many of whom suffered from treatment-resistant mood disorders. Illustrated with three representative vignettes, Grof and Fox conclude that the multi-modal nature of holotropic breathwork—which incorporates supported experiential access to the deep psyche, including its transpersonal dimensions—can significantly improve the therapeutic outcome of patients with recurrent mood disorders.

In the last essay focusing on holotropic breathwork, physicist and transpersonal psychologist William Keepin shares some of the extraordinary breakthroughs he has witnessed through his use of this therapeutic modality with leaders from such fields as the environmental movement, politics, science, social activism, and the health professions. Featuring a remarkable account of how a South African politician's experience helped her country respond to its devastating AIDS crisis, Keepin makes a convincing case for the value of holotropic breathwork as a powerful catalyst of social transformation.

The last three essays in this section report findings from second generation research in psychedelic therapy. To begin with, we hear from Christopher Bache, who shares the fruits of his experiences over a twenty-year period and seventy-three self-administered LSD sessions following Grof's protocols for psychedelic therapy. Confronted in these sessions by a seemingly endless initiatory cycle of death and rebirth, Bache proposes that, beyond the death of the personal ego, the initiatory cycle eventually involves the death (and healing through purification) of the species ego and of what he calls the "shamanic ego" of the subtler realms, and further still into "an ever-deepening spiral into the Divine."

For the last two essays of this section, we hear from two physicians whose lives and professional careers were radically transformed through their encounter with Grof. Michael Mithoefer recounts how this encounter decided his simultaneous decision to shift from emergency medicine to psychiatry and to complete the Grof training in holotropic breathwork. He shares several fundamental lessons learned from Grof's personal example of some of the skills and attitudes required for deep experiential work and how these are already being applied in the context of the recently approved clinical trials of MDMA-assisted psychotherapy for posttraumatic stress disorder (PTSD).

In the final essay of this section we hear from psychiatrist Charles Grob, how his early encounter with Grof inspired his choice of career and whose inspiration sustained him through the dark ages of ignorance of, and prejudice against, the promises evident from the first wave of psychedelic research. Picking up where Grof left off some twenty years before, Grob has been able to confirm the remarkable potential of psychedelic psychotherapy for achieving transformative outcomes for people in severe distress while approaching the end of life.

Part Three presents a group of comparative and theoretical studies that bring Grof's work into relation with figures and issues from the fields of psychology, speculative philosophy, physics, parapsychology, and religious studies. The section opens with Sean Kelly's essay on Grof and William James, an earlier pioneer in transpersonal psychology, author of the classic *Varieties of Religious Experience*, philosopher of pragmatism, and early researcher of non-ordinary states of consciousness and paranormal phenomena. Kelly explores several areas of congruence in both figures' understanding of transpersonal experiences and demonstrates the fruitfulness of Grof's theory of the perinatal process for considering the centrality of the symbol of birth in the *Varieties*. The dialogue between Grof and James also leads Kelly to propose a theoretical distinction between the concepts of the natal (actual birth), the *trans*natal (the "second" or transpersonal birth), and the *meta*natal (the deeper archetypal structure) dimensions of the perinatal process.

Moving more decidedly into the realm of speculative philosophy, John Buchanan's essay argues that the process philosophy of Alfred North Whitehead can provide a metaphysical foundation for a postmodern cosmology that answers Grof's call for a New Paradigm consistent with both the insights of ancient wisdom traditions and the findings of contemporary science. By the same token, the wealth of evidence from Grof's research into non-ordinary states of consciousness lends empirical support for Whitehead's holistic and panexperientialist vision of the cosmos.

Philosopher of science and integral theorist Ervin Laszlo begins his essay with an overview of experimental studies of psi-phenomena which point to subtle connections between individual minds or psyches, and between mind or psyche and the cosmos at large. He then reviews the much richer kinds of experience explored by Grof which also manifest these subtle connections. Seeking a coherent explanatory framework, Laszlo turns first to Jung's theory of the collective unconscious and its "psychoid" archetypes before considering a range of theories from speculative physics. The latter focuses on the idea of the "zero-point field (ZPF)" as the subtle medium and universal, holographic, and higher-dimensional ground of both psyche and cosmos.

With the next two essays we return to the theme of the perinatal. Thomas Purton reveals previously unrecognized deep correspondences between the three *guṇas* (*tamas*, *rajas*, and *sattva*) of *Sāṃkhya* philosophy and Grof's understanding of the Basic Perinatal Matrices (BPMs). Discerning these correspondences also allows Purton to propose an original reading of the *Bhagavad Gita* and to cast light on the complementary relationship between spiritual liberation, on the one hand, and psychological integration, on the other.

Next, Thomas Riedlinger presents a compelling argument for how Sartre's unintegrated mescaline experience of 1935 and the unresolved perinatal trauma which this experience catalyzed determined the guiding spirit and core assumptions of his particular brand of existentialism, one of the major philosophical movements of the twentieth century. Many of the themes and images associated with Sartre's philosophy (the absurdity of existence, the feeling of "no exit") and biography (loneliness, depression, and fantasies and hallucinations of being pursued by giant crustaceans or octopi) are typical of the phenomenology of BPM II (the first phase of labor with the cervix still closed) which Grof describes with the phrase, "Cosmic Engulfment and No Exit."

The final essay of the volume by transpersonal psychologist Jorge Ferrer brings Grof's empirical findings into dialogue with the field of contemporary religious studies. Ferrer begins by arguing that certain transpersonal experiences documented by Grof could potentially resolve one of the most controversial issues in the modern study of mysticism in deciding against the position of radical constructivism or contextualism with respect to the status of spiritual experience or knowledge. On the other hand, on the basis of his own pluralist and participatory understanding of the nature of religious or spiritual experience, Ferrer offers a friendly challenge (as do both Kelly and Buchanan on different grounds) to Grof's privileging of the neo-Advaitin type of perennial philosophy also favored by Huston Smith and others. The enormous variety and complexity of evidence Grof has collected and catalogued is itself strongly suggestive of a pluralistic and participatory spiritual ontology that may do better justice to the richness of his research.

The essays of this volume can be seen as an offering to the public, and to Stanislav Grof himself, of a bouquet of flowers that have sprung forth from the seeds of his remarkably fruitful life-work. Though by no means comprehensive, the essays are representative of the range and depth of Grof's influence on the lives and careers of countless individuals, across the whole disciplinary spectrum, and indeed at the level of the overarching paradigms and world views in which these disciplines are embedded. Doubtless, many seeds have yet to sprout, and when they do, they will further enrich the already lush garden of insights stemming from Grof's generative encounters with, and devotion to, the unbounded nature of the psyche.

* * *

The editors are grateful to John Buchanan, whose moral and financial support played a major role in making this book possible from its earliest days until its completion. Generous support along the way was also provided by Meihong Xu and Bill Melton. The Philosophy, Cosmology, and Consciousness graduate program at the California Institute of Integral Studies, with the support of President Emeritus Robert McDermott, provided an intellectually nourishing environment for work on this project, as well as a faculty home for many of the contributors, the editors, and Stanislav Grof himself. Elizabeth McAnally brought meticulous care to copyediting and formatting the various essays. Finally, the publication of this Festschrift was made possible through the sponsorship of the Multidisciplinary Association of Psychedelic Studies and Rick Doblin, and midwifed into being through the dedicated efforts of Douglas Reil and Amanda Müller of Synergetic Press.

I

EARLY WITNESSES & ALLIES

No More Horizons

JOSEPH CAMPBELL

What is, or what is to be, the new mythology? Since myth is of the order of poetry, let us ask first a poet: Walt Whitman, for example, in his *Leaves of Grass* (1855):

> I have said that the soul is not more than the body,
> And I have said that the body is not more than the soul,
> And nothing, not God, is greater to one than one's-self is,
> And whoever walks a furlong without sympathy walks to his
> own funeral, dressed in his shroud,
> And I or you pocketless of a dime may purchase the pick of
> the earth,
> And to glance with an eye or show a bean in its pod con-
> founds the learning of all times,
> And there is no trade or employment but the young man
> following it may become a hero,
> And there is no object so soft but it makes a hub for the
> wheeled universe,
> And any man or woman shall stand cool and supercilious
> before a million universes.
> And I call to mankind, Be not curious about God,
> For I who am curious about each am not curious about God,
> No array of terms can say how much I am at peace about
> God and about death.
> I hear and behold God in every object, yet I understand God
> not in the least,
> Nor do I understand who there can be more wonderful than
> myself.

> Why should I wish to see God better than this day?
> I see something of God each hour of the twenty-four, and each moment then,
> In the faces of men and women I see God, and in my own face in the glass;
> I find letters from God dropped in the street, and every one is signed by God's name,
> And I leave them where they are, for I know that others will punctually come forever and ever.[1]

These lines of Whitman echo marvelously the sentiments of the earliest of the Upaniṣads, the "Great Forest Book" (Bṛhadāranyaka) of about the eighth century B.C.

> This that people say, "Worship this god! Worship that god!"—one god after another! All this is his creation indeed! And he himself is all the gods.... He is entered in the universe even to our fingernail-tips, like a razor in a razor-case, or fire in firewood. Him those people see not, for as seen he is incomplete. When breathing, he becomes "breath" by name; when speaking, "voice"; when seeing, "the eye"; when hearing, "the ear"; when thinking, "mind": these are but the names of his acts. Whoever worships one or another of these—knows not; for he is incomplete as one or another of these.
>
> One should worship with the thought that he is one's self, for therein all these become one. This self is the footprint of that All, for by it one knows the All—just as, verily, by following a footprint one finds cattle that have been lost.... One should reverence the self alone as dear. And he who reverences the self alone as dear—what he holds dear, verily, will not perish....
>
> So whoever worships another divinity than his self, thinking, "He is one, I am another," knows not. He is like a sacrificial animal for the gods. And verily, indeed, as many animals would be of service to a man, so do people serve the gods. And if even one animal is taken away, it is not pleasant. What then if many? It is therefore not pleasing to the gods that men should know this.[2]

We hear the same, in a powerful style, even earlier, from the Egyptian

Book of the Dead, in one of its chapters, "On Coming Forth by Day in the Underworld," as follows:

> I am Yesterday, Today, and Tomorrow, and I have the power to be born a second time. I am the divine hidden Soul who created the gods and gives sepulchral meals to the denizens of the deep, the place of the dead, and heaven.... Hail, lord of the shrine that stands in the center of the earth. He is I, and I am he![3]

Indeed, do we not hear the same from Christ himself, as reported in the early Gnostic *Gospel According to Thomas*?

> Whoever drinks from my mouth shall become as I am and I myself will become he, and the hidden things shall be revealed to him.... I am the All, the All came forth from me and the All attained to me. Cleave a piece of wood, I am there; lift up the stone and you will find me there.[4]

Or again, two more lines of Whitman:

> I bequeath myself to the dirt to grow from the grass I love.
> If you want me again look for me under your boot-soles.[5]

Some fifteen years ago, I had the experience of meeting in Bombay an extraordinarily interesting German Jesuit, the Reverend Father H. Heras by name, who presented me with the reprint of a paper he had just published on the mystery of God the Father and Son as reflected in Indian myth.[6] He was a marvelously open-minded as well as substantial authority on Oriental religions, and what he had done in this very learned paper was actually to interpret the ancient Indian god Śiva and his very popular son Gaṇeśa as equivalent, in a way, to the Father and Son of the Christian faith. If the Second Person of the Blessed Trinity is regarded in his *eternal* aspect, as God, antecedent to history, supporting it, and reflected (in some measure) in the "image of God" in us all, it is then not difficult, even for a perfectly orthodox Christian, to recognize the reflex of his own theology in the saints and gods of alien worlds. For it is simply a fact—as I believe we have all now got to concede—that mythologies and their deities are productions and projections of the psyche. What gods are there, what gods have there ever been, that were not from man's imagination? We know their histories;

we know by what stages they developed. Not only Freud and Jung, but all serious students of psychology and of comparative religions today have recognized and hold that the forms of myth and the figures of myth are of the nature essentially of dream. Moreover, as my old friend Dr. Géza Róheim used to say, just as there are no two ways of sleeping, so there are no two ways of dreaming. Essentially the same mythological motifs are to be found throughout the world. There are myths and legends of the Virgin Birth, of Incarnations, Deaths and Resurrections; Second Comings, Judgments, and the rest, in all the great traditions. And since such images stem from the psyche, they refer to the psyche. They tell us of its structure, its order and its forces, in symbolic terms.

Therefore they cannot be interpreted properly as references, originally, universally, essentially, and most meaningfully, to local historical events or personages. The historical references, if they have any meaning at all, must be secondary; as, for instance, in Buddhist thinking, where the historical prince Gautama Śākyamuni is regarded as but one of many historical embodiments of Buddha-consciousness; or in Hindu thought, where the incarnations of Viṣṇu are innumerable. The difficulty faced today by Christian thinkers in this regard follows from their doctrine of the Nazarene as the *unique* historical incarnation of God; and in Judaism, likewise, there is the no less troublesome doctrine of a universal God whose eye is on but one Chosen People of all in his created world. The fruit of such ethnocentric historicism is poor spiritual fare today; and the increasing difficulties of our clergies in attracting gourmets to their banquets should be evidence enough to make them realize that there must be something no longer palatable about the dishes they are serving. These were good enough for our fathers, in the tight little worlds of the knowledge of their days, when each little civilization was a thing more or less to itself. But consider that picture of the planet Earth that was taken from the surface of the moon!

In earlier times, when the relevant social unit was the tribe, the religious sect, a nation, or even a civilization, it was possible for the local mythology in service to that unit to represent all those beyond its bounds as inferior, and its own local inflection of the universal human heritage of mythological imagery either as the one, the true and sanctified, or at least as the noblest and supreme. And it was in those times beneficial to the order of the group that its young should be trained to respond positively to their own system of tribal signals and negatively to all others, to reserve their love for at home and to project their hatreds outward. Today, however, we are the passengers, all, of this single spaceship Earth (as Buckminster Fuller once termed it), hurtling at a prodigious rate through the vast night of space, going nowhere. And are we to allow a hijacker aboard?

Nietzsche, nearly a century ago, already named our period the Age of Comparisons. There were formerly horizons within which people lived and thought and mythologized. There are now no more horizons. And with the dissolution of horizons we have experienced and are experiencing collisions, terrific collisions, not only of peoples but also of their mythologies. It is as when dividing panels are withdrawn from between chambers of very hot and very cold airs: there is a rush of these forces together. And so we are right now in an extremely perilous age of thunder, lightning, and hurricanes all around. I think it is improper to become hysterical about it, projecting hatred and blame. It is an inevitable, altogether natural thing that when energies that have never met before come into collision—each bearing its own pride—there should be turbulence. That is just what we are experiencing; and we are riding it: riding it to a new age, a new birth, a totally new condition of mankind—to which no one anywhere alive today can say that he has the key, the answer, the prophecy, to its dawn. Nor is there anyone to condemn here, ("judge not, that you may not be judged!") What is occurring is completely natural, as are its pains, confusions, and mistakes.

And now, among the powers that are here being catapulted together, to collide and to explode, not the least important (it can be safely said) are the ancient mythological traditions, chiefly of India and the Far East, that are now entering in force into the fields of our European heritage, and vice versa, ideals of rational, progressive humanism and democracy that are now flooding into Asia. Add the general bearing of the knowledge of modern science on the archaic beliefs incorporated in *all* traditional systems, and I think we shall agree that there is a considerable sifting task to be resolved here, if anything of the wisdom-lore that has sustained our species to the present is to be retained and intelligently handed on to whatever times are to come.

I have thought about this problem a good deal and have come to the conclusion that when the symbolic forms in which wisdom-lore has been everywhere embodied are interpreted not as referring primarily to any supposed or even actual historical personages or events, but psychologically, properly "spiritually," as referring to the inward potentials of our species, there then appears through all something that can be properly termed a *philosophia perennis* of the human race, which, however, is lost to view when the texts are interpreted literally, as history, in the usual ways of harshly orthodox thought.

Dante, in his philosophical work the *Convivio,* distinguishes between the literal, the allegorical, the moral, and the anagogical (or mystical) senses of any scriptural passage. Let us take, for example, such a statement as the following: *Christ Jesus rose from the dead.* The literal meaning is obvious: "A historical personage, Jesus by name who has been identified as 'Christ' (the Messiah), rose alive from the dead." Allegorically, the normal Christian reading would be: "So likewise, we too are to rise from death to eternal life." And the moral lesson thereby: "Let our minds be turned from the contemplation of mortal things to abide in what is eternal." Since the anagogical or mystical reading, however, must refer to what is neither past nor future but transcendent of time and eternal, neither in this place nor in that, but everywhere, in all, now and forever, the fourth level of meaning would seem to be that in death—or in this world of death—is eternal life. The moral from that transcendental standpoint would then seem to have to be that the mind in beholding mortal things is to recognize the eternal; and the allegory: that in this very body which Saint Paul termed "the body of this death" (Romans 6:24) is our eternal life—not "to come," in any heavenly place, but here and now, on this Earth, in the aspect of time.

That is the sense, also, of the saying of the poet William Blake: "If the doors of perception were cleansed, everything would appear to man as it is, infinite."[7] And I think that I recognize the same sense in the lines of Whitman that I have just cited, as well as in those of the Indian Upaniṣad, the Egyptian Book of the Dead, and the Gnostic Gospel of Thomas.

"The symbols of the higher religions may at first sight seem to have little in common," wrote a Roman Catholic monk, the late Father Thomas Merton, in a brief but perspicacious article entitled "Symbolism: Communication or Communion?"[8]

> But when one comes to a better understanding of those religions, and when one sees that the experiences which are the

> fulfillment of religious belief and practice are most clearly expressed in symbols, one may come to recognize that often the symbols of different religions may have more in common than have the abstractly formulated official doctrines....
>
> The true symbol [he states again] does not merely point to something else. It contains in itself a structure which awakens our consciousness to a new awareness of the inner meaning of life and of reality itself. A true symbol takes us to the center of the circle, not to another point on the circumference. It is by symbolism that man enters affectively and consciously into contact with his own deepest self, with other men, and with God... 'God is dead'... means, in fact, that symbols are dead.[9]

The poet and the mystic regard the imagery of a revelation as a fiction through which an insight into the depths of being—one's own being and being generally—is conveyed anagogically. Sectarian theologians, on the other hand, hold hard to the literal readings of their narratives, and these hold traditions apart. The lives of three incarnations, Jesus, Kṛṣṇa, and Śākyamuni, will not be the same, yet as symbols pointing not to themselves, or to each other, but to the life beholding them, they are equivalent. To quote the monk Thomas Merton again: "One cannot apprehend a symbol unless one is able to awaken, in one's own being, the spiritual resonances which respond to the symbol not only as *sign* but as 'sacrament' and 'presence.' The symbol is an object pointing to a subject. We are summoned to a deeper spiritual awareness, far beyond the level of subject and object."[10]

Mythologies, in other words, mythologies and religions, are great poems, and when recognized as such, point infallibly through things and events to the ubiquity of a "presence" or "eternity" that is whole and entire in each. In this function, all mythologies, all great poetries, and all mystic traditions are in accord; and where any such inspiriting vision remains effective in a civilization, everything and every creature within its range are alive. The first condition, therefore, that any mythology must fulfill if it is to render life to modern lives is that of cleansing the doors of perception to the wonder, at once terrible and fascinating, of ourselves and of the universe of which we are the ears and eyes and the mind. Whereas theologians, reading their revelations counterclockwise, so to say, point to references in the past (in Merton's words: "to another point on the circumference") and utopians offer revelations only promissory of some desired future; mythologies, having sprung from the psyche, point back to the psyche ("the center")—and anyone seriously turning within will, in fact, rediscover their references in himself.

Some weeks ago I received in the mail from the psychiatrist directing research at the Maryland Psychiatric Research Center in Baltimore, Dr. Stanislav Grof, the manuscript of an impressive work interpreting the results of his practice during the past fourteen years (first in Czechoslovakia and now in this country) of psycholytic therapy; that is to say, the treatment of nervous disorders, both neurotic and psychotic, with the aid of judiciously measured doses of LSD. And I have found so much of my thinking about mythic forms freshly illuminated by the findings here reported, that I am going to try in these last pages to render a suggestion of the types and depths of consciousness that Dr. Grof has fathomed in his searching of our inward sea.[11]

Very briefly, the first order of induced experience that Dr. Grof reports upon, he has termed the "Aesthetic LSD Experience." In the main, this corresponds to that which Aldous Huxley, in *The Doors of Perception,* described back in 1954, after he had swallowed and experienced the effects of four-tenths of a gram of mescalin. What is here experienced is such an astounding vivification, alteration and intensification, of all experiences of the senses that, as Huxley remarked, even a common garden chair in the sun is recognized as "inexpressibly wonderful, wonderful to the point, almost, of being terrifying."[12] Other, more profound effects may yield sensations of physical transformation, lightness, levitation, clairvoyance, or even the power to assume animal forms and the like, such as primitive shamans claim. In India, such powers (called *siddhi*) are claimed by yogis and are not supposed to have accrued to them from without, but to have arisen from within, awakened by their mystic training, being potential within us all. Aldous Huxley had a similar thought, which he formulated in Western terms, and of which I expect to have something to say a bit later on.

The second type of reaction Dr. Grof has described as the "Psychodynamic LSD Experience," relating it to an extension of consciousness into what Jung termed the Personal Unconscious, and the activation there of those emotionally overloaded contents that are dealt with typically in Freudian psychoanalysis. The grim tensions and terrified resistances to conscious scrutiny that are encountered on this level derive from various unconscious strains of moral, social, and prideful infantile ego-defenses, inappropriate to adulthood; and the mythological themes that in psychoanalytical literature have been professionally associated with the conflicts of these sessions—Oedipus complex, Electra complex, etc.,—are not really (in their references here) mythological at all. They bear, in the context of these infantile biographical associations, no anagogical, transpersonal relevancy whatsoever, but are allegorical merely of childhood desires frustrated by actual or imagined parental prohibitions and threats. Furthermore, even when traditional mythological figures do appear in the fantasies of this Freudian stage, they will be allegorical merely of personal conflicts; most frequently, as Dr. Grof has observed, "the conflict between sexual feelings or activities and the religious taboos, as well as primitive fantasies about devils and hell or angels and heaven, related to narratives or threats and promises of adults." And it will be only when these personal "psychodynamic" materials have been actively relived, along with their associated emotional, sensory, and ideational features, that the psychological "knot points" of the Personal Unconscious will have been sufficiently resolved for the deeper, inward, downward journey to proceed from personal-biographical to properly transpersonal (first biological, then metaphysical-mystical) realizations.

What Dr. Grof has observed is that, very much as patients during a Freudian psychoanalysis and in the "psychodynamic" stages of a psycholytic treatment "relive" the basal fixations (and thereby break the hold upon them) of their unconsciously rooted affect and behavior patterns, so, in leaving this personal memory field behind, they begin to manifest both psychologically and physically the symptomatology of a totally different order of relived experiences; those, namely, of the agony of actual birth—the moment (indeed, the hours) of passive, helpless terror when the uterine contractions suddenly began, and continued, and continued, and continued; or the more active tortures of the second stage of delivery, when the cervix opened, and propulsion through the birth canal commenced—continuing with an unremitting intensification of sheer fright and total agony, to a climax amounting practically to an experience of annihilation; when suddenly, release, light! The sharp pain of umbilical severance, suffocation until the bloodstream finds its new route to the lungs, and then, breath and breathing, on one's own! "The patients," states Dr. Grof, "spent hours in agonizing pain, gasping for breath, with the color of their faces changing from dead pale to dark purple. They were rolling on the floor and discharging extreme tensions in muscular tremors, twitches, and complex twisting movements. The pulse rate was frequently doubled, and it was threadlike; there was often nausea with occasional vomiting and excessive sweating.

"Subjectively," he continues, "these experiences were of a transpersonal nature—they had a much broader framework than the body and lifespan of a single individual. The experiencers were identifying with many individuals or groups of individuals at the same time; in the extreme the identification involved all suffering mankind, past, present and future." "The phenomena observed here," he states again, "are of a much more fundamental nature and have different dimensions than those of the Freudian stage." They are, in fact, of a mythological transpersonal order, not distorted to refer (as in the Freudian field) to the accidents of an individual life, but opening outward, as well as inward, to what James Joyce termed "the grave and constant in human sufferings."[13]

For example, when reliving in the course of psycholytic treatment the nightmare of the first stage of the birth trauma—when the uterine contractions commence and the locked-in child, in sudden fright and pain, is awakened to a consciousness of itself in danger—the utterly terrified subject is overwhelmed by an acute experience of the very ground of being as anguished. Fantasies of inquisitorial torture come to mind, metaphysical anguish and existential despair: an identification with Christ crucified ("My God, my God, why hast thou forsaken me?"[14]), Prometheus bound to the mountain crag, or Ixion to his whirling wheel. The mythic mode is of the Buddha's "All life is sorrowful"—born in fear and pain, expiring in fear and pain, with little but fear and pain in between. "Vanity of vanities, [...] all is vanity."[15] The question of "meaning" here becomes obsessive, and if the LSD session terminates on this note, there will generally remain a sense of life as loathsome, meaningless, a hateful, joyless inferno, with no way out either in space or in time, "no exit"—except possibly by suicide, which, if chosen, will be of the passive, quietly helpless kind, by drowning, an overdose of sleeping pills, or the like.

Passing to an intensive reliving of the second stage of the birth trauma, on the other hand—that of the tortured struggle in the birth canal—the mood and the imagery become violent, not passive but active suffering being the dominant experience here, with elements of aggression and sado-masochistic passion: illusions of horrendous battles, struggles with prodigious monsters, overwhelming tides and waters, wrathful gods, rites of terrible sacrifice, sexual orgies, judgment scenes, and so on. The subject identifies himself simultaneously with both the victims and the aggressive forces of such conflicts, and as the intensity of the general agony mounts, it approaches and finally breaks beyond the pain threshold in an excruciating crisis of what Dr. Grof has aptly named "volcanic ecstasy." Here all extremes of pain and pleasure, joy and terror, murderous aggression and passionate love are united and transcended. The relevant mythic imagery is of religions reveling in suffering, guilt, and sacrifice: visions of the wrath of God, the universal Deluge, Sodom and Gomorrah, Moses and the Decalogue, Christ's *Via Crucis*, Bacchic orgies, terrible Aztec sacrifices, Śiva the Destroyer, Kālī's gruesome dance of the Burning Ground, and the phallic rites of Cybele.

Suicides in this Dionysian mood are of the violent type: blowing out one's brains, leaping from heights, before trains, etc. Or one is moved to meaningless murder. The subject is obsessed with feelings of aggressive tension mixed with anticipation of catastrophe; extremely irritable and with a tendency to provoke conflicts. The world is seen as full of threats and oppression. Carnivals with wild kicks, rough parties with promiscuous sex, alcoholic orgies and bacchanalian dances, violence of all kinds, vertiginous adventures and explosions mark the lifestyles struck with the ferocity of this stage of the birth experience. In the course of a therapeutic session, a regression to this level may be carried to culmination in an utterly terrifying crisis of actual ego-death, complete annihilation on all levels, followed by a grandiose, expansive sense of release, rebirth, and redemption, with enormous feelings and experiences of decompression, expansion of space, and blinding, radiant light: visions of heavenly blue and gold, columned gigantic halls with crystal chandeliers, peacock-feather fantasies, rainbow spectrums, and the like. The subjects, feeling cleansed and purged, are moved now by an overwhelming love for all mankind, a new appreciation of the arts and of natural beauties, great zest for life, and a forgiving, wonderfully reconciled and expansive sense of God in his heaven and all right with the world.

Dr. Grof has found (and this I find extremely interesting) that the differing imageries of the various world religions tend to appear and to support his patients variously during the successive stages of their sessions. In immediate association with the relived agonies of the birth trauma, the usual imagery brought to mind is of the Old and New Testaments, together with (occasionally) certain Greek, Egyptian, or other pagan counterparts. However, when the agony has been accomplished and the release experienced of "birth"—actually, a "second" or "spiritual" birth, released from the unconscious fears of the former, "once born" personal condition—the symbology radically changes. Instead of mainly Biblical, Greek, and Christian themes, the analogies now point rather toward the great Orient, chiefly India. "The source of these experiences," states Dr. Grof, "is obscure, and their resemblance to the Indian descriptions flabbergasting." He likens their tone to that of the timeless intrauterine state *before* the onset of delivery: a blissful, peaceful, contentless condition, with deep, positive feelings of joy, love, and accord, or even union with the Universe and/or God. Paradoxically, this ineffable state is at once contentless and all-containing, of nonbeing yet more than being, no ego and yet an extension of self that embraces the whole cosmos. And here, I think of that passage in Aldous Huxley's *The Doors of Perception* where he describes the sense that he experienced in his first mescalin adventure of his mind opening to ranges of wonder such as he had never before even imagined.

> Reflecting on my experience [Huxley wrote], I find myself agreeing with the eminent Cambridge philosopher, Dr. C. D. Broad, "that we should do well to consider much more seriously than we have hitherto been inclined to do the type of theory which Bergson put forward in connection with memory and sense perception. The suggestion is that the function of the brain and nervous system and sense organs is in the main *eliminative* and not productive. Each person is at each moment capable of remembering all that has ever happened to him and of perceiving everything that is happening everywhere in the universe. The function of the brain and nervous system is to protect us from being overwhelmed and confused by this mass of largely useless and irrelevant knowledge, by shutting out most of what we should otherwise perceive or remember

> at any moment, and leaving only that very small and special selection which is likely to be practically useful."
>
> According to such a theory, each one of us is potentially Mind at Large. But in so far as we are animals, our business is at all costs to survive. To make biological survival possible, Mind at Large has to be funneled through the reducing valve of the brain and nervous system. What comes out at the other end is a measly trickle of the kind of consciousness which will help us to stay alive on the surface of this particular planet.... Most people, most of the time, know only what comes through the reducing valve and is consecrated as genuinely real by the local language. Certain persons, however, seem to be born with a kind of by-pass that circumvents the reducing valve. In others temporary by-passes may be acquired either spontaneously, or as the result of deliberate "spiritual exercises," or through hypnosis, or by means of drugs. Through these permanent or temporary by-passes there flows, not indeed the perception "of everything that is happening everywhere in the universe" (for the by-pass does not abolish the reducing valve, which still excludes the total content of Mind at Large), but something more than, and above all something different from, the carefully selected utilitarian material which our narrowed, individual minds regard as a complete, or at least sufficient, picture of reality.[16]

Now it strikes me as evident through all this that the imagery of mythology, stemming as it does from the psyche and reflecting back to the same, represents in its various inflections various stages or degrees of the opening of ego-consciousness toward the prospect of what Aldous Huxley has here called Mind at Large. Plato in the *Timaeus* declares that "there is only one way in which one being can serve another, and this is by giving him his proper nourishment and motion: and the motions that are akin to the divine principle within us are the thoughts and revolutions of the universe."[17] It is these, I would say, that are represented in myth. As illustrated in the various mythologies of the peoples of the world, however, the universals have been everywhere particularized to the local sociopolitical context. As an old professor of mine in Comparative Religions at the University of Munich used to say: "In its subjective sense the religion of all mankind is one and the same. In its objective sense, however, there are differing forms."

In the past, I think we can now say, the differing forms served the differing and often conflicting interests of the various societies, binding individuals to their local group horizons and ideals, whereas in the West today we have learned to recognize a distinction between the spheres and functions, on one hand, of society, practical survival, economic and political ends, and, on the other hand, sheerly psychological (or, as we used to say, spiritual) values. To return to the name, once more, of Dante: there is in the Fourth Treatise of the *Convivio* a passage in which he discourses on the divinely ordained separation of state and Church, as symbolized historically in the joined yet separate histories of Rome and Jerusalem, the Empire and the Papacy. These are the two arms of God, not to be confused; and he rebukes the Papacy for its political interventions, the authority of the Church being properly "not of this world" but of the Spirit—the relationship of which to the aims of this world is exactly that of Huxley's Mind at Large to the utilitarian ends of biological survival—which are all right and necessary too, but are not the same.

We live today—thank God!—in a secular state, governed by human beings (with all their inevitable faults) according to principles of law that are still developing and have originated not from Jerusalem but from Rome. The concept of the state, moreover, is yielding rapidly at this hour to the concept of the ecumene, i.e., the whole inhabited earth; and if nothing else unites us, the ecological crisis will. There is therefore neither any need, nor any possibility, for those locally binding, sociopolitically bounded, differing forms of religion "in its objective sense" which have held men separate in the past, giving to God the things that are Caesar's and to Caesar the things that are God's.

"God is an intelligible sphere whose center is everywhere and circumference nowhere." So we are told in a little twelfth-century book known as *The Book of the Twenty-four Philosophers*.[18] Each of us—whoever and wherever he may be—is then the center, and within him, whether he knows it or not, is that Mind at Large, the laws of which are the laws not only of all minds but of all space as well. For, as I have already pointed out, we are the children of this beautiful planet that we have lately seen photographed from the moon. We were not delivered into it by some god, but have come forth from it. We are its eyes and mind, its seeing and its thinking. And the earth, together with its sun, this light around which it flies like a moth, came forth, we are told, from a nebula; and that nebula, in turn, from space. So that we are the mind, ultimately, of space. No wonder, then, if its laws and ours are the same! Likewise, our depths are the depths of space, whence all those gods sprang that men's minds in the past projected onto animals and plants, onto hills and streams, the planets in their courses, and their own peculiar social observances.

Our mythology now, therefore, is to be of infinite space and its light, which is without as well as within. Like moths, we are caught in the spell of its allure, flying to it outward, to the moon and beyond, and flying to it, also, inward. On our planet itself all dividing horizons have been shattered. We can no longer hold our loves at home and project our aggressions elsewhere; for on this spaceship Earth there is no *elsewhere* anymore. And no mythology that continues to speak or to teach of *elsewheres* and *outsiders* meets the requirement of this hour.

And so, to return to our opening question: What is—or what is to be—the new mythology?

It is—and will forever be, as long as our human race exists—the old, everlasting, perennial mythology, in its "subjective sense," poetically renewed in terms neither of a remembered past nor of a projected future, but of now: addressed, that is to say, not to the flattery of "peoples," but to the waking of individuals in the knowledge of themselves, not simply as egos fighting for place on the surface of this beautiful planet, but equally as centers of Mind at Large—each in his own way at one with all, and with no horizons.

Letters from
Joseph Campbell to
Stanislav Grof

Joseph Campbell

136 Waverly Place
New York N.Y. –10014

SARAH LAWRENCE COLLEGE
BRONXVILLE, NEW YORK 10708

TELEPHONE
914-337-0700

March 16, 1971

Dear Dr. Grof,

I simply don't know how to thank you for the magnificent gift of your study of "Agony and Ecstasy in Psychiatric Treatment." I am reading it with fascination, marveling at the order and lucidity of your presentation, and finding every bit of it relevant and valuable to my own studies in mythology. I don't know what deity inspired you to send this to me. My thanks go to whichever it may have been, as well as to your very kind self. Your work has become my latest principal text.

All thanks again, and my very kind good wishes —

Sincerely

Joseph Campbell

P.S. When I shall have gotten further along in my reading, I shall be in touch with you again. For the present, I am progressing only slowly, because of the very heavy lecturing and teaching commitments of these next few weeks. —

Campbell File

136 Waverly Place
New York N.Y. 10014

SARAH LAWRENCE COLLEGE
BRONXVILLE, NEW YORK 10708

TELEPHONE
914-337-0700

Sept. 20, 1971

Dear Dr. Grof—

I am enclosing a copy of the last 14 pages of a book-manuscript now at the Viking Press, to appear (I am told) this spring or summer. It is a work to be named Myths to Live By, based on lectures that I gave at the Cooper Union Forum, here in New York, 1958-1971, and, as you will see, it here concludes with an extended reference to your own work and findings. I am sending ~~it~~ they on to you to ask whether you would object to the publication of such a summary. My talks were in a familiar, popular vein; but I was -- and have continued to be -- so greatly impressed and illuminated by your very important findings that I simply had to cite them as the

SARAH LAWRENCE COLLEGE
BRONXVILLE, NEW YORK 10708

TELEPHONE
914-337-0700

culminating revelation of my final lecture.

I am hoping that I may have the pleasure of meeting you in the not very distant future. Last month, at Big Sur, Dick Price spoke of you, and I had the sense of our paths approaching. Meanwhile, here is this respectful offering, in evidence of my very great admiration, and I hope that

—

you will wish to give it your *imprimatur*.

Best regards and good wishes —

Sincerely

Joseph Campbell

The Psychedelic Evidence

HUSTON SMITH

Tribute to Stanislav Grof

What do Stanislav Grof and Robert Oppenheimer have in common? A question for the parlor game, trivial pursuits? Not in my book. In my book, the answer is: both can read the *Bhagavad Gita* in Sanskrit. I cannot do that, and that is where my regard for this remarkable man begins.

It doesn't end there, though. What bonds me most securely to him is metaphysics. This may sound surprising to some readers, but his talents have many facets which allow admirers to lock into his work in many ways.

We live in a radically anti-metaphysical age to which Richard Rorty has given voice in his famous pronouncement, "there is no big picture." Not only is this untrue, I join Grof in pointing out; it does incalculable damage to the general order of intelligence in human affairs. For if there were no stable background against which perspectival truths could be assessed for their accuracy, anything would go, and blind whirl would be king. If that says it too abstractly for some readers, John Stuart Mill states the point directly: "If it were not useful to know in what order of things, under what government of the universe it is our destiny to live, it is difficult to imagine what could be considered so, for whether a person is in a pleasant or in an unpleasant place, a palace or a prison, it cannot be otherwise than useful for him to know where he is."

There is only one more point I need to make by way of introduction to the main body of my entry into this volume. Among the various metaphysical systems philosophers have come up with, the Vedanta states most clearly and succinctly what is at the heart of the worldviews that have had an enduring impact on history and remain vibrant today. And the Vedanta is the metaphysics to which both Grof and I are committed. His later works have detailed impressively how that worldview accommodates and is supported by the findings of modern science.

As my entry into this *Festschrift* to honor Grof's magisterial life's work, I reprint the appendix to my book *Forgotten Truth,* not only because it is the place where his concerns intersect my own most visibly, but more importantly because I know of no other piece of writing that presents as explicitly the empirical support that entheogens provide for the truth of the conceptual spine that underlies the history-shaping wisdom traditions. That said, I turn to the substance of my entry.

Know ten things, the Chinese say; tell nine—there is reason to question whether it is wise even to mention psychedelics in connection with God and the Infinite. For though a connection exists, it is—as in the comparable case of the role of sex in Tantra—next to impossible to speak of it without being misunderstood. It is for this reason, we suspect, that the Eleusinian mysteries were among the best-kept in history, and Brahmins came eventually to conceal, then deliberately forget, the identity of soma.[1]

If the only thing to say about psychedelics was that they seem on occasion to offer direct disclosures of the psychic and celestial planes as well as (in rare instances) the Infinite itself, we would hold our peace. For though such experiences may be veridical in ways, the goal, it cannot be stressed too often, is not religious experiences; it is the religious life. And with respect to the latter, psychedelic "theophanies" can abort a quest as readily as, perhaps more readily than, they can further it.

It is not, therefore, the isolated mystical experiences which psychedelics can occasion that lead us to add this appendix on the subject, but rather evidence of a different order. Long-term, professionally garnered, and carefully weighed, this latter evidence deserves to be called, if anything in this area merits the term, scientific. We enter it because of the ways in which, and the extent to which, this evidence seems to corroborate the primordial anthropology that Chapter 4 of *Forgotten Truth* sketched in paradigm. In contradistinction to writings on psychedelics which are occupied with experiences the mind can *have*, the concern here is with evidence they afford as to what the mind *is*.[2]

The evidence in question is not widely known, for to date, it has been reported only in a few relatively obscure journals and a book but recently off the press. At the same time, judged both by the quantity of data encompassed and by the explanatory power of the hypotheses that make sense of this data, it is the most formidable evidence psychedelics have thus far produced. The evidence to which we refer is that which has emerged through the work of Stanislav Grof.[3]

Grof's work began in Czechoslovakia, where for four years he worked in an interdisciplinary complex of research institutes in Prague and for another seven in the Psychiatric Research Institute that developed out of this complex. On coming to the United States in 1967, he continued his investigations at the Research Unit of Spring Grove State Hospital in Baltimore, Maryland. Two covering facts about his work are worth noting before we turn to its content. First, in the use of psychedelics for therapeutic and personality assessment, his experience is by far the vastest that any single individual has amassed, covering as it does over 2,500 sessions in which he spent a minimum of five hours with the subject. In addition, his studies cover another 800 cases his colleagues at Baltimore and Prague conducted. Second, in spanning the Atlantic, his work spans the two dominant approaches to psychedelic therapy that have been developed: psycholytic therapy (used at Prague and favored in Europe generally), which involves numerous administrations of low to medium doses of LSD or variant over a long therapeutic program, and psychedelic therapy (confined to America), which involves one or a few high doses in a short period of treatment.

The first thing Grof and his associates discovered was that there is no specific pharmacological effect which LSD invariably produces: "I have not been able to find a single phenomenon that could be considered an invariant product of the chemical action of the drug in any of the areas studied—perceptual, emotional, ideational, and physical."[4] Not even mydriasis (prolonged dilation of the pupils), one of the most common symptoms, occurs invariably. Psychological effects vary even more than do physiological, but the range of the latter—mydriasis, nausea, vomiting, enhanced intestinal movements, diarrhea, constipation, frequent urination, acceleration as well as retardation of pulse, cardiac distress, and pain, palpitations, suffocation and dyspnea, excessive sweating and hypersalivation, dry mouth, reddening of the skin, hot flushes and chills, instability and vertigo, inner trembling, fine muscle tremors—exceeds that of any other drug that affects the autonomic nervous system. These somatic symptoms are practically independent of dosage and occur in all possible combinations. Variability between subjects is equaled by variation in the symptoms a single subject will experience under different circumstances; particularly important from the clinical point of view are the differences that appear at different stages in the therapeutic process. All this led Grof to conclude that LSD is not a specific causal agent, but rather a catalyzer. It is, as endnote 2 indicates, an unspecific amplifier of neural and mental processes. By exteriorizing for the therapist and raising to consciousness for the patient himself material otherwise buried—and enlarging this material to the point of caricature so that it appears as if under a magnifying glass—psychedelics are, Grof became convinced, an unrivaled instrument. First, for identifying causes in psychopathology (the problem that is causing the difficulty), second, for personality diagnosis (determining the character type of the subject in question), and third, for understanding the human mind generally: "It does not seem inappropriate to compare their potential significance for psychiatry and psychology to that of the microscope for medicine or of the telescope in astronomy.... Freud called dreams the 'royal way to the unconscious.' The statement is valid to a greater extent for LSD experiences."[5]

Of the drug's three potentials, it is the third—its resources for enlarging our understanding of the human mind and self—that concerns us in this book. The nature of man has been so central to our study that even flickers of light from Grof's work would make it interesting. That the light proves to be remarkably clear and steady makes it important.

We come at once to the point. The view of man that was outlined in *Forgotten Truth* presented him as a multilayered creature, and Grof's work points to the same conclusion. As long as the matter is put thus generally, it signals nothing novel, for existing depth psychology—psychiatry, psychoanalysis—says the same; the adjective "depth" implies as much, and metaphors of archaeology and excavation dot the writings of Freud, Jung, and their colleagues. The novelty of Grof's work lies in the precision with which the levels of the mind it brings to view correspond with the levels of selfhood the primordial tradition describes.

In chemo-excavation, the levels come to view sequentially. In this respect, too, images of archaeology apply: surface levels must be uncovered to get at ones that lie deeper. In psychedelic (high-dose) therapy, the deeper levels appear later in the course of a single session; in psycholytic (low-dose) therapy, they surface later in the sequence of therapeutic sessions. The sequences are parallel, but since the levels first came to Grof's attention during his psycholytic work in Prague, and since that earlier work was the more extensive, covering eleven of the seventeen years he has been working with the drugs, we shall confine ourselves to it in reporting his experimental design.

The basic study at Prague covered fifty-two psychiatric patients. All major clinical categories were represented, from depressive disorders through psychoneuroses, psychosomatic diseases, and character disorders to borderline and clear-cut psychoses in the schizophrenic group. Patients with above-average intelligence were favored to obtain high-quality introspective reports; otherwise, cases with dim prognosis in each category were chosen. Grof himself worked with twenty-two of the subjects, his two colleagues with the remainder. The number of psycholytic sessions ranged from fifteen to one hundred per patient, with a total of over 2,500 sessions being conducted. Each patient's treatment began with several weeks of drug-free psychotherapy. Thereafter the therapy was punctuated with doses of 100 to 250 micrograms of LSD administered at seven- to fourteen-day intervals.

The basic finding was that "when material from consecutive LSD sessions of the same person was compared it became evident that there was a definite continuity between these sessions. Rather than being unrelated and random, the material seemed to represent a successive unfolding of deeper and deeper levels of the unconscious with a very definite trend."[6]

The trend regularly led through three successive stages preceded by another which, being less important psychologically, Grof calls a preliminary phase. In this opening phase, the chemical works primarily on the subject's body. In this respect, it resembles what earlier researchers had called the vegetative phase, but the two are not identical. Proponents of a vegetative phase assumed that LSD directly caused the manifold somatic responses patients typically experience in the early stages of psychedelic sessions. We have seen that Grof's more extensive evidence countered this view. Vegetative symptoms are real enough, but they vary so much between subjects and for a single subject under varying circumstances that it seems probable that they are occasioned more by anxieties and resistances than by the chemical's direct action. There is also the fact that they are far from confined to early phases of the LSD sequence. These considerations led Grof to doubt that there is a vegetative phase per se. The most he is prepared to admit is that the drug has a tendency at the start to affect one specific part of the body: its perceptual and particularly its optical apparatus. Colors become exceptionally bright and beautiful, objects and persons are geometrized, things vibrate and undulate, one hears music as if one were somehow inside it, and so on. This is as close as the drug comes to producing a direct somatic effect, but it is sufficient to warrant speaking of an introductory phase which Grof calls aesthetic.

With this preliminary phase behind him, the subject begins his psycholytic journey proper. Its first stage is occupied with material that is psychodynamic in the classical sense: Grof calls it the psychodynamic or Freudian stage. Experiences here are of a distinctly personal character. They involve regression into childhood and the reliving of traumatic infantile experiences in which Oedipal and Electra conflicts and ones relating to various libidinal zones are conspicuous; first and last, pretty much the full Freudian topography is traversed. The amount of unfinished business this layer of the self contains varies enormously; as would be expected, in disturbed subjects, there is more than in normals. But the layer itself is present in everyone and must be worked through before the next stratum can be reached. "Worked through" again means essentially what psychiatry stipulates: a reliving not only in memory but in emotion of the traumatic episodes that have unconsciously crippled the patient's responses. Freud and Breuer's hypothesis that insufficient emotional and motor abreaction during early traumatic episodes produces a "jamming" of affect that provides energy thereafter for neurotic symptoms is corroborated, for when patients in the course of a number of sessions enter into a problem area to the point of reliving it completely and integrating it into consciousness, the symptoms related to that area "never reappear," and the patient is freed to work on other symptoms.

This much was in keeping with Grof's psychiatric orientation; it came as "laboratory proof of the basic premises of psychoanalysis."[7] But there, that model gave out. For the experiences that followed, "no adequate explanation can be found within the framework of classical Freudian psychoanalysis."[8]

Negatively, the new stage was characterized by an absence of the individually and biographically determined material that had dominated the sessions theretofore. As a result, the experiential content of this second stage was more uniform for the population than was the content of the first. We have already cited Grof's contention that LSD is not so much an agent that produces specific effects as it is an amplifier of material that is already present, and in the first stage, the enlarging process worked to magnify individual differences: "the sessions of patients belonging to various diagnostic categories were characterized by an unusual inter- and also intra-individual variability."[9] In the second stage, the process was reversed. With the magnifying glass still in place, variations receded. "The content seemed to be strikingly similar in all of the subjects."[10]

This is already important, for the emergent similarity suggests that the subjects were entering a region of the mind which they shared in common, a region that underlay the differing scrawls their separate biographies had incised upon it. As to content, "the central focus and basic characteristics of the experience on this level are the problems related to physical pain and agony, dying and death, biological birth, aging, disease and decrepitude"[11]—Buddha's First Noble Truth, Grof somewhere observes, and three of the Four Passing Sights that informed it. Inevitably, he continues, "the shattering encounter with these critical aspects of human existence and the deep realization of the frailty and impermanence of man as a biological creature, is accompanied by an agonizing existential crisis. The individual comes to realize through these experiences that no matter what he does in his life, he cannot escape the inevitable: he will have to leave this world bereft of everything that he has accumulated, achieved and has been emotionally attached to."[12]

Among the phenomena of this second stage, the theme of death and rebirth recurred so frequently that it sent Grof to a book he had heard of in his psychiatric training but had not studied, it having been written by a psychoanalytic renegade, Otto Rank. It bore the title *The Trauma of Birth*, and to use Grof's own word, he was "flabbergasted" to find how closely the second-stage psycholytic experiences conformed to it. He and his colleagues fell to calling the second stage perinatal or Rankian.

During the weeks through which the stage extends, the patient's clinical condition worsens. The stage climaxes in a session in which the patient experiences the agony of dying and appears to himself actually to die: "The subjects can spend hours in agonizing pain, with facial contortions, gasping for breath and discharging enormous amounts of muscular tension in various tremors, twitching, violent shaking and complex twisting movements. The color of the face can be dark purple or dead pale, and the pulse rate considerably accelerated. The body temperature usually oscillates in a wide range, sweating can be profuse, and nausea with projectile vomiting is a frequent occurrence."[13] This death experience tends to be followed immediately by rebirth, an explosive ecstasy in which joy, freedom, and the promise of life of a new order are the dominant motifs.

Outside the LSD sequence, the new life showed itself in the patients' marked clinical improvement. Within the sequence, it introduced a third experiential landscape. When Grof's eyes became acclimated to it, it appeared at first to be Jungian—Jung being the only major psychologist to have dealt seriously and relatively unreductionistically with the visions that appeared. Later, it seemed clearer to refer to the stage as transpersonal. Two features defined this third and final stage. First, its "most typical characteristics ... were profound religious and mystical experiences."[14] "Everyone who experientially reached these levels developed convincing insights into the utmost relevance of spiritual and religious dimensions in the universal scheme of things. Even the most hardcore materialists, positivistically-oriented scientists, skeptics and cynics, uncompromising atheists, and anti-religious crusaders such as the Marxist philosophers, became suddenly interested in spiritual search after they confronted these levels in themselves."[15]

Grof speaks of levels in the plural here, for the "agonizing existential crisis" of the second stage is already religious in its way: death and rebirth are ultimates, or none exist. The distinguishing feature of the third stage is not, strictly speaking, that it is religious but that it is (as Grof's words indicate) mystically religious—religious in a mode in which (a) the whole predominates over the part, and (b) within the whole evil is rescinded. This connects with the stage's other feature, its transpersonal aspect, which was so pronounced as to present itself in the end as the logical candidate for the name by which the stage should be designated. A trend toward transpersonal experiences, that is, one occupied with things other than oneself, had already shown itself in stage two. Suffering, for example, which in the first stage presented itself in the form of recollected autobiographical traumas, had in the second stage taken the form of identifying with the suffering of others, usually groups of others: famine victims, prisoners in Nazi concentration camps, or mankind as a whole with its suffering symbolized archetypally by Christ on his cross, Tantalus exposed to eternal tortures in Hades, Sisyphus rolling his boulder incessantly, Ixion fixed on his wheel, or Prometheus chained to his rock. Likewise with death, already by stage two, "the subjects felt that they were operating in a framework which was 'beyond individual death.'"[16] The third stage continues this outbound, transpersonal momentum. Now the phenomena with which the subject identifies are not restricted to mankind or even to living forms. They are cosmic, having to do with the elements and forces from which life proceeds. And the subject is less conscious of himself as separate from what he perceives. To a large extent, the subject-object dichotomy is itself transcended. So much for description of the three stages. Now to interpretation and explanation.

Grof was and is a psychiatrist. Psychiatry is the study and practice of ontogenetic explanation: it accounts for present syndromes in terms of antecedent experiences in the life history of the individual. Freud had mined these experiences as they occur in infancy and childhood, but Grof's work had led to regions Freud's map did not fit. Clearly, as a psychiatrist, Grof had nowhere to turn for explanations save further in the same direction—further back. His very methodology forced him to take seriously the possibility that experiences attending birth and even gestation could affect ensuing life trajectories.

Taking his cues from *The Trauma of Birth* while emending it in important respects, Grof worked out a typology in which second- and third-stage LSD experiences are correlated with four distinct stages in the birth process: (a) a comfortable, intrauterine stage before the onset of labor; (b) an oppressive stage at labor's start when the fetus suffers the womb's contractions and has "no exit" inasmuch as the cervix has not opened; (c) the traumatic ensuing stage of labor during which the fetus is violently ejected through the birth canal; and (d) the freedom and release of birth itself. (b) and (c) seemed to Grof to vector the second or Rankian stage in the LSD sequence. In the reliving of (b), the oppressiveness of the womb is generalized and the entire world, existence itself, is experienced as oppressive. (c), when relived—the agony of labor and forced expulsion through the birth canal—produces the experience of dying: traumatic ejection from the only life-giving context one has known. The rebirth experience in which the Rankian stage climaxes derives from reliving the experience of physical birth (d) and paves the way for the ensuing transpersonal stage. The sense of unshadowed bliss that dominates this final stage taps the earliest memories of all: before the womb grew crowded, when the fetus blended with its mother in mystic embrace (a).

Even in bare outline, Grof's hypothesis is plausible, and when fleshed out with the case histories and experiential accounts that gave rise to it (material that is fascinating but which space precludes our entering here), it is doubly so. When subjects in their Rankian stage report first suffocation and then a violent, projective explosion in which not only blood but urine and feces are everywhere, one is persuaded that revived memories of the birth process play at least a part in triggering, shaping, and energizing later-stage LSD experiences. The question is: Are these the only causes at work? As we have noted, in the psychiatric model of man, once the Freudian domain has been exhausted, there is nowhere to look for causes save where Rank did, and Grof does: the ego, driven back to earlier and yet earlier libido positions, finally reenters the uterus. In the model of man that was sketched in *Forgotten Truth*, however, things are different. There the social and biological history of the organism is not the sole resource for explanation. "The soul that rises with us...

> Hath had elsewhere its setting,
> And cometh from afar:
> Not in entire forgetfulness...
> But trailing clouds of glory do we come....

From whence? "From God," Wordsworth tells us, and we agree. When he adds in the line that follows: "Heaven lies about us in our infancy!" we again concur; as the celestial plane, it envelops our souls not only in their infancy but always. More proximately, however, it is the intermediate or psychic plane from which we stem. Whereas in the psychiatric perspective body is basic, and explanations for mental occurrences are sought in body's endowments or history. In the primordial psychology, body represents a kind of shaking out of what has condensed on the plane of mental phenomena that exist prior to body and are more real than body. We are back at the point *Forgotten Truth* made in the context of dreams: it is not so much that we dream as that we are dreamed, if we may use this way of saying that the forces that come to the fore in our dreams pull the strings that govern our puppet existences. They do not govern them entirely—man is man, not manikin—but to say that they govern them is closer to the truth than is the epiphenomenal view in which body pays the piper and calls the tunes that dreams play out.

Thus, to Grof's finding that later stages in the LSD sequence conform to the stages of the birth process to a degree that warrants our saying that they are influenced by those stages, we add: influenced only, not caused. To a greater degree the experiences of these stages put the subject in direct touch with the psychic and archetypal forces of which his life is distillation and product. Birth and death are not physical only. Everyone knows this, but it is less recognized that physical birth and death are relatively minor manifestations of forces that are cosmic in blanketing the manifest world, the terrestrial and intermediate planes combined. Buddhism's *pratitya-samutpada* (Formulation of Dependent Origination) says profound things on this point, but all we shall say is that when a psychic quantum, germ of an ego, decides—out of ignorance, the Buddhists insert immediately—that it would be interesting to go it alone and have an independent career, in thereby distinguishing itself from the whole, and setting itself in ways against the whole, the ego shoulders certain consequences. Because it is finite, things will not always go its way, hence suffering in its manifold varieties. And the temporal side of the self's finitude ordains that it will die—piecemeal from the start as cells and minor dreams collapse, but eventually in its entirety. Energy is indestructible, however, so in some form there is rebirth. Confrontation of these principal truths in their transpersonal and trans-species scope and intensity is the basic stuff of later-stage LSD experience. Biological memory enters, but conceivably with little more than a "me too"—I too know the sequence from the time I was forged and delivered.

Spelled out in greater detail, the primordial explanation of the sequence would run as follows. Accepting LSD as a "tool for the study of the structure of human personality; of its various facets and levels," we see it uncovering the successively deeper layers of the self which Grof's study brings to light. Grof's psychiatric explanation for why it does so is that "defense systems are considerably loosened, resistances decrease, and memory recall is facilitated to a great degree. Deep unconscious material emerges into consciousness and is experienced in a complex symbolic way."[17] Our explanation shifts the accent. Only in the first stage are the defense systems that are loosened—ones that the individual ego builds to screen out painful memories. For the rest, what is loosened are structures that condition the human mode of existence and separate it from modes that are higher: its corporeality and compliance with the spatio-temporal structures of the terrestrial plane. The same holds for the memory recall that LSD facilitates. In the first stage, it is indeed memory that is activated as the subject relives, directly or in symbolic guise, the experiences that had befallen it, but in later stages what the psychiatrist continues to see as memory—an even earlier, intra-uterine memory—the ontologist (short of invoking reincarnation) sees as discovery: the discovery of layers of selfhood that are present from conception but are normally obscured from view. Likewise, with the "peculiar double orientation and double role of the subject" that Grof describes. "On the one hand," he writes, the subject "experiences full and complex age regression into the traumatic situations of childhood; on the other hand, he can assume alternately or even simultaneously the position corresponding to his real age."[18] This oscillation characterizes the entire sequence, but only in the first stage is its not-immediate referent the past. In the later sessions, that which is not immediate is removed not in time but in space—psychological space, of course. It lies below the surface of the exterior self that is normally in view.

The paradigm of the self that was sketched in *Forgotten Truth* showed it to be composed of four parts: body, mind, soul, and Spirit. Working with spatial imagery, we can visualize LSD as a seeing-eye probe that penetrates progressively toward the core of the subject's being. In the early sessions of the LSD sequence, it moves through the subject's *body* in two steps. The first of these triggers peripheral somatic responses, most regularly ones relating to perception, to produce the aesthetic phase. The second moves into memory regions of the brain where, Wilder Penfield has posited, a complete cinematographic record of everything the subject has experienced lies stored. That the events that were most important in the subject's formation are the ones that rush forward for attention stands to reason. We are into the first of the three main stages of the psycholytic sequence, the psychodynamic or Freudian stage.

Passage from the Freudian to the Rankian stage occurs when the chemicals enter the region of the mind that outdistances the brain and swims in the medium of the psychic or intermediate plane. The phenomenological consequences could almost have been predicted:

1. Biographical data—events that imprinted themselves on the subject's body, in this case, the memory region of his brain—recede.
2. Their place is taken by the "existentials"—conditioning structures—of human existence in general. The grim effect of this stage could be due in part to memories of the ordeals of gestation and birth, but the torment, the sense of the wistfulness and pathos of a suffering humanity and indeed life in all its forms, derives mainly from the fact that the larger purview of the intermediate plane renders the limitations (*dukkha*) of the terrestrial plane more visible than when the subject is immersed in them.
3. In the death and rebirth experience that climaxes this phase, Rankian factors could again cooperate without precluding causes that are more basic. The self had entered the intermediate plane through the soul's assumption of—compression into—mind; as the Hindus say, the *jiva* assumed a subtle body. Now, in the reversal of this sequence, mind must be dissolved (die) for soul to be released (reborn).

The sense of release from the imprisoning structures of mind signals the fact that the probe has reached the level of soul. The phenomenological consequences are the ones Grof's subjects reported in the transpersonal stage, the main ones being the following:

1. Whereas in the Rankian stage "there...was...a very distinct polarity between very positive and very negative experience,"[19] experience is now predominantly beatific, with "melted ecstacy" perhaps its most reported theme. Subjects "speak about mystic union, the fusion of the subjective with the objective world, identification with the universe, cosmic consciousness, the intuitive insight into the essence of being,

the Buddhist nirvanam, the Vedic samadhi, the harmony of worlds and spheres, the approximation to God, etc."[20]

2. Experience is more abstract. At its peak it "is usually contentless and accompanied by visions of blinding light or beautiful colors (heavenly blue, gold, the rainbow spectrum, peacock feathers, etc.)"[21] or is associated with space or sound. When its accouterments are more concrete, they tend to be archetypal, with the archetypes seeming to be limitless in number. The celestial plane which the soul inhabits is, we recall, the plane of God and the archetypes. The distinction between the two, which if fleshed out would result in an ontology of five tiers instead of four,[22] is for purposes of simplification and symmetry being played down in the present book.
3. The God who is almost invariably encountered is single and so far removed from anthropomorphism as to elicit, often, the pronoun "it." This is in contrast to the gods of the Rankian stage which tend to be multiple, Olympian, and essentially enlarged titans.

Beyond the soul lies only *Spirit*, an essence so ineffable that when the seeing eye strikes it, virtually all that can be reported is that it is "beyond" and "more than" all that had been encountered theretofore.

Up to this point we have noted Grof's empirical findings and compared the way they fit into his Rank-extended psychiatric theories on the one hand and into the primordial understanding of man on the other. It remains to point out how the findings of seventeen years affected his own thinking.

Engaged as he was in "the first mapping of completely unknown territories,"[23] he could not have foreseen where his inquiry would lead. What he found was that in "the most fascinating intellectual and spiritual adventure of my life [it] opened up new fantastic areas and forced me to break with the old systems and frameworks."[24] The first change in his thinking has already been noted: the psycholytic sequences showed the birth trauma to have more dynamic consequences than Grof and his strictly Freudian associates had supposed. This change psychoanalysis could accommodate, but not the one that followed. "I started my LSD research in 1956 as a convinced and dedicated psychoanalyst," he writes. "In the light of everyday clinical observations in LSD sessions, I found this conception untenable."[25] Basically, what proved to be untenable was "the present...gloomy...image of man, which is to a great extent influenced by psychoanalysis."[26] This picture of man,

> that of a social animal basically governed by blind and irrational instinctual forces...contradicts the experiences from the LSD sessions or at least appears superficial and limited. Most of the instinctual tendencies described by psychoanalysis (incestuous and murderous wishes, cannibalistic impulses, sadomasochistic inclinations, coprophilia, etc.) are very striking in the early LSD sessions; these observations are so common that they could almost be considered experimental evidence for some of the basic assumptions of psychoanalysis. Most of them, however, appear in the sessions for only a limited period of time. This whole area can be transcended [whereupon] we are confronted with an image of man which is diametrically opposed to the previous one. Man in his innermost nature appears then as a being that is fundamentally in harmony with his environment and is governed by intrinsic high and universal values.[27]

This change in anthropology has been the solid effect of psychedelic evidence on Grof's thinking. In psychoanalytic terms, if Freud discovered the importance of infantile experience on ontogenetic development and Rank the importance of the experience of birth itself, Grof's discoveries carry this search for ever earlier etiologies—in psychoanalytic theory, earlier is synonymous with stronger—to its absolute limit: his optimistic view of man derives from discovering the influence and latent power of early-gestation memories; memories of the way things were when the womb was still uncongested and all was well. Beyond this revised anthropology, however, Grof has toyed with a changed ontology as well. Endowments that supplement his psychiatric competencies have helped him here: he has an "ear" for metaphysics and an abiding ontological interest. These caused him to listen attentively from the start to his subjects' reports on the nature of reality, and in one of his recent papers, "LSD and the Cosmic Game: Outline of Psychedelic Cosmology and Ontology,"[28] he gives these reports full rein. Laying aside for the interval his role as research psychiatrist, which required seeing patients' experiences as shaped by if not projected from early formative experiences, in this paper, Grof turns phenomenologist and allows their reports to stand in their own right. The view of reality that results is so uncannily like the one that has been outlined in *Forgotten Truth* that, interlacing paraphrases of passages in Grof's article with direct quotations from it, we present it here in summary:

> The ultimate source of existence is the Void, the supracosmic Silence, the uncreated and absolutely ineffable Supreme.
>
> The first possible formulation of this source is Universal Mind. Here, too, words fail, for Mind transcends the dichotomies, polarities, and paradoxes that harry the relative world. Insofar as description is attempted, the Vedantic ternary—Infinite Existence, Infinite Intelligence, Infinite Bliss—is as adequate as any.
>
> God is not limited to his foregoing, "abstract" modes. He can be encountered concretely, as the God of the Old and New Testaments, Buddha, Shiva, or in other modes. These modes do not, however, wear the mantle of ultimacy or provide answers that are final.
>
> The phenomenal worlds owe their existence to Universal Mind, which Mind does not itself become implicated in their categories. Man, together with the three-dimensional world he experiences, is but one of innumerable modes through which Mind experiences itself. The "heavy physicality" and seemingly objective finality of man's material world, its space-time grid and laws of nature that offer themselves as if they were *sine qua non* of existence itself—all these are in fact highly provisional and relative. Under exceptional circumstances man can rise to a level of consciousness where he sees that taken together they constitute but one of inumerable sets of limiting constructs Universal Mind assumes. To saddle that Mind itself with these categories would be as ridiculous as trying to understand the human mind through the rules of chess.
>
> Created entities tend progressively to lose contact with their original source and the awareness of their pristine identity with it. In the initial stage of this falling away, created entities maintain contact with their source and the separation is playful, relative, and obviously tentative. An image that would illustrate this stage is that of waves on the ocean. From a certain point of view they are individual entities; we can speak of a large, fast, green, and foamy wave, for example. At the same time it

is transparently evident that in spite of its relative individuation the wave is part of the ocean.

At the next stage created entities assume a partial independence and we can observe the beginnings of "cosmic screenwork." Here unity with the source can be temporarily forgotten in the way an actor on stage can virtually forget his own identity while he identifies with the character he portrays.

Continuation of the process of partitioning results in a situation in which individuation is permanently and for all practical purposes complete, and only occasionally do intimations of the original wholeness resurface. This can be illustrated by the relationship between cells of a body, organs, and the body as a whole. Cells are separate entities but function as parts of organs. The latter have even more independence, but they too play out their roles in the complete organism. Individuation and participation are dialectically combined. Complex biochemical interactions bridge provisional boundaries to ensure the functioning of the organism as a whole.

In the final stage the separation is practically complete. Liaison with the source is lost and the original identity completely forgotten. The "screen" is now all but impermeable; radical qualitative change is required for the original unity to be restored. Symbol of this might be a snowflake, crystallized from water that has evaporated from the ocean. It bears little outward similarity to its source and must undergo a change in structure if reunion is to occur.

Human beings who manage to effect the change just referred to find thereafter that life's polarities paradoxically both do and do not exist. This holds for such contraries as spirit/matter, good/evil, stability/motion, heaven/hell, beauty/ugliness, agony/ecstasy, etc. In the last analysis there is no difference between subject and object, observer and observed, experiencer and experienced, creator and creation.

In the early years of psychoanalysis, when hostility was shown to its reports and theories on account of their astonishing novelty, and they were dismissed as products of their authors' perverted imaginations, Freud used to hold up against this objection the argument that no human brain could have invented such facts and connections had they not been persistently forced on it by a series of converging and interlocking observations. Grof might have argued equally: to wit, that the "psychedelic cosmology and ontology" that his patients came up with is as uninventable as Freud's own system. In fact, however, he does not do so. In the manner of a good phenomenologist, he lets the picture speak for itself, neither belittling it by referring it back to causes that in purporting to explain it would explain it away, nor arguing that it is true. As phenomenologists themselves would say, he "brackets" his own judgment regarding the truth question and contents himself with reporting what his patients said about it.

> The idea that the "three-dimensional world" is only one of many experiential worlds created by the Universal Mind... appeared to them much more logical than the opposite alternative that is so frequently taken for granted, namely, that the material world has objective reality of its own and that the human consciousness and the concept of God are merely products of highly organized matter, the human brain. When closely analyzed the latter concept presents at least as many incongruences, paradoxes and absurdities as the described concept of the Universal Mind. The problems of finity versus infinity of time and space; the enigma of the origin of matter, energy and space; and the mystery of the prime impulse appear to be so overwhelming and defeating that one seriously questions why this approach should be given priority in our thinking.[29]

Perception and Knowledge

Reflections on Psychological and Spiritual Learning in the Psychedelic Experience

FRANCES E. VAUGHAN

Little controlled research has been done with psychedelics. But my own experience, coupled with my observation of hundreds of clients, students, and acquaintances who have used LSD in both controlled and uncontrolled settings, has convinced me that we have much to learn from an appropriate investigation of this powerful mind-altering chemical. The dearth of research has not lessened the impact of psychedelic experiences on people's lives and on the culture at large. Psychology, in general, has failed to keep pace with personal explorations in altered states of consciousness, many of them induced by LSD or similar psychedelic substances.

In the past decade, transpersonal psychology has emerged as that branch of psychology specifically concerned with the study of human consciousness. It attempts to expand the field of psychological inquiry to include such human experiences as those induced by psychedelics, as well as similar states attained through the practice of meditation or other disciplines. As a transpersonal psychologist, I have been particularly interested in the study of consciousness as it pertains to psychological health and well-being. My clinical practice is devoted to facilitating human growth and development, often on the border between psychological and spiritual domains.

My personal introduction to LSD occurred under optimum conditions. In his book, *LSD Psychotherapy*, Dr. Stanislav Grof observes that normal people benefit most when participating in a supervised psychedelic program and that the experience can move them in the direction of self-actualization.[1] My own experience supports this view. As a subject in early LSD research, I was thoroughly screened and well prepared. I also had an opportunity to talk with other subjects who felt they had benefited. My first session was a profound and overwhelming mystical experience. Subsequent sessions seemed less important but served as reminders of insights gained in the initial one.

A most striking feature of my psychedelic experience was the noetic quality of consciousness as it expanded from its usual perceptual range to a vast contextual awareness that recognized the relativity of all perception in space/time. I find the term re-cognize particularly appropriate, since the knowledge that was suddenly revealed to me under LSD seemed to be remembered rather than learned. I was awed by the vast range of consciousness, yet felt that I was simply uncovering what I had always known, i.e., the truth which had previously been hidden behind a veil of relative unconsciousness. As the illusory, changeable nature of ordinary reality became increasingly clear, I also realized how a normally constricted perceptual framework permits one to see only a fraction of reality, inevitably distorted to suit personal projections and presuppositions.

During the experience, I felt I understood what mystics throughout the ages have claimed to be the universal truth of existence. I had an academic background in philosophy and comparative religion, but I realized that mystical teachings had now taken on an added dimension. My perception seemed to have shifted from a flat, two-dimensional intellectual understanding of the literature, to a three-dimensional sense of immersion in the mystical reality.

The perennial philosophy and the esoteric teachings of all time suddenly made sense. I understood why spiritual seekers were instructed to look within, and the unconscious was revealed to be not just a useful concept, but an infinite reservoir of creative potential. I felt I had been afforded a glimpse into the nature of reality and the human potential within that reality, together with a direct experience of being myself, free of illusory identifications and constrictions of consciousness. My understanding of mystical teachings, both Eastern and Western, Hindu, Buddhist, Christian, and Sufi alike, took a quantum leap. I became aware of the transcendental unity at the core of all the great religions and understood for the first time the meaning of ecstatic states.

I now felt I had some direct experience of the ineffable realms of union with God, and I discovered that my dissatisfaction with conventional religion was not due to the death of God, as some theologians proclaimed, but rather to impoverished concepts of God currently in vogue. Whether one spoke of God, the Void, or the Self, Being, Bliss, or Consciousness, did not matter, for words were far removed from the experience. They were only fingers pointing to the moon; they bore little resemblance to the depth of realization that became available when I let go of my preconceptions about the nature of the universe. As far as I knew, such insights into the nature of consciousness had only been attained by rare individuals, many of them advanced practitioners of spiritual disciplines.

The worldview that made most sense of this experience was clearly a mystical one. Neither the subjective nor the objective pole of experience could encompass the totality. The possibility of transcending boundaries between self and other, the illusory nature of the ego, the interdependence of opposites, the relative nature of dualism and the resolution of paradox in transcendence became clear. All mental content was simply the play or the dance of life, and what could be known about consciousness became the focus of my attention. Psychodynamic material that came into awareness seemed irrelevant. My own personal drama was no more significant than light playing on a movie screen. Even feelings of joy, ecstasy, and liberation in letting go of attachments were less important than the insight and sense of knowing, or remembering, inexpressible truth. "Know the truth, and the truth shall make you free" were the words that seemed best to capture the nature of my experience. I felt free to be exactly who I was, free of fear and social constraints, and filled with love and compassion for all beings.

Although many of the insights that flooded my awareness were forgotten, many remained to influence my life. I felt I could see how much human suffering is self-imposed, how our beliefs shape our reality, and what it means to awaken to the realization that life is a dream of our own making. The dreamlike quality of existence, the unreality of past memories and future fantasies, and the acceptance of the interrelatedness of all things were insights subsequently confirmed as I learned more of the perennial teachings of both Eastern and Western contemplative traditions.

I also gained a new appreciation for the Christian teaching of forgiveness. I saw how our own condemnation injures us, and how our difficulty in forgiving ourselves for imagined imperfections contributes to neurotic guilt and anxiety. Not only did I feel forgiven for being just as I was, I saw that, in reality, there was nothing to forgive. This seemed to remove the obstacles to the experience of love. I felt an extension of love and forgiveness to all beings everywhere.

The subjective nature of time also became starkly apparent. My Newtonian worldview was sufficiently shaken to make it relatively easy for me to accept some of the more apparently nonsensical propositions of the new subatomic physics, when they later came to my attention. Likewise, parapsychological phenomena no longer seemed incomprehensible. The fact that we could not explain part of our human experience in the existing paradigm seemed to indicate that the paradigm needed re-examination rather than to justify dismissal of the evidence.

For the first time, I understood the meaning of "ineffable." There seemed to be no possibility of conveying in words the subjective truth of my experience. A veil had been lifted from my inner vision and I felt able to see, not just images or forms, but the nature of truth itself. The doors of perception were so cleansed, they seemed to vanish altogether, and there was only infinite being. Krishnamurti's characterization of truth as a pathless land seemed an appropriate description of this domain.

I felt that I had now experienced the grace of God. Truly, I had been given a gift of infinite worth. I could understand why human beings throughout history have relentlessly pursued truth and sought enlightenment. I knew now why some felt impelled to sit in caves for years trying to become enlightened, why some were willing to die for ideals, and why suffering was endured. If asceticism was perceived as a means of attaining this state of oneness, I could understand why a person might choose it. I understood that the essence of my being was identical with the timeless essence of every living thing, that formlessness was the essence of form, that the whole universe was reflected in every psyche and that my separateness was only an illusion, a dream from which I had, in this moment, fully awakened.

As I faced old fears and watched the tricks of my mind, I became increasingly aware of my ability to choose my subjective state. Consciousness seemed infinitely plastic. I could choose to focus the lens of attention on anything. Barriers and resistance had dissolved, and fears had disappeared along with them. In that moment, I knew that I had nothing to fear. Only the creations of my own mind and my own thought forms could threaten me, and I could see them as if in a lucid dream, parading through the field of awareness. I was free to either attend to them or let them pass, choosing instead to experience more fully the bliss of pure being, just being present to my experience of the moment, with no added fantasy or distraction.

The affective tone of my experience was pure love. After the barriers dissolved, I could feel the depth of my love for life itself, and for my husband and children. They seemed perfect just as they were, yet I did not need them and therefore felt no fear or possessiveness. Life itself was enough. I, too, was complete and acceptable just as I was. Old feelings of inadequacy and uncertainty had vanished.

My aesthetic sensibilities were profoundly enhanced, not only during the few hours of the session, but afterwards as well. The effect has lasted over a period of fifteen years. My appreciation of music, art, nature, and human beings has continued to grow since that time. I remember being particularly struck by the joy of hearing music as I never had heard it before. I could laugh at my old self image, which included "not being musical." I was deeply moved by each piece of music that was played. As I listened without distraction, each one evoked a different aspect of my psyche, and at the center of each was the perfect still point of pure being where one could experience union with God.

I gained a new appreciation of my own capacity for choice and the role of consciousness in creating experience. For the first time, I saw the possibility of taking responsibility for my own experience. I also felt I was truly participating fully in the universal human condition. All of my experience, including the experience of separateness and aloneness, was something I had in common with all human beings. Although my personal history and the events of my life were unique, the underlying unity of life became starkly evident. The forms of expression and experience were diverse, but the underlying qualities of being were universal.

I also felt a reduction in nonspecific anxiety, and a greatly diminished fear of death. As the illusory nature of many of my worries and fears became apparent, I became more trusting and accepting of myself and more willing to enter into unfamiliar situations and take risks in exploring new creative endeavors. As I was released from feelings of neurotic guilt and inadequacy, my increased ability to relax also contributed to enhanced sexual enjoyment. My appreciation of life itself and of the simple tasks of everyday living was also profoundly enhanced. I found myself more open in my intimate relationships and better able to give and receive love without fear.

I also became aware of a desire to be of service in the world, to make some contribution to humanity through my work. At the same time, I felt more able to tolerate paradox and ambiguity. The recognition of the interdependence of opposites has since become a useful therapeutic tool in my practice: I often think of psychological growth as a balance and synthesis of opposites. In working with others to heal internal splits and conflicts, enabling them to take increasing responsibility for their own lives and well-being, I have had many opportunities to appreciate the importance of this capacity.

The effects of this experience seemed to me equivalent to what I might have expected from several years of insight therapy. I had been able to see through, and let go, of many constricting patterns of thought and behavior that previously seemed automatic and beyond conscious control. Some of the far-reaching effects appeared immediately in my personal life. For several months after this experience, I remained in a semi-euphoric state in which I experienced being in love all the time. Everything in my life seemed to be exactly as it was supposed to be. Everything was all right. None of the small things I used to get upset about seemed to matter any more. I was experiencing a state of inner peace and serenity that allowed me to cope more effectively with everything I needed to do, while I felt in touch with a sense of divinity within.

This period of my life coincided with what seemed to be a time of new hope for humankind. The flower children in San Francisco were happily rebelling against the old order, and a better future seemed within reach. A sense of euphoria was in the air; the more sordid side of psychedelia became apparent only as time went by. My interest in understanding the experience led me to graduate school to study psychology, but I soon found that Western psychological models could not accommodate it. Yet, I knew I was not unique. Many other people were reporting similar experiences. Eastern consciousness disciplines seemed to offer the best maps of this inner world, and they also offered instruction for attaining such states without the use of chemicals. Now I could hear, as if for the first time, the depth of the wisdom in their teachings and in the mystical doctrines of all ages and all cultures. As I sought for words to express my own inner experience, I gained a new appreciation for those individuals who had attempted to communicate their own insights in writing or art. I also became interested in understanding intuitive ways of knowing; many years later I wrote a book about the development of intuition entitled *Awakening Intuition*.[2]

My intellect was eager to incorporate what I had learned into working psychological models. I saw a need to formulate new psychological theories that could encompass such experiences. Among Western psychologists, only Carl Jung had addressed transpersonal experiences. He wrote, "[T]he fact is that the approach to the numinous is the real therapy and inasmuch as you attain to the numinous experience you are released from the curse of pathology."[3] That was apparently true of my experience, but it later became clear that a psychedelic experience in and of itself was not necessarily therapeutic. The popularity of psychedelics increased greatly, but few of their users achieved the therapeutic benefits I had experienced.

In his extensive research on LSD psychotherapy, Stanislav Grof noted that transpersonal experiences occur only rarely in early sessions of psychedelic therapy but are quite common in advanced sessions.[4] Grof has provided a detailed map of the death/rebirth experience which he found to be therapeutic for many of his subjects. The experience of ego death may be liberating and ecstatic, as it was for me, but it may also be terrifying to a person who is unprepared. However, under appropriate, carefully controlled conditions, a subject may be enabled to surmount the difficulties encountered in letting go of limiting self-identifications.

Phenomenologically, personal accounts of drug-induced mystical experiences may be indistinguishable from spontaneously occurring mystical experiences. In either case, the effects may or may not last. The glimpse of a larger reality that such experience affords may change a person's life if he or she chooses to integrate it. If, however, the experience is repressed, denied, or invalidated, it may only contribute to exacerbating existential guilt and anxiety. When a person is not able to stabilize such glimpses into transcendent reality and incorporate them into existing belief systems, they can certainly disrupt the ordinary adjustment of everyday life.

Transpersonal psychology has attempted to formulate a conceptual framework for such experiences since they obviously are not going to go away.[5] Although psychedelics have been restricted, the public continues to experiment, and research continues to lag far behind. Moreover, the striking parallels between such experiences and those described by mystics raise many questions for mental health professionals. In the transpersonal domain, where psychological and spiritual growth are one, psychedelics appear to be powerful tools for the investigation of consciousness; they could enable us to expand our understanding of the human mind and the nature of creative consciousness. A willingness to question our assumptions and to keep an open mind with respect to potential benefits and potential hazards is essential.

For the past ten years, I have been practicing transpersonal psychotherapy and training therapists to work in this area. The lack of serious study in the field of psychedelic drugs has unfortunately restricted their use to uncontrolled personal experimentation. The dearth of research is clearly a drawback when therapists are so often called upon to handle situations where clients have been involved in uncontrolled experimentation. Although many people in our culture have taken psychedelics, few therapists are capable of assessing, evaluating, and integrating psychedelic experiences in a useful way. Psychedelics, like any powerful tool, may be used skillfully for the benefit of humanity, or unskillfully, to the detriment of those whose ignorance leads to abuse.

As we search for ways of understanding the possibly infinite resources of human consciousness, I suggest that the potential of psychedelics as tools for learning should not be ignored. Today, when the survival of our planet is at stake, there is an urgent need to work responsibly in every facet of human endeavor. By refusing to tread where fools rushed in, we may be turning away from significant learning about human experience and how the mind works. People of differing views and persuasions must join together in exploration of the universals of psychological health and well-being and work to find ways of facilitating experiences that foster growth toward wholeness for everyone.

The Psychology of Birth, the Prenatal Epoch, and Incarnation

Ancient and Modern Perspectives

RALPH METZNER

Through his research and writing, Stanislav Grof has made ground-breaking contributions in three areas of modern psychology: psychotherapy with LSD and other psychedelics; the impact of birth patterns on subsequent psychological development, including psychopathology and psychedelic experiences; and the understanding of how transformative realizations can emerge out of crisis experiences when birth patterns are taken into account. Grof's explorations have led him into far-reaching reconsiderations of the metaphysical foundations of the Western scientific worldview and to the extension of transpersonal psychological perspectives to the perennial questions of death and rebirth, problems of karma and good vs. evil, and humanity's place in the greater cosmos.

I propose, in this essay, to focus on Grof's central theoretical contribution, or perhaps I should say "discovery," namely, the four-fold sequence of "basic perinatal matrices" (BPMs), and the impact of this formulation in understanding psychedelic states and on the development of the field of pre- and perinatal psychology. From there, I want to examine the role and significance of the prenatal epoch in human psychological life more generally, and present, in schematic form, an integrative heuristic model for understanding prenatal formative influences and the process of incarnation. Finally, I comment on some related perspectives emerging from Jewish Wisdom Literature and the Tibetan Buddhist Bardo teachings.

I first met Grof in 1965, when I was living in Millbrook, NY, in an experimental community called the Castalia Foundation, founded by Timothy Leary, Richard Alpert, and their associates. Grof was a visiting psychiatrist from Czechoslovakia, one of the pioneers of the *psycholytic* approach to psychotherapy prevalent in Europe. In this method, the patient/client received gradually increasing low doses of LSD over a series of psychoanalytically oriented therapy sessions. This was different from the *psychedelic* approach, developed in Canada and the US, which involved a single high dose session, most often used with alcoholics. Our group, at Harvard and in Millbrook, was not actually involved in psychotherapy, though one prisoner "behavior change" project had been carried out at Harvard. Rather, our work was dedicated to the applications of psychedelics in psychology, creativity, the arts, personal growth, and spirituality; we were also involved in an ongoing series of small group experiments in communal living. My impression of Stan Grof was that of a man in the grip of a powerful mystical vision: he spoke intensely about his realization, with LSD, of the Vedantist teachings of the oneness of *Atman* and *Brahman,* the individual Self and the cosmic Self, or Godhead. This language was somewhat different from the language of Tibetan Buddhism with which I was more familiar from working on our adaptation of the *Tibetan Book of the Dead*. However, I felt a strong affinity for Grof the fellow consciousness explorer, like me a native European; our professional paths and interests have continued to evolve in an amazingly parallel fashion.

My next major encounter with Grof's work occurred about a decade later, in the mid-1970s, when he was scholar-in-residence at the Esalen Institute and invited me to be a guest teacher in several of the month-long seminars that he and his wife Christina were then conducting. I was in the midst of a ten-year immersion in a Western esoteric school, the School of Actualism founded by Russell Schofield, that taught yogic methods of working with light-fire energies, anciently known as *Agni Yoga* (from *Agni*, the Vedic Fire Deity), and had stopped working with psychedelics personally, in part due to a health crisis. Grof had been involved in research giving psychedelic experiences to people dying from cancer, extending the work on religious experience begun by Walter Pahnke at Harvard, that the latter referred to as "experimental mysticism." Pahnke tragically died in a scuba diving accident in 1971; Grof and his colleagues continued their studies at the Maryland Psychiatric Research Center for a few more years (since LSD had been made a controlled substance on Schedule I in 1964, legitimate medical research projects gradually dwindled over the next decade). During those years, Grof began writing a book on the dying project[1] as well as *Realms of the Human Unconscious*, which described case histories from his LSD psychotherapy practice in Prague and outlined his model of the stages of birth.[2]

Reading Grof's books was a revelation for me; for the first time, I felt I understood certain aspects of psychedelic experiences I had years before, particularly some painful, hellish visions I had never been able to interpret. I had learned early on that Freud's psychoanalytic theories clearly did not touch the psychedelic realms. The Tibetan Buddhist model of three *bardos* one went through between death and rebirth was metaphysically illuminating but did not really describe the typical course of a psychedelic experience; it was clearly written for highly trained lama-monks who had spent a lifetime practicing one-pointedness and absorbing the elaborate iconography of *Vajrayana* Buddhism. I had found the theories and body-oriented therapy methods of Wilhelm Reich very apposite in helping to understand the energy-flow and energy-blockage experiences that can occur in these altered states of consciousness. Grof's model, on the other hand, gave a detailed description of the content of certain experiences, tracing them to the qualitative template or matrix provided in one or another of the four BPMs.

Let me give a personal example. Here are some extracts from my description of a high-dose psilocybin session that occurred when we were still at Harvard:

> As I looked around the room, I saw great bands of moving streams of energy particles traversing the space, passing through and between myself and the other people. We all seemed to be part of these moving, ever-changing bands of energy. They were familiar to me from other mushroom sessions when I had seen them as luminous vibrating filigree networks. But this time, the intensity frightened me. As my fear level increased, the energy bands congealed and stopped moving; they took on a greyish hue, like prison bars. All at once, I felt immobilized and trapped, like a fly in a gigantic metallic spider's web.

> I couldn't even talk and explain what was happening to me; my voice felt paralyzed.
>
> Everyone seemed to be frozen into immobility by these metallic web-cages. I felt my mind was paralyzed too. I couldn't think or understand what was happening. I couldn't tell whether what I was experiencing was real or a drug-induced hallucination (an experience psychiatry refers to as "derealization").
>
> Somehow, I managed to reach Timothy Leary on the telephone and asked him to come over and help us....While waiting for Tim to arrive, I was holding on to sanity with the thought that when he got here, he would free us all from this monstrous spider's web we were all caught in, which also had the effect of making us speechless. Or, at least it felt like I couldn't hear, say or understand anything. I felt completely dehumanized, not even like a biological organism, more like a mechanical puppet or device.
>
> I was so relieved when Tim came through the door—I could see him moving freely through the sticky web of grey steel bands. But then, after a few minutes, as I watched in horror, he too became trapped in them: his movements slowed down, became mechanical, robot-like, and I plunged into despair as I realized he too was helplessly caught....I began to wish I could be killed in order to get out of this hell world. After a couple of hours of objective time and a hellish eternity in subjective time, the intensity of the experience began to diminish.[3]

The following are Grof's descriptions of typical LSD experiences from the perinatal realm of BPM II, the stage of the birth process when contractions have started, but the cervix has not yet opened thus no movement is possible for the fetus.

> The activation of this matrix results in a rather characteristic spiritual experience of "no exit' or "hell." The subject feels encaged in a claustrophobic world....This experience is characterized by a striking darkness of the visual field and by ominous colors....Another typical category of visions related to this perinatal matrix involves the dehumanized, grotesque, and bizarre world of automata, robots and mechanical gadgets...or of a meaningless "honky tonk" or "cardboard"

> world. Agonizing feelings of separation, alienation, metaphysical loneliness, helplessness, hopelessness, inferiority and guilt are standard components of BPM II....An interesting variety of the second perinatal matrix seems to be related to the very onset and the initial stages of the delivery. This situation is experienced in LSD sessions as an increasing awareness of an imminent and vital danger or as cosmic engulfment...intensification of this experience typically results in the vision of a gigantic and irresistible whirlpool, a cosmic maelstrom sucking the subject and his world relentlessly to its center...typical symptoms involve extreme pressures on the head and body....[4]

I understood then how such experiences are not so much "birth memories," but rather subjective experiences of extreme bodily sensations that have the same qualitative characteristics as the "matrix" from the corresponding birth stage. In another early psilocybin session, I was triggered by a chance interaction with a friend in the session into feeling guilty about what I had said. Next, I found myself in a medieval torture dungeon, where I was being beaten to a bloody pulp by men with gigantic clubs. While the experience of hallucinating torturing guilt was explicable in terms of the interpersonal dynamics, the grotesque intensity of these hellish visions was not. Here is what Grof says about the dynamics of BPM III, where the cervix has opened, and now there is movement—intense, powerful pushing and thrusting. "The most important characteristic of this pattern is the atmosphere of a titanic struggle, frequently attaining catastrophic proportions. The intensity of painful tension reaches a degree that appears far beyond what any human can bear...sadomasochistic orgies...unbridled murderous aggression...tortures and cruelties of all kinds, mutilations and self-mutilations."[5]

Having observed hundreds of sessions in individuals or groups, with psychedelic or other forms of activation and amplification, I can attest that such experiences typical of BPM stages II and III can be readily identified by an observer, even if the subjects themselves do not initially make such identification with birth memories. The objective facts of one's birth, if available, can then often be correlated with the qualitative, subjective experience.

In his theoretical writings on the birth matrices, Stan Grof pointed out that he was, in a sense, reviving the long-neglected theories of Otto Rank,

Freud's disciple who had emphasized the trauma of birth as an important source of anxiety-laden memories and fantasies in the unconscious. Rank's observations were accepted by Freud himself, but never really incorporated into the psychoanalytic canon, possibly due to a widely held prejudicial belief that the sensory life of infants begins at birth and not before. Rank did point out, in his book, *The Myth of the Birth of the Hero*, that world mythology contains numerous stories that reflect aspects of the birth experience. Grof, in his concept of birth matrices, has extended this view much further—giving examples of numerous mythic and legendary themes correlated with the different birth matrices, which are often also experienced by persons in psychedelic experiences. In his exploration of reflections of early somatic experiences in mythology, Grof's approach is much more like Jung's views on archetypes—those deep structures of the collective unconscious responsible for such universally occurring images as the hero, the Great Mother, or Wise Old Man. In the specificity with which his mythic amplifications are anchored in somatic experience, Grof's theories go far beyond what Rank or Jung were able to do.

I do not wish, in the framework of this essay, to enter into the discussion of whether Grof has proven that traumatic birth experiences cause or bring about later image clusters; or that in certain LSD experiences one is actually re-experiencing somatic memories of birth stages. There are complicated questions in epistemology and the philosophy of science involved here. The causal question can equally be raised in relation to Freudian theories of developmental fixations being connected to later character neuroses. Do the childhood experiences "cause" the later patterns of fantasy and behavior?

My own inclination is to regard developmental models, including Grof's, as *heuristic analogies*: one can observe qualitative analogies between the experiential birth matrix and subsequent experiences in altered states of consciousness. Such analogies offer a framework for understanding that is useful to the person undergoing a profound transformation of consciousness. As Gregory Bateson pointed out,[6] analogies or metaphors, contrary to what many believe, are pervasive in the natural and social sciences and should not be judged by the standards of a causal-deterministic explanation. Grof's postulate is that the birth process, with its four distinct stage patterns, is a useful and quite specific analogy to psycho-spiritual transformations of all kinds. In some ways, it is reminiscent of Ernst Haeckel's principle that *ontogeny repeats phylogeny*—which, although rejected by science as a causal explanatory law, nevertheless offers valuable insight and understanding.

Grof's work on the perinatal matrices and their reflections in psychedelic states as well as adult psychopathology, had a seminal impact on the development of the field now known as "pre- and perinatal psychology and health." Grof has made an important contribution to this field, but in his own theoretical work, apparently moved more into further exploration of the transpersonal dimensions of human life, with perinatal memory experiences seen as providing a kind of access channel or transition phase.

Developments in Pre- and Perinatal Psychology

Over the last three decades of the 20th century, other researchers developed this field of pre- and perinatal psychology further, showing how not only birth itself, but the entire experience of the embryo and fetus have a profound impact on subsequent human physical and psychic health and well-being, as well as on existential attitudes and spiritual aspirations. An association was formed—APPPAH (Association for Pre- and Perinatal Psychology and Health) that publishes a journal and organizes conferences. The following extract from the Mission Statement of the APPPAH gives a good sense of the underlying philosophical assumptions shared by this eclectic group of researchers, which includes psychologists, psychiatrists, brain scientists, obstetricians, midwives, childbirth educators, and others:

> The APPPAH is dedicated to the in-depth exploration of the psychological, emotional and social development of babies and parents from preparation for pregnancy through the postpartum period. We find that pre- and perinatal experiences have a profound impact on health and human behavior. Life is a continuum which starts not, as is commonly thought, at birth, but at conception. We see mother and child as fundamentally interconnected, and understand that babies are sentient beings who are best nurtured when their mothers are provided with a supportive and healthy prenatal environment in which their choices are respected. A strong and loving pre- and perinatal foundation can encourage human bonding and sensitivity to others, with lasting benefits. Womb ecology becomes world

> ecology as the seeds of peace—or violence—are sown by parents, educators and caregivers.[7]

It would go far beyond the limits of this essay to survey the whole field of pre- and perinatal psychology, which, in my view, represents as significant an extension of the paradigm boundaries of mainstream psychology as do transpersonal psychology and ecopsychology. I will just mention two main lines of development: (1) the psychology of the prenate, and (2) the impact of pre- and perinatal trauma.

(1) The founder of "radical psychiatry," R.D. Laing, had published an early impressionistic study of the experiential dimensions of the entire prenatal period from conception on, and its reflection in myths of the hero's journey.[8] Psychological studies of the perceptions, feelings, thoughts, and communication abilities of the unborn child, as reported in works by Thomas Verny and David Chamberlain, have revealed a vast world of prenatal experience every bit as rich and differentiated as the experiential world of the neonate and infant.[9] Jane English, in a work entitled *Different Doorway*, has described, based on her own experience and interviews with others, the uniquely different psychology of Caesarean-born individuals.[10] Many, though not all, of such findings are derived from hypnotherapeutic regression work with adults, who can often be brought to remember details of their pre- or perinatal experience that can then also be verified by other means. The San Francisco obstetrician and hypnotherapist David Cheek, for example, found that he could, using the ideo-motor signaling method, regress adults to their birth experience, the reported physical/medical details of which he would then verify by checking his files—having personally delivered them perhaps 25 years before.[11]

(2) The impact of prenatal and perinatal trauma has been studied extensively by a fairly large number of therapists, using some variation of hypnotic, imagistic, and breathing methods to access and resolve such traumas. I think it would be fair to say that agreement, perhaps amounting to consensus, exists in this community that the impact of pre- and perinatal shock and trauma is every bit as important, if not more so, than traumata in subsequent infancy and childhood and that the resolution of prenatal traumata can facilitate and accelerate the resolution of later neurotic and character disorders. I will return to this point below, with a suggestion as to why this may be so. Various non-drug methods of working therapeutically with

pre- and perinatal trauma have been described in articles in the *Journal of Prenatal and Perinatal Psychology and Health,* as well as in several chapters in the two-volume *Regression Therapy: Handbook for Professionals,* edited by Winafred Lucas.[12]

I will mention, as examples of this line that I am personally familiar with, only the work of pioneering therapists William Emerson and Ray Castellino. Emerson studied with the English theologian and psychiatrist Frank Lake, who had originally done research on the use of small doses of LSD for facilitating pre- and perinatal regression, and had switched to group breath and body therapies, when LSD work became impossible. Emerson subsequently developed astonishingly profound and effective methods of working, not only with adults, but with children and even infants in their mother's laps, to resolve birth difficulties. Emerson's methods and approach to the prenatal period are described in publications and video tapes available from Emerson Training Seminars;[13] and in the book *Remembering Our Home,* coauthored with Christian spirituality writers Sheila Linn, Dennis Linn, and Matthew Linn. In this extraordinary volume, the authors describe the stages of prenatal development and the possible trauma impacts at each stage, as Emerson has come to understand them, and compare these with the prototypical healing birth stories given in the Biblical account of the birth of Jesus Christ.[14]

I have personally attended several of Emerson's training workshops, as well as one with Ray Castellino, who has further developed highly innovative procedures for therapeutic work with babies and their parents.[15] Coming from my experiences with guided divinations, both with and without psychoactive amplification, for myself and with others, I would have to say that I found Emerson's and Castellino's descriptions of the *experience* of prenatal (and neonatal) development, and the role of trauma and shock at each stage, convincing in an almost revelatory way. As with Grof's analysis of the perinatal stages, I saw how many experiences occurring in altered states, with psychedelics, deep body work, or experiential therapies, which are reported by the experiencer in terms of somatic sensations and feelings in different areas of the body (e.g., heat, pressure, fear, pain, tension, release, etc.), can be understood as memories of the prenatal epoch, if the therapist is trained to recognize them and can provide a *psychic-energetic container* for them—though not necessarily a verbal interpretation. In other words,

I could *confirm* many of their observations through my own recollected prenatal memories as well as my observations in work with clients. I had developed, in my therapeutic regression work, ways of connecting with one's own conception, as an event equally impactful as birth. Three strands or tracks of memory imprints can be identified when regressing consciously to conception: *first,* the mother's and father's familial-genetic inheritance, genetic strengths and weaknesses, that are pre-programmed into the unicellular embryo beginning a human life journey at that moment; *second,* the personality, attitudes, sexual conditioning and existential orientation of each of the parents is imprinted into the embryonic being at conception, and later, throughout the prenatal period—and can be identified as such. How could it be otherwise? The *third* strand, described in detail in Emerson's work, are the *cellular memories* of the subjective experience of sperm and ovum; their origin (ovulation, spermatogenesis) and climactic, fusion encounter can also all be recollected in regressive memory journeys. And further, that not only conception but also *discovery, implantation, placenta formation,* and other key events in embryogenesis, have recognizable impacts in later character formation, and when traumatic, leave "hot spots" in specific locations in the adult body that can be worked with in particular ways.

In conclusion, I would like to comment on the theoretical question as to the reason for the importance of working with prenatal and perinatal defensive patterns—those resulting from shock and trauma, as well as those that form a kind of matrix or template for "normal" personality or character development. Wilhelm Reich, the pioneering psychoanalyst whose work on character "armoring" became the inspiration for most of the numerous forms of body, breath, and movement-oriented forms of psychotherapy developed in the late 20th century, had proposed that throughout the formative years, repeated patterns of muscular tension in the developing body actually congeal into a kind of characteristic postural attitude that functions both defensively and expressively. *The muscular armor and the character armor are functionally equivalent,* was his summary formulation.[16] Working on dissolving armoring through breath and movement helps dissolve the neurotic character defenses. The metaphor of muscular armoring can be extended to birth patterns, since the musculature of the fetal body is clearly and totally engaged in that process (though less so in Caesarean births). However, the question arises as to how perinatal "armoring" patterns are established in cases of normal, easy, or untraumatic birth; and how patterns are imprinted at conception and during the prenatal period.

I suggest that the imagery of *skins, clothing,* or *coverings* (or *koshas*—sheaths, as in the Indian Yoga tradition) makes a better analogy than armor for the ensemble of characteristic protective/defensive complexes that we receive in our formative years of development and carry with us throughout life. For one thing, it's an organic rather than a mechanical metaphor. Another advantage of the clothing or covering metaphor is that it provides us with an image of the goal of bio-energetic psychotherapy, whereas it's difficult to conceive how one would function in the world if all one's armoring is dissolved or removed, it makes sense that one's character clothing or coverings could become comfortable, flexible and appropriate, rather than rigid and hard, like metal. In my discussion of the metaphors of armor, clothing, veils and the like, in *The Unfolding Self,* I quoted the following passage from Meister Eckhart: "*A man has many skins in himself, covering the depths of his heart. Man knows so many things, but he does not know himself. Why, thirty or forty skins or hides, just like an ox's or bear's, so thick and hard, cover the soul.*"[17]

A poetic expression of this same metaphor is found in some lines from William Blake's aphoristic *Auguries of Innocence*:

> Joy and Woe are woven fine,
> A clothing for the soul divine.
> Under every grief & pine
> Runs a joy with silken twine.

I've come to appreciate this analogy of skins or coverings of the psyche in thinking about prenatal imprints: I think of them as a series of body-clinging wetsuits, built into cellular tissues. The coverings and skins imprinted at conception and in the prenatal period are the first and earliest, and since all subsequent cells in the body are generated from that fertilized and implanted ovum, these earliest ones are the most deep-seated and pervasive. It therefore makes sense that when we resolve them, loosen them, make them flexible and comfortable, all subsequent superimposed coverings and armorings would also be more easily dissolved or resolved.

Incarnational Choice: The Journey of the Prodigal Soul

Stanislav Grof, in his analysis of experiences in LSD psychotherapy and holotropic breathwork, found himself increasingly impressed by the prevalence of transpersonal elements in the imagery and thought processes of persons reliving perinatal traumas. Similarly, some of the psychologists working with hypnotic regression to the prenatal period and conception found themselves confronted by imagery and thoughts that seemed like memories from a time before conception, before the beginnings of biological life. In other words, these experiences relate to the *level of soul*—that mysterious being or essence that "comes down," as it were, and incarnates in human form (personality and body). William Emerson identified two distinct feeling-complexes, coming from pre-conception, that can have long-lasting effects in a person's life until resolved. One he called "divine homesickness," a feeling of wanting to return back to the heaven world from which one came; and the other "divine exile," a feeling of having been abandoned or exiled by God.[18] He adds, "However literally or symbolically one interprets these reports, they do correlate with how people experience their lives. Thus, although I do not believe that God actually throws anyone out, people who perceive God as having done this to them are likely to live their lives as exiles until their perception of God is healed."[19]

It is clear that we are here leaving the realm of commonly accepted psychological theory, "evidence" as commonly understood, conventional Christian religious teaching, and even the Western worldview. The theme of human life as exile and alienation, at a basic existential spiritual level, is, however, very prevalent in the Gnostic literature of the first few centuries of the Common Era.[20]

A growing body of literature from birth professionals and psychotherapists deals with this pre-conception level of soul incarnation, collecting stories of spiritual communications between parents and unborn children, and memories of the soul's pre-conception existence.[21] My own statements here, though dealing with transpersonal or spiritual realms, are empirically based, in that they are made on the basis of my decade-long experiential study in a contemporary Western mystery school, as well as in shamanic, yogic, and alchemical traditions with a variety of teachers. They are further based on my observations and those of others, both individually and in groups, in deep non-ordinary states of consciousness, over the past forty years. I have also found confirmatory observations in the writings of reincarnation therapists who have increasingly focused on the "interlife" period between death and rebirth;[22] in the writings of some highly experienced clairvoyant psychics,[23] and, to my surprise, in some strands of Western and Eastern esoteric teachings, to which I will return below.

I have come to understand that patterns in our psychic make-up that seemingly come from a time "before" conception, actually come from a "place" in consciousness, outside of time and space. I call this place the *soul communion* or *soul council*—where the decision is made to incarnate, in total agreement between the three souls of mother, father, and child, and sometimes the presence of other ancestors and elders as well. There is unconditional love and acceptance of one another, and a deep knowing of the prime purpose or mission of one's life. In addition, there may be a knowing of other roles played by these souls in other lives—the one who is now child was perhaps father or spouse. When people tune in to this place of soul communion and know-feel-sense that they have, with high spiritual awareness, *chosen* to be here, on Earth, in this particular life for a particular purpose (or purposes)—this can be a life-changing realization that puts all other striving and frustration in perspective. I believe this is the moment of choice and freedom pointed to by the famous Zen koan: What was your original face before you were born? In other words, where were you facing, what was your original intention? Koans are not meant to be answered—just asked, repeatedly. The act of asking opens up awareness to higher purpose. My translation of this koan for our time is: *What was—and is—your soul's vision for this incarnation?*

This place of soul communion, out of which comes the choice to incarnate, is "pre-conception" when seen from the perspective of ordinary linear time; hence we can get there by regressive remembering. But it is also still present now, until the end of our present incarnation; hence we can get there by direct divination, and then bring that knowing-feeling-sensing awareness "down" through and into the mental, emotional, perceptual, and physical personality systems (a process known as *soul infusion*). The highly skilled medical intuitive Caroline Myss calls the agreement made between souls the *sacred contract,* relating major types of soul purpose to Zodiac signs.[24] She says, "a Sacred Contract is an agreement your soul makes before you are born." Sylvia Browne describes the process of choice as the designing of a *chart,* featuring major life events and circumstances in a kind of advanced scenario.[25]

In addition to the shared knowing and acceptance of the soul's purpose, there is a second knowing that can be asked for from this council of souls and elders at the moment of incarnational choice: and that is the question of why the soul chose this particular family, this mother and father. In individual therapy, this can be an astonishingly revealing and liberating question to ask. I will give an example from my own process to illustrate the kind of insight that can occur when reflecting on the question of how choosing one's parents might reflect some aspect of the soul's mission. My parents came from extremely divergent backgrounds—in class, education, and nationality. My father was an intellectual German publisher with a doctorate in political science; my mother came from a working-class family in the Scottish lowlands, who didn't have a college education, and who didn't speak the German language. They met and married when both were working at the League of Nations in Geneva, reflecting their shared passion for cross-national, cross-boundary communication and peacemaking. My brothers and I grew up living in Hitler's Germany, at a time when our parents' respective countries were at war with one another. When I was an adolescent, I used to think I should become a diplomat. I've come to understand that my life purpose does have to do with the crossing of boundaries and resolving of differences—between worlds of consciousness, more than nations (though the two are also linked).

Sometimes, the soul's choice of parents will seemingly reflect a working out of some karmic entanglement or indebtedness from another existence. A client may say, "Choose my parents? Never in a thousand years would I choose those people." I have to remind them we are speaking of soul, not personality. And sometimes, it seems as if the soul chose a deliberately difficult situation to be born into, to learn from. A woman I worked with found that her mission was to "learn to love," and she had chosen a family in which her mother hated her with great cruelty. She had chosen to learn to love in a situation where love was hardly manifested to her as a child. *Souls love challenges,* apparently; they often do not take an easy path. Perhaps this is because the greater the difficulties, the greater the learning.

The soul, all three souls (of mother, father, and child) know, at that level, that once we incarnate in the time-space world of Earth conditioning, all bets are off, so to speak—everything, all knowledge of our origin and mission, may be forgotten, even the existence of the soul may be denied or buried in the unconscious personality layers of mother and father. That's the challenge, the risk, the testing, and the learning. As soon as the first step into biological form is taken, at conception, the veils of forgetting, the sheaths of conditioning, come in and layer on during the entire prenatal epoch, culminating with great intensity at the trauma of birth, especially if this occurs with little or no consideration for the consciousness and spiritual nature of the child. The processes of conditioning continue of course, during infancy and childhood, the generally acknowledged formative years, in which inherited predispositions are combined with learned reaction patterns. This understanding of how the processes of conditioning apply layers of unconscious defensive reaction patterns, starting from the time of conception, and therefore already quite developed at the time of birth, is consistent with the teachings of both Vedanta and Buddhism, which claim that the default condition of the human being at birth is unconsciousness *(avidya)*. This Sanskrit term, literally "not-knowing," and sometimes misleadingly translated as "ignorance" or "delusion," actually refers to the non-consciousness by the neonate of their true spiritual origin, as a soul, a child of God.

And yet, at every moment of this whole process, the possibility of remembering our true origin, re-connecting to our soul and its purpose arises. The consciously conceived Indigo children, we are told, often speak spontaneously of their purpose in life, and of memories from before this life. These soul-memories tend to fade progressively with age. Emerson and his colleagues tell the beautiful story of a little girl who kept asking to be alone with her newborn sibling. At first, the parents were worried that she might harm the baby out of sibling jealousy. They finally agreed to the girl's request, but listened via intercom from the next room. After a period of silence, they heard their daughter say to the baby, "Tell me about Heaven. I'm beginning to forget."[26] In my divination ritual workshops, I've observed that sometimes, tuning in to the memory of one's earliest experience of the bonding-gazing between newborn infant and mother can lead directly to the recognition and remembrance of the communion of two souls. On the other hand, some people, when asked to recall the first bonding-gazing experience, will only remember sensing anxiety, coldness or even hatred in the mother's eyes (which is a traumatic experience that then needs to be healed).

In his poem *Intimations of Immortality from Recollections of Early Childhood*, William Wordsworth expresses beautifully the spiritual origin of the soul and the progressive, but not complete, forgetting of the formative years:

> Our birth is but a sleep and a forgetting
> The Soul that rises with us, our life's Star
> Hath had elsewhere its setting,
> And cometh from afar:
> Not in entire forgetfulness,
> And not in utter nakedness,
> But trailing clouds of glory do we come
> From God, who is our home;
> Heaven lies about us in our infancy!

A summary overview of what we have learned from pre- and perinatal psychology, as well as traditional esoteric teachings, about the process of incarnation and the prenatal period is shown in the chart "Prenatal Formative Influences—Predispositions (*Samskaras*)." *Samskaras* (from *sam*- "together," and *kri*—"make") are associative patterns, markings, tracks or traces, that predispose us to think, feel, and perceive in particular ways. For example, genetic predispositions to certain illnesses are *samskaras,* as are genetic resilience, or health. There are also familial, ancestral and ethnic traditions, attitudes, talents, and gifts. The chart also shows how physical environmental factors (e.g., climate, health of the mother), social factors (e.g., the family's presence in war or impoverished conditions), and familial factors (e.g., the father's leaving, or a grandparent dying) can play a significant role in affecting prenatal development. Finally, the planetary transits (studied by Richard Tarnas with Stan Grof for their impact on transformations of consciousness) can be seen as an extension of ecological relational factors beyond the Earth. Their symbolic indicator role in the crucial perinatal period is, of course, the foundation of the ancient divination art of astrology.

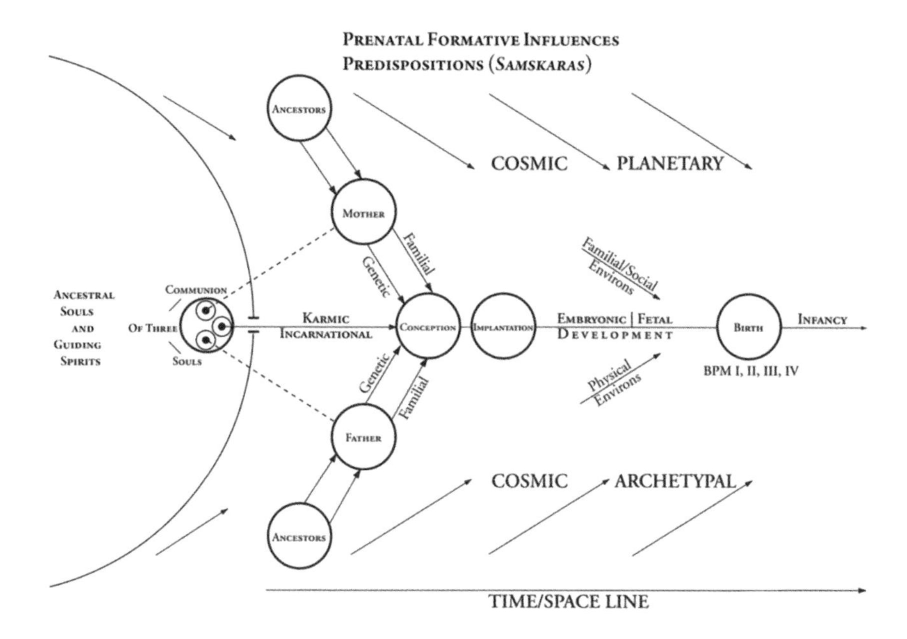

The Prenatal Epoch in the Jewish Wisdom Literature: The Angel Lailah

In the vast literature of Jewish mystical teaching stories referred to as *midrashim*, there are references to the activities of an "angel of conception," also called "midwife of souls," which illuminate the points made here about the process of incarnation. I first heard about this angel Lailah in an article on prenatal psychology by Thomas Armstrong, quoting from Howard Schwartz's collection of such tales, called *Gabriel's Palace*;[27] Michael Ziegler also referred me to some related passages in *The Book of Legends*.[28] In correspondence with Daniel Matt, translator of the most recent edition of the medieval *Zohar—The Book of Splendor*,[29] he related to me another variant of this story, which emphasizes the androgyny of the soul. Daniel Matt also pointed out that the name *Lailah* means "Night" in Hebrew, and that this angel is elsewhere in the Zohar identified with Gabriel. In the Gospel stories, Gabriel is the messenger of the Annunciation, announcing the conception of John to Zachariah[30] and of the Christ child to Mary.[31]

The two older versions of the story emphasize the reluctance or resistance of the soul to entering into human life, which seems to contradict the notion of incarnational choice. It is, however, consistent with the observation that, associated with the memory of our original abode, there is often a kind of divine homesickness or sense of alienation.

> Among the angels there is one who serves as midwife of souls. This is Lailah, the angel of conception. When the time has come for conception, Lailah seeks out a certain soul hidden in the Garden of Eden and commands it to enter the seed. The soul is always reluctant, for it still remembers the pain of being born, and it prefers to remain pure. But Lailah compels the soul to obey, and that is how new life comes into being.
>
> While the infant grows in the womb, Lailah watches over it, reading the unborn child the history of its soul. All the while a light shines upon the head of the child, by which it sees from one end of the world to the other. And Lailah shows the child the rewards of the Garden of Eden, and the punishments of Gehenna.[32]

So here is the image of the soul reluctantly leaving its heavenly abode, and being sent on a mission into the world. The soul is assisted by angelic guide, who tells it about its previous incarnations, and provides it with divine foreknowledge, as well as a scenario of likely consequences of good and bad actions. From our present understanding, the only amendment I would suggest to this vision is that the guidance from Lailah even precedes the actual conception (entering into the seed) and then continues during the prenatal period.

In another version of the story, we are told that God predetermines the characteristics and gifts we bring with us, but the incarnating soul chooses between right and wrong. The "angel in charge of spirits" functions as both guide, instructor, and protector:

> Before the formation of the embryo in its mother's womb, the Holy One decrees what it is to be in the end—male or female, weak or strong, poor or rich, short or tall, ungainly or handsome, scrawny or fat, humble or insolent. He also decrees what is to happen to it. But not whether it is to be righteous or wicked, a matter He places solely in the man's power. He beckons the angel in charge of spirits…and the angel goes and brings the spirit to the Holy One.…The Holy One makes the

> spirit enter the drop against its will. Then the angel returns and has the spirit enter the mother's womb. Moreover, two angels are designated for the spirit to guard it, so that it will not leave the embryo or fall out of it. There a lamp is lit over its head, and it is able to look and see from world's end to world's end.[33]

The Zohar's version of this extraordinary tale does not mention any reluctance on the part of the soul to enter into form, but does make it clear that the soul is originally androgynous, and only becomes polarized as male or female on the descent into human form. The two then become fused into unity at conception (coupling) and become one body.

Daniel Matt interprets the coming together of male and female as referring to the interpersonal meeting of male and female soul mates.[34] However, the mention of "right and left," which is the energetic polarization of the human body in the esoteric traditions of both yoga and alchemy, suggests that it is the intrapsychic androgyny of male and female human beings that is being referred to here.

One of the most fascinating details of the conception story involving Lailah occurs only in the version reported by Schwartz, said to have originated in the Babylonian period around the 9th century BCE. After using the light above the head to show the unborn the rewards of the Garden of Eden and the punishments of Gehenna (as cited above), "When the time has come to be born, Lailah extinguishes the light and brings forth the child into the world, and as it is brought forth, it cries. Then Lailah lightly strikes the newborn above the lip, causing it to forget all it has learned. And that is the origin of this mark, which everyone bears."[35] The other variant of the story also mentions that the birth is painful, and this pain is the cause of the soul forgetting its origin. This is certainly consistent with the observations from the studies of the effects of birth trauma—that it causes massive amnesia. However, according to folklore, touching with the finger on the philtrum above the upper lip is a gesture we make when trying to remember something. It occurred to me that the gesture of Lailah, the angel that guides the soul into this world (and from this world at death), should really be considered a gesture to help us remember our origin when we are *in extremis:* in other words, the *philtrum is a point of remembrance, not forgetting.*

In relating this observation to a friend and acupuncturist Susan Fox, she pointed out to me that in traditional Chinese medicine, this point at the philtrum, above the upper lip just below the nose, is Governor Vessel .26, and is considered a point of "consciousness," to be stimulated to revive someone unconscious. The Governor Vessel meridian is the central *yang* channel which comes up the back from the tailbone, over the top of the head, the nose, and then ends under the lip; the central *yin* Conception Vessel comes up from the pubic bone and terminates at the lower lip. Thus, this point, GV.26, is said to establish the connection between Heaven (*yang*) and Earth (*yin*) and is also called "man's middle," since man (the human) is situated between Heaven and Earth. According to the *Manual of Acupuncture,* stimulating this point "restores consciousness and calms the spirit." Among its indications are "sudden loss of consciousness, coma, acute and chronic childhood fright, ...mania-depression, epilepsy, (and) wasting and thirsting disorder." The authors say, "when the harmonious interaction of *yin* and *yang* is lost and they begin to separate, there is loss of consciousness (death being the ultimate manifestation of this separation).... Renzhong DU-26 was indicated for resuscitation, and it is the single most important acupuncture point to revive consciousness and re-establish *yin-yang* harmony."[36]

With this unexpected confirmation from a completely different and independent system of subtle energy anatomy, I feel there is some support for interpreting the action of Lailah, the guiding angel of incarnation and conception, as setting up for us a gesture of conscious remembering, for reminding us our true nature as souls. One could practice it that way.

Phenomenological Parallels between Birthing and Dying

A cartoon published in the newsletter of the APPPAH shows twin fetuses hanging side by side in the womb vessel prior to birth. One of them asks: "Is there life after birth?" The other one replies, "We don't know. No one has come back to tell us." The joke points to the curious parallelism between birthing and dying, as well as our attitudes to them. Though mainstream science and medicine have long denied the reality of consciousness before birth, this is now gradually gaining acceptance due to the work of researchers in pre- and perinatal psychology. The possibility of consciousness before conception, and of consciousness after death, the traditional realms of the soul and of spirit, pose even more radical challenges

to the worldview of materialist science—though not perhaps to common understanding and beliefs.

In his investigations of the phenomenology of perinatal memories in psychedelic and holotropic states, Stan Grof observed that the imagery of birthing and dying seemed to be deeply intermingled, especially in experiences related to the second and third stages of birth (BPM II and III). In *Psychology of the Future,* he pointed out how the experiential association of birth and death imagery has a quite realistic basis in the physical facts of the birth process.

> The intimate connection between birth and death in the unconscious psyche makes eminent sense. It reflects the fact that birth is a potential or actual life-threatening event. The delivery brutally terminates the intrauterine existence of the fetus. He or she "dies" as an aquatic organism and is born as an air-breathing, physiologically and even anatomically different form of life. And the passage through the birth canal is, in and of itself, a difficult and possibly life-threatening event. Various complications of birth...can further increase the emotional and physical challenges associated with this process. The child and the mother can actually lose their lives during delivery, and

children might be born blue from asphyxiation, or even dead and in need of resuscitation.[37]

Grof's statements probably express a virtual consensus among birth researchers and psychotherapists working in the field of perinatal regression therapy. I have myself made the same observation numerous times in my own self-exploration and altered states divination work with individuals and groups. What in the literature on psychedelic states is sometimes expressed as "ego-death," i.e., the sense that "I," myself, am dying, can occur in the context of perinatal regression. You realize then that, in fact, at birth (even relatively non-traumatic birth), the fetal self dies, and an infant self is born.

Buddhist, Hindu, and other esoteric teachings of reincarnation and rebirth speak of a sequence of death, an intermediate phase in the afterlife, and then rebirth. Elaborate cartographies exist in many cultures, so-called Books of the Dead, describing the landscape of the afterlife; and Grof has made extensive studies of this literature, relating it to experiences of shamanic initiation and death-rebirth mysteries in non-ordinary states.[38] In some ways, the close phenomenological association of birthing and dying is perhaps even more intimate than as envisioned in these texts. We see here a *simultaneity of birthing and dying*: the passage through the birth canal, in birth or rebirth process, is at once the death of the fetal self and the birth of the neonate self. In the Jewish mystical tales of Lailah, discussed above, this angel is not only the guide and escort for souls destined to be conceived and born, but also for souls that are dying and returning to the heavenly home from which they originated. "Lailah is a guardian angel who watches over us all of our days. And when the time has come to take leave of this world, it is Lailah who leads us to the World to Come."[39]

The Prenatal Epoch in Tibetan Buddhism: The Bardo of Seeking Rebirth

Undoubtedly, the best-known text in world religious literature concerning the afterlife and the process of reincarnation is the *Bardo Thödol*, popularly known as *The Tibetan Book of the Dead*. This book, in a translation by Lama Kazi Dawa-Samdup and W.Y. Evans-Wentz, with introductions by C.G. Jung and Lama Anagarika Govinda,[40] served as the text which Leary, Alpert, and myself used as the basis for a metaphorical manual for the psychedelic experience.[41] The original *Bardo Thödol* is attributed to the legendary 8th century Indian Buddhist adept Padmasambhava, who brought Buddhism to Tibet. According to Buddhist scholar Robert Thurman, who has published a more recent translation and commentary,[42] the title of the work should more accurately read *The Great Book of Natural Liberation Through Understanding in the Between*. "Between" is Thurman's preferred translation of the term *bardo,* which has also been called "intermediate state," and corresponds, in my view, to what some might call a "state of consciousness." A state of consciousness is a division of time *between* two transition points—for example, "sleep" is the state between falling asleep and waking up.

Tibetan Buddhism teaches that there are six bardo states that we all experience, three between birth and death, and three between death and rebirth in the subsequent incarnation. It is with the latter three that the *Bardo Thödol* mostly concerns itself—giving detailed and explicit instructions on how people, both highly proficient yogic practitioners and ordinary people with no particular yogic aptitude, can be helped to find their way through the confusing and terrifying afterlife states and come to make the most favorable kind of rebirth possible. The *Book of Liberation through Understanding the Bardo States* teaches that liberation from the *samsaric* round of conditioned existence can occur in, or from, any of the bardo states, *if* we understand and remember the teachings, recognize the bardo state we are in, and choose the most enlightened conscious option available to us. It is for this reason that Buddhist teachers refer to it not only as a book of preparation for dying, but really a profound guidebook for both living and dying. Included in the *Bardo Thödol* is a set of poetic meditations for "waking up" in each of the six in-between states, called the "Root Verses of the Six Bardos." There is also a text called the *Six Yogas of the Bardos*, for example, the yoga for the *bardo of dreams* includes instructions for developing what Western dream researchers call lucid dreaming.

Although the *Bardo Thödol* does not explicitly mention the prenatal epoch as such, we can find in the teachings of the after-death bardo states some suggestive parallels with findings from the work of prenatal and past-life regression therapists and highly developed intuitives.[43] As discussed above, the work of Emerson and others (myself included) has uncovered what appear to be cellular memories of conception and perceptions of the father and mother at the time of conception; past life therapists speak of a series of learning experiences undergone by the soul during the *interlife* period (and remembered in deep trance states); and a choice point I call "soul council" at which the decision to incarnate is made, according to one's intentions, choosing the parents and circumstances of one's incarnation or rebirth.

The teachings of the *Bardo Thödol,* in outline, are that immediately at death, in the *bardo of the moment of dying,* highly proficient meditators who can maintain one-pointed concentration will be able to attain liberation through the "Clear Light." Most people are not able to concentrate, get caught in fear and confusion and enter then into the second phase, called the *bardo of the experiencing of reality,* in which there are elaborate visions of "peaceful and wrathful deities," depicted in the fantastic iconography of Tibetan Buddhism. The deceased are repeatedly reminded, by his or her attendant, not to be overwhelmed by either the heavenly or the hellish visions, but to remember that they are all projections of one's own mind. Due to lack of training and/or preparation on the part of most ordinary people, the bardo traveler, after repeatedly lapsing into unconsciousness, then finds himself in the third phase, the *bardo of seeking rebirth,* in which he wanders about seeking to orient himself again to ordinary existence. Thurman's translation calls this phase the "existence between."

In this phase, the bardo traveler is repeatedly admonished to remember where he is and that his thoughts and intentions will profoundly affect the kind of experience he/she may have. He is told that he is not in his ordinary body, but a "mental body," or "*bardo* body," or "desire body," which can't be killed, but which can fly, pass through walls, and have all kinds of supernormal capacities. In other words, he/she is *in* what are also called, in esoteric traditions, the "intermediate planes" (including astral, mental, and so forth). The deceased is then reminded of the six possible worlds of *samsara* (existence) into which he might find him or herself drifting, carried along by the karmic propensities of their previous existence. They are advised to avoid the hell worlds, and the worlds of *pretas* (hungry ghosts) and *asuras* (demi-gods), but, if unavoidable, to go with the heavenly worlds of benign spirits (*devas*) or the human world—considered to be offering the best opportunities for liberation.[44] "If thou art to be born on a higher plane, the vision of that higher plane will be dawning upon thee."[45] And, "in whatever continent or place thou art to be born, the signs of that birthplace will shine upon thee then."[46]

Then follow a series of instructions on how to "close the womb-door," the point here being to delay the rebirth as long as possible, so that one can prolong one's stay in the higher planes and avoid being sucked into unfavorable birth forms by one's karmic propensities *(samskaras)*. The first method of closing the womb-door is to remember that you are in this *bardo* of rebirth and focus on positive intentions: "holding in mind one single resolution, persist in joining up the chain of good karma...this is a time when earnestness and pure love are necessary."[47] The second method of closing the womb-door is used as the deceased has visions of "males and females coupling;" he/she is to think of them as a divine Father-Mother pair and withhold from joining them.

The third, fourth, and fifth methods of closing the womb-door involve further ways of dealing with the visions of a man and woman copulating. I suggest that *this is the vision of one's own conception,* which can be reached in prenatal regression work, and which here is arrived at from the other side, as it were. The *Bardo Thödol* then says that if the voyager feels an attraction to the female and aversion to the male, he will be reborn as a male; and if attraction to the male and aversion to the female, she will be born as a female.

Commenting on the apparent Oedipal complex being described, C.G. Jung, in his introduction to the Evans-Wentz translation, pointed out that this is the point where Western (Freudian) psychology connects with Buddhist rebirth teachings "as if from below."

If even after using the various methods of preventing or postponing rebirth by meditating with conscious intention on light, and on one's chosen deity, one is still drawn into a womb for birth, the deceased is given instructions for "choosing of the womb-door." First, there are "premonitory visions of the place of rebirth"[48]—the continents in four directions are described, where one might be born. There are renewed instructions for going for the world of *devas,* and avoiding the hell world or *preta* world. "Since thou now possesseth a slender supernormal power of foreknowledge, all the places of birth will be known to thee, one after another. Choose accordingly."[49] One might still choose a supernormal birth into a paradise realm, or if that is not possible, the deceased are advised to use their foresight to choose a human birth in an area in which religion prevails.

> Since it is important to direct the wish, direct it thus: I ought to take birth as a universal emperor; or as a Brahmin; or as the son of an adept in *siddhic* powers; or in a spotless hierarchical line; or in the caste of a man who is filled with religious faith; and, being born so, be endowed with great merit so as to be able to serve all sentient beings. Thinking thus, direct thy wish, and enter into the womb. At the same time, emit thy gift-waves of grace upon the womb thou art entering, transforming it into a celestial mansion.[50]

To summarize, the instructions for the most favorable kind of birth are: to delay the return from the light- and wisdom-filled heavenly worlds as long as possible, and when the time comes, which you know by seeing the acts of conception between men and women, to choose a birth family where the likelihood of coming into contact with the *dharma* teachings are greatest. The ending of the interlife period is the beginning of the bardo of rebirth—the decision is made to reincarnate, in a blending of karmic tendencies and conscious choice, and conception takes place in a fleshly human womb. This *rebirth phase then ends with the actual physical birth*, nine months later, when we start cycling through the three bardos of waking life, dreaming, and meditating. In conclusion, below is my version of the Root-Verse for the bardo of rebirth:

> Now, as the bardo of rebirth dawns upon me,
> I will hold one-pointedly to a single wish –
> To continue directing intention with a positive outlook.
> Delaying the return to Earth-Life as long as possible.
> I will concentrate on pure energy and love,
> And cast-off jealousy while meditating on the Guru
> Father-Mother.

Journeys Beyond Space and Time

Fritjof Capra

I met Stan Grof in early 1977, two years after the publication of my first book, *The Tao of Physics*.[1] The book had been an unexpected success, and I was invited to give lectures and seminars to professional organizations from many disciplines and had discussions with people from all walks of life.

Surprisingly, there were many psychologists and psychotherapists among them who told me that the kind of paradigm shift I discussed in *The Tao of Physics*—the shift from a mechanistic to a holistic worldview—was happening also in their field. In many of these discussions, the name Stan Grof was mentioned, and it was often suggested to me that I should meet this man who was an important figure in the human potential movement and entertained ideas about science and spirituality that were close to my own. So, I became very curious about Stan and was delighted when I received an invitation to a small gathering in a private home in the Bay Area where Stan would present a summary of his work.

When I met Stan at that gathering, I was very surprised. I had always heard people refer to him as "Stan Grof," and it never occurred to me that his full name is Stanislav. I expected to meet a Californian psychologist, and when I shook hands with him, I realized to my great surprise that not only was he European, but that he came from a cultural background very close to my own. His native Prague and my native Vienna are a mere hundred miles apart, and our countries share a long common history during which the two cultures blended considerably. Meeting Stan, therefore, felt somewhat like meeting a distant cousin, which created an immediate bond that would later turn into a close friendship.

During the months following our first meeting, I visited Stan several times at the Esalen Institute where he lived. We had many substantial discussions and soon planned to give a series of joint seminars, which we called "Journeys Beyond Space and Time." What we had in mind was an outer journey into the realms of subatomic physics and an inner journey into the realms of the unconscious.

The seminars were very successful and generated great excitement. I would begin by presenting the perception of reality that emerged in atomic and subatomic physics, where the solid objects of our everyday experience dissolve into energy patterns, where particles can travel backward and forward in time, and where these energy patterns, moreover, are intrinsically dynamic, the whole universe being engaged in a cosmic dance of energy. Stan would follow by describing very similar perceptions emerging in psychedelic, transpersonal, and mystical experiences, and we would then compare the worldviews emerging from these adventures.

In doing so, we were able to compare three perspectives that lead to essentially the same vision of reality: the mystical experience, the transpersonal experience, and the experience of physicists in those subatomic experiments. What the physicists discovered to their great surprise and shock was that, at the subatomic level, there are no separate objects. A subatomic particle, like an electron or a proton, for a physicist, is not an object but a set of relationships that reach outward to other things; and these other things turn out to be again sets of relationships, and you never end up with any "things" at all; you always end up with relationships. The world of subatomic physics is a world of patterns of relationships, a world of processes. Today, this realization has become part of a general paradigm shift in the sciences and in society at large. We no longer see the world as a machine made of some kind of basic building blocks. We are beginning to understand it as a network of inseparable patterns of relationships. That fundamental shift from seeing the world as a machine to understanding it as a network, which in science occurred first in physics, has now occurred also in biology, medicine, economics, and in various other sciences. In all these fields, a new conception of life has emerged.

During the last forty years, I developed a synthesis of this new understanding of life; a conceptual framework that integrates four dimensions of life: the biological, the cognitive, the social, and the ecological. I presented summaries of this framework as it evolved in several books, beginning with a first outline in *The Turning Point*, published in 1982, to which Stan contributed significantly.[2] My final synthesis was published in 2014 in a textbook titled *The Systems View of Life* and coauthored by Pier Luigi Luisi.[3] From its early formulation in *The Turning Point*, I have called my synthesis "the systems view of life" because it requires a new kind of thinking—thinking in terms of relationships, patterns, and context. In science, this is now known as systemic thinking, or "systems thinking."

One of Stan's many great contributions has been to give this new conception of life, this new perception of reality, an experiential content. When people have transpersonal experiences, seeing the world in terms of patterns of relationships is not an intellectual exercise. It is a deep and highly emotional experience. Stan has helped people from all walks of life to have these experiences and to integrate them into their lives.

Stan's influence on my work in the late 1970s and early 1980s was enormous. When I branched out beyond physics into various other fields, I soon realized that I could not explore the emerging new paradigm in those fields on my own, because I had no expertise in biology, medicine, psychology, or economics. I would not even know where to start. So I adopted a technique—and this has stayed with me until today—of learning things from people through dialogues. I developed a talent for seeking out and recognizing individuals who were experts in their fields, systemic thinkers, and who shared my general worldview and values, and Stan was one of them.

I actually realized that right away when I first met Stan. At that gathering in 1977, he presented the cartography of the unconscious he had developed, which allowed him to integrate various schools of psychology and psychotherapy.[4] These schools were often contradictory, but Stan saw them addressing different levels of the human unconscious. So, I realized on that evening with great excitement that Stan not only knew a lot about psychology; he also had a framework to integrate these various schools of psychology. For me, this was tremendously helpful, and there was more to it.

My basic premise for *The Turning Point*, the book I was working on at the time, was that the sciences in the eighteenth and nineteenth centuries had all modeled themselves after Newtonian physics; they had adopted a mechanistic outlook. In the early twentieth century, the mechanistic worldview was replaced by a holistic (or systemic) worldview in physics, and in the late twentieth century, a similar change of paradigms was emerging in the other sciences.

In psychology, it is quite easy to see how behaviorism—one of the main schools of twentieth century psychology—had a mechanistic outlook. Instead of studying consciousness, behaviorists more or less eliminated consciousness and concentrated purely on behavior, analyzing behavior in very mechanistic terms, in terms of reflexes, and so on. That is fairly easy to recognize, but to see how Freud modeled psychoanalysis after Newtonian physics is much more subtle; and that is what Stan helped me to do. Stan practically tutored me in the basics of psychoanalysis, and he showed me how Freud's theory in very subtle ways was also modeled after Newtonian physics.

Furthermore, the whole notion of mental illness that Stan developed was very influential on my thinking and has remained so even today. Stan distinguishes between two types of human experiences. One is our everyday experience of reality—which we may call a Cartesian experience, where our environment is made up of separate objects; time flows in a linear way, and there are clear cause-and-effect relationships, and so on. Then, we may also have transpersonal experiences, where there are no separate objects but only webs of relationships; where everything is moving, vibrating, dancing, and there are no strict cause-and-effect relationships. So human beings can have these two types of experiences; and, according to Stan, mental illness occurs when people cannot integrate those two experiences properly. This means that the way to recognize mental illness is not by the contents of the experience, but rather by the inability of a person to integrate experiences.

This is very revolutionary and very important for therapy because still today, most psychiatrists try to suppress their patients' psychotic experiences with drugs. What Stan says is that you should not suppress them. On the contrary, you should allow the experiences to emerge and come to a conclusion and integration. Stan sees symptoms of mental illness as "frozen" experiences, which need to be "unfrozen." He would do this initially with LSD and then later on with other techniques, such as holotropic breathing.

When I first learned about Stan's view of mental illness, this was very exciting to me because around the same time, I also had many discussions with R. D. Laing, the Scottish psychiatrist who developed a revolutionary approach to the treatment of mental illness and who had the same view as Stan. He expressed it slightly differently, but it was exactly the same idea.[5] Laing had an activist approach, protesting against the habits of psychiatric hospitals of constraining patients against their will and not recognizing them as authentic human beings. So, for me, this was a very exciting time in the late 1970s and early 1980s, when Stan and I had many in-depth discussions and saw each other frequently.

In those days, Stan often invited me to a very special seminar that he and his wife Christina organized at Esalen. It was a month-long seminar, widely known simply as "the Grof monthlong." During four weeks, a group of about two dozen participants lived together in Esalen's "Big House" and interacted with a string of outstanding guest lecturers who would come for two or three days each, often overlapping and interacting with one another. The unique feature of the Grof monthlongs was that Stan and Christina offered the participants not only intellectual enrichment through stimulating and challenging discussions, but also experiential contact with the ideas discussed through art, meditative practice, ritual, and other nonrational modes of cognition. A special feature was sessions of holotropic breathing, just called "Grof breathing" at the time. The participants lived together during that month as a learning community, almost like a tribe. I participated in many of these month-long seminars. This was an extraordinary time for me, a time of both personal and intellectual growth.

Since the 1970s and 1980s, when I had my formative discussions with Stan, there have been several major advances in the life sciences, and my synthesis of the systems view of life has evolved accordingly. One of the main innovations was the development of complexity theory, technically known as "nonlinear dynamics." It is a new nonlinear mathematics that allows scientists for the first time to describe and model the complexity of living systems mathematically.[6]

Another major innovation has been the so-called Santiago theory of cognition, which sees mind as a process of cognition and identifies it closely with the very process of life. In this view, mind and life are inseparably connected, and the relationship between mind and body is one between process and structure.[7]

Both of these radical innovations were developed in the 1970s but became widely known only much later. They do not negate anything Stan and I discussed during that time. In particular, the fact that Stan's work has added an experiential dimension to the systems view of life is still fully valid.

A major discovery in complexity theory, and in my view its most important discovery, is a process known as emergence. It is well-known that living systems generally remain in a stable state, even though energy and matter flow through them and their structures are continually changing. But every now and then such an open system will encounter a point of instability where there is either a breakdown or, more frequently, a spontaneous emergence of new forms of order.

This spontaneous emergence of order at critical points of instability, often referred to simply as "emergence," is one of the hallmarks of life. It has been recognized as the dynamic origin of development, learning, and evolution. In other words, creativity—the generation of new forms—is a key property of all living systems. Life constantly reaches out into novelty.

This process of emergence is exactly what Stan describes when he says that the symptoms of mental illness are frozen experiences, and that healing takes place when you unfreeze them. When that happens, a new state of being emerges from the completion and integration of these transpersonal experiences. This means that Stan's revolutionary view of mental illness is now supported by a theoretical model from one of the most advanced branches of contemporary science.

Psyche and Cosmos

Planetary Correlations with Extraordinary States of Consciousness

RICHARD TARNAS

So it comes to pass that, when we pursue an inquiry beyond a certain depth, we step out of the field of psychological categories and enter the sphere of the ultimate mysteries of life. The floorboards of the soul, to which we try to penetrate, fan open and reveal the starry firmament.
—Bruno Schulz

Background of Research

After many intensive years practicing psychotherapy using LSD and other psychedelic substances, first in Prague and later in Maryland, Stanislav Grof moved to Esalen Institute in Big Sur, California, in the fall of 1973 to work on the series of books that would summarize his clinical findings. Shortly after he arrived at Esalen, I joined him to work under his supervision on my doctoral dissertation on LSD psychotherapy. The move to Esalen turned out to be long-term and pivotal for both of us. During the greater part of the 1970s and 1980s, Stan served as Esalen's Scholar-in-Residence, organizing and teaching major month-long seminars with many distinguished scholars, while I, first as a staff member and later as Esalen's director of programs and education, collaborated with him on the research I will describe below. In 1993, we both joined the faculty of the California Institute of

Integral Studies in San Francisco, where we have taught graduate seminars in the Philosophy, Cosmology, and Consciousness program during the past three decades.

At the start of our work together at Esalen, we were especially interested in the radical variability of psychedelic experiences, a phenomenon widely observed but not well understood. From the beginning, researchers had noted a seeming paradox in which two individuals of similar clinical status could ingest the same substance, the same number of micrograms, in the same clinical setting, yet undergo extremely different experiences. One subject might have an experience of deep spiritual unity and mystical transcendence, while another who had received the identical substance and dosage might confront a state of metaphysical panic or bottomless despair that seemingly allowed no hope for release. Moreover, the same person at different times could have strikingly different psychedelic experiences from opposite ends of the experiential spectrum.

This variability also took another form, in which different individuals seemed to be constitutionally prone to encounter certain enduring constellations of related experiences—specific complexes, emotionally charged biographical memories, perinatal matrices, transpersonal events—in an evolving way through multiple psychedelic sessions, reflecting dominant elements or qualities appearing again and again in their life's personal journey. Each individual seemed to have his or her own characteristic set of enduring themes that, in the course of time, could take variable forms with either positive or negative inflections at multiple levels of consciousness, often during the same session.

Stan and his colleagues in Prague and Baltimore had long sought a reliable way to predict the nature and outcome of psychedelic sessions, hoping to find tools that would be helpful for anticipating how different individuals might react to psychedelic therapy and whether they would benefit from it. Yet years of research on the problem had been unsuccessful, as none of the standard psychological tests—the MMPI (Minnesota Multiphasic Personality Inventory), POI (Personal Orientation Inventory), TAT (Thematic Apperception Test), Rorschach Test, Wechsler Adult Intelligence Scale, and others—proved to have any predictive value for this particular purpose. For at least the second form of variability, involving the same person taking the same substance at different times, such an outcome was understandable, since retesting individuals with a standard psychological test typically does not change the results. In general, if one tests the same person today and tests again a month from today, the results will not significantly change, whereas if a subject takes LSD today and then takes the same dose next month, the session could be altogether different.

Nevertheless, given the extraordinary intensity of psychedelic experiences, the possibility of being able to anticipate even in general terms how different individuals might respond to such therapy, perhaps even how the same individual might respond at different times, inspired the hope that a useful method might someday be found. While we were not fully aware of it at the time, several decades earlier, C. G. Jung had opened up another possible approach to this variability in psychological experience. On the basis of lengthy studies of various esoteric systems, he came to regard astrology as providing a unique window for understanding the qualitative dimension of time, and specifically the archetypal dynamics at work at any particular time, including that of birth. He believed that time was not merely quantitative, an ongoing neutral or homogeneous continuum, but rather possessed an intrinsically qualitative dimension.

More surprisingly, he came to regard the qualitative dimension of time as intrinsically connected in some undetermined manner to the positions of the Sun, Moon, and planets relative to the Earth. As he wrote in *Memories, Dreams, Reflections*, "Our psyche is set up in accord with the structure of the universe, and what happens in the macrocosm likewise happens in the infinitesimal and most subjective reaches of the psyche."[1] In his later years, Jung came to employ the archetypal analysis of birth charts as a regular aspect of his analytical work with patients. Yet, given the intellectual climate of his time, and indeed of our own time to some extent, one can well appreciate his reluctance to make more public the extent of his use of astrology. He had perhaps already pushed the envelope of twentieth-century intellectual discourse as far as might reasonably be expected.

During the years of Stan's and my residence there, Esalen Institute was widely known as an experimental educational center where an unusually diverse range of perspectives and transformational practices was explored—Eastern and Western, ancient and modern, psychological, somatic, philosophical, scientific, shamanic, mystical, esoteric. Of all those perspectives and practices, astrology was perhaps the last one that Stan or I would have imagined seriously investigating. In modern intellectual culture, astrology has served as a kind of gold standard of superstition, that to which one compares something if one wants to emphasize how ludicrous it is, beneath serious discussion. In spite of our skeptical starting point, in early 1976, prompted by a suggestion from an Esalen seminar participant who had extensively studied astrology, we decided we should at least examine the evidence for possible correlations. The participant, an artist named Arne Trettevik, was especially focused on the study of planetary "transits," the ongoing movements of the planets from day to day, year to year, as they move into specific geometrical alignments with respect to an individual's birth chart. He studied the ways in which transits seemed to correspond with the different kinds of experiences people undergo in the course of life—periods especially marked by personal happiness or depression, for example, or falling in love, entering a new phase of life, sudden crises, and so forth. After hearing Stan's lectures, he had suggested that planetary transits might be similarly relevant for understanding the kinds of experiences people would have in the powerful states of consciousness catalyzed by psychedelic substances.

Stan in turn mentioned the idea to me, and subsequently Trettevik showed us how to calculate birth charts as well as transits, using the necessary reference works such as a planetary ephemeris, a world atlas with time-zone references, and the required mathematical tables. This was still in the years before personal computers were available, so each birth chart and transit calculation had to be done by hand. We also secured several standard astrological reference works that set out the characteristic meanings for various planetary combinations and their alignments as measured in celestial longitude along the ecliptic (for example, Saturn opposite the Sun, or Jupiter conjoining the Moon).[2] Because Stan and I both had records for our own LSD sessions over the years, including dates and major themes, we were able to compare our actual experiences retrospectively with the descriptions in the astrological texts of what kinds of events and experiences were supposedly likely to take place during the concurrent transits.

Initial Correlations

Much to our astonishment, we were highly impressed by both the quality of the correlations and their consistency. What we had experienced in our sessions during those transits seemed to be archetypally intensified versions of the more common life experiences that were generically described in the astrological texts. For example, based on the specific planets and alignment involved, the text might indicate that the period of a particular planetary transit was potentially an appropriate time for expanding one's intellectual horizons, learning new perspectives, or traveling to a distant country and discovering a new culture. It might indicate a period of potentially increased spiritual insight, or in the case of other transits, increased tensions and frustrations within one's career and life direction, or the emergence of problematic family issues. One transit might be described as coinciding with greater accident proneness and risk-taking impulses, while another might be characterized as indicating a greater potential for heightened anger or aggressiveness, depression, or generalized anxiety. These astrological text descriptions of more common circumstances and emotions proved helpful for us in gaining a sense of what underlying archetypal energies might be at work in each case.

In fact, I was very much struck by how the underlying archetypal nature of the astrological paradigm was apparent even in the many astrological texts that did not use a Jungian vocabulary or reflect a conscious relationship with the Platonic tradition or with its own deeper esoteric roots in which some form of the archetypal perspective was central. Each planet was understood as bearing an underlying cosmic association with a particular archetypal principle, which could express itself multivalently in diverse inflections and in different dimensions of life—psychologically, circumstantially, interpersonally, physically, spiritually, artistically, professionally, and so forth—but always with a clear connection to the essential nature of that archetypal complex. The correlations were not concretely predictive, but rather archetypally predictive.

Based on our records of experiences during such transits, it seemed that LSD sessions typically catalyzed more intense, often perinatal or transpersonal versions of the more common states and themes, the ordinary ups and downs of life, described in the standard astrological texts. During the psychedelic session, one might experience a sudden opening of consciousness to a much vaster view of reality, deep insight into another culture's religion or mythology, mystical awakening, or spiritual rebirth; or conversely, powerful states of cosmic aloneness, a sudden confrontation with the ruthless inevitability of human mortality, or an eruption of collective aggression and fear such as that activated within an entire nation at war. One factor that made the correlations much easier to recognize than expected was the fact that in psychedelic states, the archetypal qualities constellated during the session tended to be unmistakable because of their relative intensity—for example, not only feeling constrained or oppressed by one's particular life circumstances but undergoing a deep experiential identification with all people who have ever been imprisoned, enslaved, or oppressed. Most surprisingly, those archetypal qualities were intelligibly correlated with the individual's natal chart and current transits. On occasion, the experiential intensity within the psychedelic session could take the form of a direct experience of the archetypal dimension that underlay both the more ordinary conditions and the collective transpersonal experiences, with the particular mythic figures or archetypal powers encountered closely matching the specific principles that the astrological tradition associated with the relevant natal and transiting planets.

After this initial examination of our own sessions, we turned our attention to a larger range of individuals and their experiences, beginning with the fifty or sixty long-term members of the Esalen community who asked to have their own transits calculated and interpreted. Then we expanded the research to include the ongoing stream of scores of new and returning seminar participants who came to Esalen week after week. The institute was in fact an ideal laboratory for such research as thousands of people traveled there each year with the specific intention of pursuing deep self-exploration and potentially transformative experiences. Esalen was at that time a kind of epicenter of psychospiritual experiment and consciousness research. We therefore had available a substantial and continually growing database to work with. In addition to these current cases, we had access to case histories, session dates, and birth dates of a number of Stan's patients and subjects from previous years.

I should mention that, despite our initial impression that the evidence showed remarkable correlations, our understanding of the nature of those correlations arrived in halting stages that produced significant revisions in our first tentative conclusions. Over time, our grasp of the evidence underwent a definite evolution. We initially noticed a very general division in which transits involving certain planets and alignments seemed to coincide with easier, more smoothly resolved sessions, while other transits seemed to coincide with more difficult sessions that ended without a resolution. Then, more specific observations emerged concerning sessions that brought dramatic psychological and spiritual breakthroughs in comparison with others that remained stuck in agonizing "no-exit" situations. Eventually, through much trial and error, it became apparent that the initial impression of simple binary patterns masked a far more complex interplay of multiple transiting and natal factors that were involved in the fuller range of psychedelic experiences in their extraordinary diversity.

Correlations with Perinatal Experiences

An especially surprising finding from the early period of research concerned a remarkably robust correlation between the four basic perinatal matrices (BPMs) and the four outer-planet archetypes as described in the standard astrological texts. On the one hand, the complex phenomenology of each BPM had first been observed in psychedelic session reports by Stan in the mid-1960s, by which point he had also recognized the connection between these four dynamic constellations of experience and the successive stages of biological birth. On the other hand, working within a completely distinct tradition of research and interpretation going back many centuries, astrologers had gradually arrived at a strong consensus about the meanings of Saturn (the outermost planet known to the ancients), and in more recent centuries, Uranus, Neptune, and Pluto (discovered by telescope in the modern era). Fairly early in our research, I had noticed an apparent one-to-one general correspondence between experiences reflecting the four perinatal matrices and coinciding transits involving the four slower-moving outer planets.[3] To our amazement, upon closer reading of the astrological texts, it became apparent that every single feature of the four BPMs closely fit the widely accepted astrological meanings of the four outer planets. Because the perinatal category of correlations is typical of the kinds of archetypal correspondences we subsequently found involving the wider range of psychedelic experiences we researched, I will take a moment here to indicate the correspondences involved, comparing the phenomenology for each matrix as set forth in Stan's work with the standard planetary meanings delineated in the astrological literature. I will begin with BPM IV, the first perinatal matrix for which I noted this pattern.

The fourth perinatal matrix is associated both biologically and archetypally with the emergence from the birth canal and the moment of birth. It is reflected in experiences of sudden breakthrough, unexpected liberation, release from constriction and imprisonment, radiant brilliance of vision and understanding, awakening to a sense of deeper meaning and purpose in life, being flooded by intensely bright light, sudden intellectual and spiritual illumination, the feeling of being reborn after a long and dangerous passage, and so forth. In its negative aspect, when activated but uncompleted, BPM IV can take the form of manic inflation, restless impatience, eccentric ideation accompanied by a sense of unprecedented personal genius, insatiable craving for excitement, and compulsive hyperactivity.

Having observed a correlation between BPM IV experiences and major transits of Uranus, I was very much struck by how fully the set of symbolic meanings universally attributed to Uranus by contemporary astrologers coincides with the phenomenology of BPM IV. The astrological Uranus is typically described as the principle of sudden change, of unexpected openings and awakenings, creative breakthroughs and inventiveness, the brilliance of inspiration and achievement, sudden illumination, and flashes of insight. It is also associated with the impulse toward freedom, rebellion against constraints and the status quo, tendencies towards eccentric or erratic behavior, instability, restless unpredictability, the drive towards novelty and the new, the unexpected, the disruptive, the exciting, and liberating.

By contrast, the second perinatal matrix is associated with the difficult perinatal stage of uterine contractions when the cervix is still closed. BPM II is typically expressed in experiences of claustrophobic constriction, images of imprisonment and hell, physical and emotional pain, helpless suffering and victimization, fear of dying, states of intense shame and guilt, depression and despair, feelings of "no exit" in Sartre's sense, existential alienation and meaninglessness, being trapped in a perspective in which all that exists is mortal life in a disenchanted material world with no deeper meaning or purpose.

In this case, I noticed how often the planet Saturn was involved in transits that coincided with BPM II states. And again, the set of symbolic meanings long attributed by the astrological tradition to the planet Saturn closely matched the BPM II phenomenology: constraint, limitation, contraction, necessity, hard materiality, the pressures of time, the weight of the past, strict or oppressive authority, aging, death, the endings of things; judgment, guilt, trials, punishment; separation, negation, opposition, difficulty, problems, deprivation, defeat; the labor of life, the workings of fate, karma, the consequences of past action, pessimism, sorrow, and melancholy.

However, whereas in the three other cases, both positive and negative sides of the astrological principle involved seemed to be expressed in the large range of potential experiences related to each perinatal matrix, in the case of BPM II, only the negative and problematic features of the Saturn archetype were evident. Subjects under the sway of the second perinatal matrix seem to experience everything as if through an encompassingly negative filter that allows no positive or redeeming dimension of life to be evident. Only in retrospect, after the perinatal process has unfolded and been at least to some extent resolved and integrated, does the experience of BPM II come to be seen in a different light with new meaning. Then the positive dimension of contraction, separation, loss, suffering, the encounter with death, and so forth becomes altogether evident in the concrete manifestation of biological birth or spiritual rebirth, in the experience of having joined the realm of the "grateful dead" because one is now happily reborn. The crushing defeat of the old identity and form of reality is seen as making possible a deeper wisdom that has known both sides of life, and that can embrace pain and loss as necessary for a more profound mode of being. The realities of aging and of mortality itself are seen in a new way, permitting the emergence of the positive qualities that are also traditionally associated with the Saturn archetype — wisdom, maturity, groundedness, patience, endurance, inner equilibrium, capacity for the long view, a deeper understanding, the soul-deepening of suffering and loss, the harvest of time and experience.

In the case of BPM III, Stan and I were especially struck by the uncannily similar, indeed virtually identical, sets of meanings correlating that matrix with the astrological Pluto. The phenomenology of the third perinatal matrix is unusually diverse and brings together a unique constellation of extremely intense experiences. In terms of the stages of biological birth, it is associated with the propulsion of the baby through the birth canal with the cervix fully dilated. Experientially, one finds a powerful convergence of experiences involving titanic elemental energy of volcanic proportions, intense arousal of sexual libido and aggression, enormous discharge of pent-up energies, dramatic experiences involving violent struggle, life-and-death danger, bloody biology, war, scenes of immense destruction, descent to the underworld, demonic evil, sadomasochism, pornographic sexuality, degradation and defilement, scatology, sewers and decay, purifying fire or pyrocatharsis, elemental transformation, ritual sacrifice, orgiastic bacchanalia, and the paradoxical merging of agony into ecstasy. In general, BPM III represents overwhelmingly intense elemental energies within a cathartic, transformational crucible that culminates in the experience of death and rebirth.

Given these several distinctive themes converging within one perinatal matrix, we found the consistent coincidence of BPM III experiences with transits of Pluto especially extraordinary, for the descriptions of the many-sided principle of Pluto by contemporary astrologers encompassed precisely the same convergence of diverse themes: elemental intensity, depth, and power; that which compels, empowers, and intensifies whatever other archetypal principle it combines with, sometimes to overwhelming and catastrophic extremes; a dominant concern with survival, sexuality, or power, the lower chakras; the primordial instincts both libidinal and aggressive, destructive and regenerative; the volcanic, cathartic, eliminative, transformative, ever-evolving; the biological processes of birth, sex, and death, the cycles of death and rebirth; breakdown, decay, and fertilization. Characteristic features of BPM III sessions include the violent purging of repressed energies, power struggles, and situations of life-and-death extremes. Events that are titanic, potent, and massive are typical. Subjects describe the powerful forces of nature emerging from its chthonic depths both within and without, the intense, fiery underworld and underground in many senses (geological, psychological, sexual, urban, political, criminal, demonic, mythological). They make reference to Freud's primordial id, "the broiling cauldron of the instincts," and to Darwin's ever-evolving nature and the biological struggle for existence.

As with the other perinatal matrices, subjects often had direct experience of specific mythic deities when reaching the deeper dimensions of that matrix. In the case of BPM III, the mythic figures encountered tended to be the same as those brought up in astrological texts describing the nature of the Pluto archetype: deities of destruction and regeneration, descent and transformation, death and rebirth such as Dionysus, Hades and

Persephone, Pan, Priapus, Medusa, Lilith, Inanna, the volcano goddess Pele, Quetzalcoatl, Kundalini activation and the Serpent power, Shiva, Shakti, Kali.

Finally, a similar set of parallels was apparent in examining the coincidence of the very different range of BPM I experiences with transits of Neptune. The first perinatal matrix is associated with the prenatal condition immediately prior to the beginning of the birth process: experiences of the amniotic universe, floating oceanic sensations, the melting of boundaries, a porous relationship to the environment, lack of differentiation between inner and outer, embryonic experiences multidimensionally blended with aquatic, interstellar, galactic, and cosmic experiences. Here also are found experiences of mystical unity, spiritual transcendence, the dissolution of material reality and of the separative ego. There is typically a sense of merging with the womb, with the mother, with other persons or beings, with all life, as well as with the divine, giving access to other ontological dimensions beyond consensus reality, and a transcendence of time and space. Experiences of idyllic nature such as lush tropical islands or timeless childhood play in beautiful meadows or on seashores can merge into experiences of cosmic unity, oceanic ecstasy, and images of Paradise. In its negative aspect, BPM I is associated with experiences that involve a disorienting loss of boundaries, the dissolving of a stable identity or reality structure, susceptibility to delusional thinking, feeling enveloped by a threatening atmosphere filled with invisible dangers and subtly infecting influences, and fetal experiences of a toxic womb merging into experiences of drug poisoning, psychic contamination, or oceanic pollution.

Remarkably, astrologers have long symbolically associated the planet Neptune with experiences having a spiritual, transcendent, or mystical character; with the subtle and intangible, the unitive, timeless, immaterial, and infinite; and with all that transcends the limited world of matter, time, and concretely empirical reality. Neptune is connected with states of psychological fusion, physical and psychological permeability, and the longing for the beyond. It also has a symbolic association with water, the sea, streams and rivers, mist and fog, with liquidity and dissolution of any kind. Relevant here is Romain Rolland's term "the oceanic feeling," which he described as a religious sensation of eternity and being one with the world as a whole, which Freud in turn associated with very early infantile experience, merging with the mother and so forth.[4] Negatively, Neptune is described in terms of perceptual and cognitive distortions, projection, inability to distinguish the inner world from the outer, tendencies towards illusion and delusion, *maya*, deception and self-deception, disorganization, escapism, intoxication, addiction, and vulnerability to toxic drug reactions, infections, and contamination.

With respect to all four perinatal matrices, what especially struck us was the two-fold nature of the correlations: On the level of the comparative study of symbol systems, the fact that two entirely distinct interpretive traditions, psychology and astrology, could have independently formulated four fundamental sets of qualities and meanings which so closely corresponded to each other—point for point, matrix to archetype—was certainly surprising on its own terms. But apart from these clear parallels of meaning, the fact that the timing of subjects' experiences of each perinatal matrix in psychedelic sessions actually coincided so consistently with transits involving the very planet that bore the matching astrological character seemed to us astounding.

These perinatal correspondences that emerged into view early in our research became greatly complexified as time passed and we gained a greater understanding of how the major geometrical alignments involving two or more planets (the major planetary "aspects" such as conjunction, opposition, and square) played out in natal charts and transits. For example, Neptune transiting a person's natal Sun seemed to play out differently than if it transited the Moon, even though there were common "Neptunian" features in both instances. In the case of Neptune transits to Saturn, the coinciding experiences tended to be in certain respects virtually opposite those of Neptune transits to Jupiter, though again they had essential underlying features in common that reflected in different ways the archetype associated with Neptune. Though there is not space in this brief overview to describe the various archetypal inflections that typically accompanied a particular planet's alignments with each of the other planets, we found that the differences in each case seemed to be directly related to the archetypal qualities associated with the second planet involved in the transit—drawing out, as it were, different potentialities with respect to the same planet, in this case, Neptune.

Each planetary combination seemed to involve a mutual activation of the two archetypal principles involved, with each archetype infusing and inflecting its specific nature through the other, with each archetype thereby shaping the expression of the other and creating a live compound of the two. Moreover, though there was an underlying archetypal commonality, different individuals seemed to experience the same transit of a specific planet to a natal planet differently depending on how that natal planet was situated relative to the other planets in the birth chart, or depending on what other transits were occurring simultaneously. Nothing happened in a vacuum. Everything was always situated in and shaped by a unique context, not only biographical and circumstantial, cultural and historical, but also at the archetypal level, reflecting the larger complex of planetary factors involved at any given time.

As another example of these complexities, in full-fledged BPM II experiences a specifically perinatal level of content in psychedelic sessions generally seemed to involve the presence of Saturn in alignment with one of the three outer planets Pluto, Neptune, or Uranus. A Saturn transit on its own, for example, in alignment with the Sun, Moon, or one of the inner planets, tended to coincide with more common life experiences reflecting the various themes associated with the Saturn archetype (a separation, illness, career difficulty, or period of moderate depression). By contrast, full-scale experiences of BPM II during psychedelic sessions were more likely during Saturn transits that involved Pluto or Neptune, each bringing out specific inflections of the second matrix reflecting the relevant archetype: Saturn with Pluto, for example, being more likely to coincide with experiences of helpless suffering in the face of extreme cruelty or titanically intense contractions with no release, while Saturn with Neptune was found more in the case of confrontations with the meaninglessness of mortal life, the dark night of the soul, loss of spiritual meaning, suicidal despair, or fears of insanity. Transits involving Saturn together with Uranus were more associated with experiences involving the sudden confrontation with death, unexpected fall from grace, or the sudden collapse of previously secure structures, whether of identity or of reality itself. Uranus, Neptune, and Pluto all seemed to have a more emphatically transpersonal character, while Saturn represented more of a symbolic threshold between the personal and transpersonal, and between life and death. These four outer planets together appeared intimately connected with the perinatal domain of the psyche: the nexus point of the personal and transpersonal, the biological and spiritual, the biographical and the cosmic.

Moreover, all of the above combinations of planets in transit could unfold in ways that moved through the perinatal depths into very different healing inflections of the same archetypal principles: Stan's dictum that the full affective and somatic experience of a difficult emotion is "the funeral pyre of that emotion"[5] was relevant to understanding how a given archetypal complex could evolve from its most challenging forms to highly positive expressions. Thus, after the integration of traumatic or otherwise problematic unconscious material, including BPM II experiences of hellish torment, Saturn-Pluto transits were observed in connection with a new capacity for unflinching courage in the face of death and danger, a new willingness to engage life's gravest realities and bear great burdens, to mobilize immense energy with sustained effort and determination for prolonged periods of time, as in the titanic hard labor of birth. Similarly, experiences during Saturn-Neptune transits could unfold into heightened compassion for the suffering of others and sustained practical efforts to alleviate that suffering, a new capacity for sacrifice and renunciation of personal attachments in the service of one's spiritual aspirations, or working to bring spiritual ideals into concrete manifestation in a practical, focused, and disciplined way.

All these complexities reflected the intrinsic multivalence of archetypes, their carrying of a far greater range of interconnected meanings than a simplistic grid of keywords or phrases could ever convey. It also gradually became clear that the archetypes that we observed in perinatal experiences and that corresponded with the specific transiting and natal planets seemed to exist at a supraordinate level with respect to the perinatal dimension of the psyche. This supraordinate status became apparent when we noticed a remarkable category of correlations involving the COEX systems, which informed not only the personal unconscious but also the transpersonal, with the perinatal level often serving as an experiential gateway between the two realms.

Correlations with COEX Systems

In the course of his early work in the 1960s using psycholytic therapy, before he had recognized the perinatal matrices, Stan observed the existence of certain dynamic constellations of emotionally charged memories that shared similar affective and somatic qualities: the COEX systems, or systems of condensed experience. In serial sessions involving low to medium doses of LSD, these COEX systems gradually emerged as various memories from different periods of life that were thematically related came to the surface, eventually often converging in a cluster and becoming condensed into a powerful multidimensional experience. Such retrieved memories in earlier sessions were typically based on more recent events and experiences, while in later sessions they reached deeper into the unconscious to early childhood and infantile experiences that were thematically related to the more recent memories.

Different individuals tended to carry their own particular sets of COEX systems, both negative ones such as various experiences of abandonment, shame, or claustrophobic constriction, and positive ones such as diverse experiences of nourishing love, joyful triumph, or expansive awakening. Careful analysis of individuals' birth charts and their many transits in the course of their lives suggested that these COEX systems corresponded in striking ways with major planetary alignments in the natal chart whose archetypal meanings were directly relevant to the themes of the emerging COEX. Even more striking was the fact that the timing of the major events that contributed to the COEX system in the course of life, such as the death of a parent, a numinous childhood experience, or a romantic awakening, consistently coincided with major transits crossing the specific planetary configurations in the natal chart related to that COEX. In turn, the emergence into consciousness and integration of such a COEX during a psychedelic session tended to take place when the relevant natal configuration was again undergoing a major transit. The evidence suggested that such transits of a particular COEX-related natal configuration could be experienced either as a further magnification of the unconscious COEX system, thus increasing its psychological power, or as an opportunity to bring it more fully to consciousness and, in the case of a negative COEX, release the jammed energies and painful emotions associated with the original trauma.

As Stan has frequently discussed, he found that the COEX systems associated with various biographical experiences turned out to be more deeply rooted in the unconscious in one of the four perinatal matrices, whose rich and complex phenomenology contained in prototypical form the elementary themes of virtually all COEX systems. Biographical experiences of abandonment such as the tragic loss of a parent during childhood, a devastating romantic rejection, or a wrenching divorce were all thematically connected with each other but also had common roots in the perinatal experience of the primordial loss of the maternal womb and organismic separation from the mother. By contrast, later experiences of unexpected personal achievement at school or sports in one's youth, of major professional success in one's career during one's adult years, or more generally of sudden joy after overcoming great obstacles found deeper roots in the experience of sudden successful emergence from the constricting birth canal.

Yet, over time, in more advanced stages of therapy and self-exploration, the COEX systems proved to be rooted at even deeper levels of the unconscious than the perinatal, as evidenced in ancestral, historical, collective, karmic, and phylogenetic experiences. What in earlier sessions might have seemed to be an acutely personal psychological issue or biographical theme specific to the individual could be discovered to be grounded in underlying familial patterns stretching back across many generations, or in vivid experiences connected to an earlier historical era, to an entire culture, or to another species of life. In the case of negative COEX systems, integration of these deeper transpersonal levels of a COEX system was often necessary before deep healing and release from the underlying traumatic syndrome could occur. In the case of positive COEX systems, connecting with deeper transpersonal sources such as mystical unity with the Godhead or the all-nourishing embrace of the Great Mother Goddess could provide an especially powerful healing experience. At the core of each such COEX system, we consistently found a particular archetypal principle or complex whose thematic character informed and interconnected the biographical, perinatal, and transpersonal dimensions of the COEX.

This finding closely resembled Jung's understanding of the archetype as constituting the core of every psychological complex, but the COEX system added a dynamic temporal dimension in which major events and experiences from different periods of life, from the stages of birth, and from various prenatal, historical, and other transpersonal levels of the psyche can accumulate and constellate into an integrated system that can be experientially accessed in extraordinary states of consciousness. The concept of the archetypal complex developed in Jungian and archetypal psychology provided us with a nuanced understanding of the various archetypal principles with their rich array of mythological and esoteric meanings and interrelationships.[6] In turn, Stan's concept of the COEX system offered a more precisely delineated view of the multilayered dynamic constellations of memories and events lodged in the deep psyche, rooted in biographical, perinatal, fetal, ancestral, karmic, historical, phylogenetic and other transpersonal levels. Over time these COEX systems seem to accumulate greater and greater psychic and somatic charge, like a snowball going downhill, drawing into themselves more events and experiences that magnify the inherited psychosomatic structures and impulses until they are made conscious and integrated. The overall picture that emerged from this synthesis is depicted in the diagram below.

Systems of Condensed Experience (COEX)
and Levels of Consciousness

Ordinary ego consciousness
Personal unconscious
Biographical memories
Birth chart
BPM I
BPM II
BPM III
BPM IV
Perinatal memories
Karmic memories
Ancestral memories
Transpersonal unconscious
Phylogenetic memories
Collective memories
Mythological experiences
Archetypal dimension

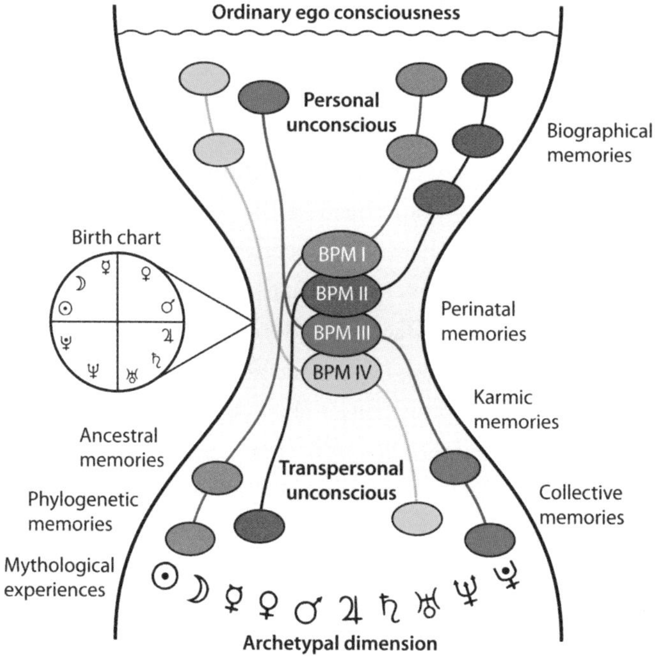

Two important implications of our research findings can be mentioned here, both of which are suggested in the above diagram. One is the supraordinate role of the archetypes in relation to all three levels of consciousness—biographical, perinatal, and transpersonal—informing the differentiated dynamic constellations and matrices of experience at each level, and thematically unifying them within the multivalent complexes of meaning carried by each archetype or combination of archetypes. The archetypal forms seem to serve as general organizing principles of the deep psyche, much as described in Jungian and archetypal psychology (as well as in Platonic and Neoplatonic philosophy). In turn, these supraordinate principles express themselves within the more differentiated psychodynamic architecture that emerged from psychedelic therapy, transpersonal consciousness research, and other holotropic experiences, which Stan has articulated in his "cartography of the unconscious" in his many books and lectures.

The second implication worth noting here is the unexpected correspondence between Stan's finding of the crucial psychological importance of birth and astrology's focus on the planetary positions at birth. One might say that both birth and the birth chart in some sense mediate access to archetypal and transpersonal dimensions. More specifically, the careful study of the birth chart and the reliving of birth and encounter with death in perinatal experiences of death and rebirth both appear to provide powerful ways for individuals to access more directly and consciously the deeper archetypal and transpersonal dimensions that inform their lives and influence their present state of consciousness. Both the perinatal level of the psyche and the astrological natal chart seem to represent a gateway, a *via regia,* opening up consciousness to the depths of the unconscious.

The above diagram can in fact be read in both directions, from the top down and from the bottom up. In long-term serial sessions of experiential psychotherapy and self-exploration, a characteristic sequence is a movement from more recent experiences that share certain underlying qualities to successively earlier and earlier experiences of a similar emotional or somatic character from youth, childhood, and the pre-Oedipal and infantile periods of life; then a significant deepening to the perinatal level and the death-rebirth complex of experiences; with this in turn connecting with and opening up to a vast range of transpersonal experiences in the collective unconscious. Beyond and in a sense surrounding and informing all the above is the archetypal realm, in some manner associated with the vast cosmos and starry sky. The revelation of such a domain is depicted in Plato's well-known Allegory of the Cave in book seven of the *Republic* (514a–520a).

Reading the above diagram in the other direction: After experiencing the larger spectrum of experiences across these many levels, the individual often gains insight into how various factors from the larger transpersonal domain—ancestral, karmic, historical, and so forth—seem to have translated themselves into specific powerful aspects of the birth experience. Here the vector of the diagram can be seen as moving upward from the transpersonal to the personal. For example, an experience of death by hanging in a previous life can morph into a birth this lifetime in which the umbilical cord is wrapped around the neck causing near-asphyxiation, which in turn can be experienced as unfolding within the postnatal life in various forms such as suffering severe difficulty in breathing during an episode of infantile diphtheria or whooping cough, or being aggressively choked as a child by an older sibling or a bully in a fight. In both directions, the perinatal can be seen as the point of convergence between the transpersonal and personal.

Over the years of research, with further analysis and the wider range of data that emerged during the 1980s, 1990s, and 2000s, we gained a somewhat different understanding of the role of the perinatal in relation to the rest of the unconscious in therapeutic and transformational work. Rather than serving as a mandatory threshold through which all individuals inevitably pass in the course of their journeys into the deep psyche, we found that a person could potentially undergo a powerful transpersonal experience, such as a deep identification with Gaia or the entire Earth community, or what seemed to be an event from another historical era or a memory from a past life, without necessarily having undergone the biographical-perinatal-transpersonal sequence observed by Stan in many patients and subjects during the 1960s and early 1970s. Instead, an individual might access any level at any time, depending not only on the catalyzing method (psychedelic session, breathwork, kundalini yoga, gestalt therapy, spiritual emergency, meditation, etc.), but also on the setting, the stage of therapy or self-exploration, and the specific psychedelic medicine used and dosage level. Perhaps other less knowable factors enter in as well, such as the spontaneous unfolding of the individual's inner healing intelligence, the *telos* of individuation in Jung's sense, the holotropic movement towards wholeness in Stan's sense, perhaps even karma or grace.

What proved to be key across all these variables, however, was the archetypal character of the experience, which consistently tended to coincide with specific natal and transiting planetary alignments, and which could express itself at any level, whether biographical, perinatal, or transpersonal. With extraordinary consistency, the dominant qualities of any particular psychedelic session, holotropic experience, spiritual emergency/emergence, or therapeutic turning point could be discerned in archetypal terms and correlated with the natal chart and transits.

World Transits

I have so far been discussing correlations involving individuals' birth charts and personal transits. After the initial years of research in which we focused on the lives and experiences of individuals involved in deep self-exploration, psychotherapy, psychedelic experimentation, and various other transformational practices, I increasingly turned my attention to the study of major cultural and historical figures. I was curious, for example, what transits Freud had when on July 24, 1895, "The secret of dreams was revealed" to him, as he later put it, and he grasped how the unconscious symbolically expressed itself through the dream; or what convergence of transits Jung had during the crucial 1913-18 period of his life when he underwent a powerful descent into his own unconscious that brought forth the flood of images and ideas with which he would work for the rest of his career. I was curious what transits were taking place when Galileo first turned his telescope to the heavens in 1609-10 and glimpsed the new Copernican universe that he helped open up to modern understanding. Or what transits Rosa Parks had in December 1955 when she refused to get up from her seat on the segregated bus in Montgomery, Alabama and catalyzed the US civil rights movement. Or what transits Beethoven had when he wrote the *Eroica* Symphony and revolutionized classical European music—or, by contrast, what he had when he first came to the tragic realization that he was becoming incurably deaf, unable to hear his own music. With each of these major biographical and cultural turning points, and hundreds more like them, I discovered the same consistency and archetypal precision of the planetary correlations as we had been finding in the psychotherapeutic and psychedelic research.

Gradually, however, another level of understanding opened up that recontextualized the findings so far considered. As the five slower-moving outer planets, from Jupiter through Pluto, orbit around the Sun along with the Earth, they gradually move into and then out of major alignments with each other relative to the Earth—conjunctions, oppositions, and so forth—in ongoing cycles. Depending on the planets and orbital speeds involved, some of these periods of cyclical alignments last longer and happen more rarely, such as the Uranus-Pluto conjunction that encompassed the entire period of the 1960s and early 1970s, while others are shorter in duration and happen more frequently, such as the Jupiter-Uranus cycle whose conjunctions and oppositions each last about fourteen months and happen approximately every seven years. I found that the periods encompassed by these world transits were consistently marked by a remarkable convergence of major historical events, cultural movements, and public figures in many countries and areas of human activity, all of which reflected a shared zeitgeist whose archetypal character corresponded with the planets aligned with each other during that time with respect to the Earth.

These historical patterns were both synchronic and diachronic in nature, a dual form of archetypal patterning that was strikingly consistent throughout the larger body of historical evidence. The *synchronic* patterns took the form of many events of the same archetypal character occurring simultaneously in different cultures and individual lives in coincidence with the same planetary alignment—simultaneous revolutionary movements or major waves of artistic creativity occurring independently in separate countries and continents, or multiple scientific breakthroughs achieved at the same time by different scientists working entirely independently of each other. The *diachronic* patterns reflected the fact that events taking place during one planetary alignment had a close archetypal and often historically causal association with the events that occurred during preceding and subsequent alignments of the same two planets in a manner that suggested a distinct unfolding cycle. The relevant periods were thus linked to each other not only because they shared the same archetypal character but also by virtue of their unfolding historical and causal connections from one cycle to the next. The associated historical trends and cultural movements seemed to undergo a sharply intensified or accelerated development during each successive period in what appeared to be a continuously unfolding but cyclically "punctuated" spiralic evolution.

Because I have already published an extensive account of such historical correlations between planetary cycles and archetypal patterns in my book *Cosmos and Psyche*,[7] I will not discuss them further here except as they proved relevant for understanding individual experiences in therapeutic, psychedelic, and holotropic contexts. For as I deepened my analysis of individual transformational experiences over the decades, I became aware that the overarching archetypal dynamics reflected in the world transits appeared to provide a kind of meta-context that encompassed and informed the specific archetypal dynamics reflected in an individual's personal transits. For example, the Uranus-Pluto conjunction of the 1960s and early 1970s with its characteristic quality of intensified elemental energy and revolutionary transformation seems to have provided the archetypal context for the powerful emergence of perinatal experiences that Stan observed and formulated at that time. The entire collective field had a perinatal intensity as well as perinatal quality that expressed itself in the LSD sessions in a manner that seemed to come directly from a larger archetypal source.

By contrast, the long Uranus-Neptune conjunction from the mid-1980s to the end of the century provided a different archetypal context, reflected in such archetypally relevant phenomena as the widespread use of MDMA or Ecstasy with its characteristic stimulation of numinous merging experiences in group settings like the countless raves occurring throughout the world beginning in the late 1980s; the increasing participation in ayahuasca rituals not only in indigenous South American settings but in North American and European societies, which was part of a more general widespread engagement with the sacred ritual use of vision plants; and the rapid spread of holotropic and other forms of breathwork and deep meditation techniques. Reports from many psychedelic and holotropic experiences at this time made clear that individuals were accessing various transpersonal dimensions without necessarily first passing through the titanic upheavals and breakthroughs of the perinatal domain. Similarly appropriate in archetypal terms during this time of increasing globalization and internet connectivity was the increasing dissolution of boundaries between different cultural and religious traditions. This dissolving of boundaries occurred not only at the collective level of multicultural interaction and a resulting creative religious syncretism, but also at an entirely interior individual level whereby subjects in extraordinary states of consciousness reported having spontaneous religious and mythological experiences and insights from cultural traditions entirely outside the compass of their previous knowledge, suggesting that the collective psyche was undergoing an unprecedented internal globalizing process apart from the more literal one taking place in the external world.

In addition, other major world transits of briefer duration during these decades, such as the several Saturn-Neptune or Jupiter-Uranus alignments, coincided with still other major archetypal tendencies in individual experiences and extraordinary states of consciousness. Moreover, it became clear that with both world transits and personal transits, the faster-moving transits (those involving Mars, Venus, Mercury, Sun, or Moon) seemed to synchronistically "trigger" or catalyze the specific timing of events and experiences associated with the longer, more powerful transits of the slower-moving outer planets. Finally, there was the important ongoing issue of multiple transits happening simultaneously that were often of very different archetypal quality, and on occasion virtually polar opposite in nature. Only gradually did we gain a sense for how to synthesize and weigh the relative importance of these multiple transiting and natal factors as they were expressed in psychedelic and holotropic sessions, and in individual lives more generally.

The Issue of Causality

From our first encounter with the evidence of potential planetary correlations, Stan and I were confronted with the theoretical difficulty of imagining how the physical planets, of greatly varying mass and orbiting at widely varying distances from the Earth, could exert an influence on not only external events in human history and biography but the interior realities of private human experience. It was difficult to imagine any physical factor, at least as conventionally understood, that could serve as a plausible source or medium of the observed correlations. Very early in our research, Stan suggested that a more likely explanation for what we were seeing is that the universe has woven into its very fabric a meaningful coherence between the macrocosm and the microcosm. Instead of a Cartesian-Newtonian form of linear causality involving some kind of physical factor, such as electromagnetic radiation, the nature of the correspondences suggested more of an intrinsic synchronistic orchestration between planetary movements in the heavens and archetypal patterns in human experience. Indeed, the concept of synchronicity had been invoked by Jung on several occasions as a possible explanation for why astrology worked in spite of modern assumptions that it should not.[8]

Today, after several decades of research, I believe that the range of correspondences between planetary positions and human existence is too vast and multidimensional—too clearly ordered by structures of meaning rather than physically measurable forces, too suggestive of creative intelligence, too pervasively informed by aesthetic patterning, too symbolically multivalent, too experientially complex and nuanced, and not least, too responsive to human participatory inflection—to be explained by straightforward material factors alone. A more plausible and comprehensive explanation of the available evidence points to a conception of the universe as a fundamentally interconnected whole, informed by creative intelligence and pervaded by patterns of meaning and order that extend through every level. This would represent, as Jung suggested, a cosmic expression of the principle of synchronicity. It also parallels the Hermetic axiom, "As above, so below." In this perspective, the planets do not "cause" specific events any more than the hands on a clock "cause" a specific time. Instead, the planetary positions seem to be *indicative* of the cosmic state of archetypal dynamics at that time. The Neoplatonic philosopher Plotinus expressed a world conception along this line in the *Enneads*:

> The stars are like letters which inscribe themselves at every moment in the sky.... Everything in the world is full of signs.... All events are coordinated.... All things depend on each other; as has been said, "Everything breathes together."[9]

There is, however, a sense in which causality does appear relevant in this context, and this is in the form of *archetypal* causation, comparable to Aristotle's concepts of formal and final causation. While the movements of the physical planets may bear a synchronistic rather than mechanistically causal connection with a given human experience, one could say that the experience is in a certain sense being constellated—patterned, shaped, colored, affected, impelled, drawn forth—by the relevant archetypes, and in this sense it may be appropriate to speak, for example, of Saturn (as archetype) as "influencing" one in a specific way, as "governing" certain kinds of experience, and so forth. But while the archetype may be *a* cause, I would not consider it *the* cause, as archetypal factors are always acting in complex recursive relationship with human agency, level of consciousness, cultural context, concrete circumstance, interpersonal field, genetic inheritance, past actions, and many other possible factors.

The Nature of Archetypes

The evidence of planetary correlations with human experience centers on the multidimensional principle of *archetypes.* When Jung brought the concept of archetypes into contemporary discourse in his recognition of certain universal constants that structure the deep levels of the human psyche, he was influenced by his own experience as well as by Kant's critical philosophy of *a priori* forms and Freud's mythically framed theory of the instincts. In choosing the term "archetype," he was deliberately employing a vocabulary drawn from the Platonic philosophical tradition. And in the background of both the Jungian and Platonic perspectives was the ancient mythological experience of gods and goddesses, essentially personified expressions of the equally numinous Platonic Forms and Jungian archetypes.

To simplify a complex historical development, in the course of which the cultural focus evolved from myth to philosophy to psychology, one could say that the classical Platonic tradition gave philosophical articulation to the primordial mythic vision of powerful essences or beings that both informed and transcended human life. In turn, while Plato understood the transcendent Forms or Ideas to be the fundamental structuring principles of an ensouled cosmos, Jung understood the archetypes to be the fundamental structuring principles of the human psyche. These important distinctions reflected the long epistemological and cosmological evolution that took place in Western thought during the past twenty-five hundred years, gradually differentiating psyche from cosmos, subject from object, human from non-human, and leading to the modern disenchantment of the world within which, or out of which, depth psychology emerged over a century ago.

On the basis of his long study of synchronicities, however, Jung came to the conclusion that the archetypes could not justifiably be localized within human subjectivity but instead seemed to inform both psyche and world, serving as an underlying unitive principle. In this sense, the later development of Jung's archetypal theory more closely approached the Platonic view, though with a greater psychological emphasis and a fuller recognition of both the fluid multivalence and the shadow dimension of the archetypes. Jung's later thinking is also consistent with the many archetypal experiences reported in the psychedelic literature, which suggest that archetypes can express themselves as psychological forms, as cosmic principles, or as mythic beings, in a kind of multidimensional continuum.

Contrary to the disenchanted modern world view, the evidence of systematic planetary correlations with the archetypal patterns of human experience suggests that the cosmos is a living, ever-evolving matrix of being imbued with meaning and purpose, within which the human psyche is embedded as a co-creative participant. In Jungian terms, the evidence points to the possibility that the collective unconscious is in some way embedded in the universe itself, whereby the planetary motions reflect at a macrocosmic level the unfolding archetypal dynamics of human experience. In Platonic terms, the evidence seems to reflect the existence of an *anima mundi* informing the cosmos, a world soul in which the human psyche participates as a microcosm of the whole. In mythic terms, the evidence indicates a continuity with the world views of the great archaic civilizations, such as ancient Mesopotamia and Egypt, with their awareness of an intimate connection between the gods and the heavens that both inspired and structured their religious and social life, astronomical observations, and monumental architecture.

In retrospect, humanity's long evolution of consciousness and world views seems to have been informed by an evolution in how the archetypal domain has been perceived and theorized, as well as how it was eventually negated and then rediscovered in new forms. In the course of that evolution, and especially in its modern disenchanting stages, there occurred a gradual and then finally decisive differentiation of an autonomous self and a strengthening of human agency. In a further dialectical unfolding, more recent developments in archetypal theory and experience have emphasized the participatory and multivalent nature of the archetypes. This emerging perspective both recognizes the underlying power of archetypes while giving the human being a greater co-creative as well as co-responsible role in their expression. This has led to the possibility of a new form of human relationship to the *anima mundi* that permits and even thrives on the simultaneous existence of autonomy and embeddedness. Yet paradoxically, the disenchantment of the universe and radical separation of human consciousness from the whole may have been the precondition for both the alienation that helped precipitate the crisis of modern consciousness and the forging of an autonomous modern identity capable of reengaging the *anima mundi* in a newly participatory manner. The journey of depth psychology from the 1880s to the present, from Freud to Grof, so to speak, would have been neither possible nor necessary without the long cosmological and existential evolution that preceded it.[10]

Returning to the planetary correlations with psychedelic experiences: Only as I came to fully recognize the multidimensional and multivalent nature of archetypes—their formal coherence and consistency that could nevertheless give rise to a plurality of meanings and possible expressions—did the extraordinary elegance of the planetary correlations become discernible. Any particular manifestation of a given archetype can be "positive" or "negative," creative or destructive, admirable or base, profound or trivial. The archetypes associated with specific planetary alignments are equally apt to express themselves in the interior life of the psyche as in the external world of concrete events, and often both at once. Closely linked yet entirely opposite polarities contained within the same archetypal complex can be expressed in coincidence with the same planetary configuration. A person undergoing a particular transit can be on either the acting or the receiving end of the relevant archetypal gestalt, with altogether different consequences. Of these many related possibilities, which mode actually occurs does not seem to be observable in the birth chart or planetary alignments per se. Instead, the archetypal principles at work in these correlations appear to be dynamic yet radically indeterminate in their multivalent nature. Though they represent enduring forms or essences of complex meaning, and are clearly discernible underlying the flux and diversity of the observed phenomena, they are also both fundamentally shaped by many relevant circumstantial factors and co-creatively modulated and enacted through human will and intelligence.

Because of this combination of dynamic multivalence and sensitivity to particular conditions and human participation, I believe that, contrary to its traditional reputation and employment, such an astrology is best regarded, as mentioned before, not as concretely predictive but as archetypally predictive. Compared with, for example, some forms of intuitive divination with which astrology has often been systematically conjoined, the focus of an archetypal astrology that is reflective of the evidence we have studied is not the prediction of specific outcomes but rather the precise discernment of archetypal dynamics and their complex unfolding in time. I believe that such an understanding shines a light on numerous long-standing issues surrounding astrology, such as the question of fate versus free will, the problem of identical planetary configurations coinciding with concretely different though archetypally parallel phenomena, and the fundamental inadequacy of statistical tests for detecting most astrological correlations.

The planetary correlations appear to offer a uniquely valuable form of insight into the dynamic activity of archetypes in human experience—indicating which ones are most operative in a specific instance, in what interactive combinations, during which periods of time, and as part of what larger patterns. In providing such a perspective, archetypal astrology can be seen as essentially continuing and deepening the depth psychology project: to make conscious what is unconscious, to help free the conscious self from being a puppet of unconscious forces (as in blind acting out, inflated identification, projection, self-sabotage, drawing towards one as "fate" what is repressed or unconscious, and so forth). Its study can mediate a heightened quality of communication and coordination between consciousness and unconscious factors, with "the unconscious" now suggestive of considerably larger dimensions than originally conceived—less exclusively personal and subjective, more cosmically embedded, less a literal entity, more a mode of being. Astrology provides this mediation, however, not by spelling things out in a literal, concretely predictive manner, but rather by disclosing intelligible patterns of meaning whose very nature and complexity—multivalence, indeterminacy, sensitivity to context and participation, and a seemingly improvisatory creativity—are precisely what make possible a dynamically co-creative role for human agency in participatory interaction with the archetypal forces and principles involved.

One of the most provocative patterns of correlations that emerged shone a light on the notorious diversity of psychological theories and systems that have been proposed by the major psychologists from Freud on. The specific character and spirit of each theory, the particular vision of the psyche articulated by a given psychologist such as Freud or Jung, Melanie Klein or Wilhelm Reich, Marie-Louise von Franz or James Hillman, was consistently reflected in the birth chart of that psychologist. It was, as Stan often said, as if each psychologist were a mouthpiece for the archetypal dynamics depicted in their birth charts, which shaped their psychological imagination in a particular way. Each theorist rendered his or her definite theory of the infinitely complex psyche, which remained an ever-evolving, indeterminate mystery that no theory could ever finally capture and comprehend. This correlation was true not only with respect to the general emphasis of the theory—as with Freud's more Pluto-dominated birth chart and his more Plutonic paradigm of the psyche as driven by the biological instincts of sex and aggression, compared with Jung's more Neptune-dominated chart and his more Neptunian psychological vision with its greater appreciation of the spiritual, esoteric, imaginal, and numinous dimensions of human experience. It was also true for the many more specific aspects and nuances of their theories and practice, which were reflected in their various other major natal planetary aspects. Moreover, the biographical timing of their discoveries and their evolving perspectives closely matched the major transits they underwent at those times.

We recognized that these patterns were evident with respect to us researchers as well, which helped contextualize our own findings and perspectives. We could see how, as a general principle, the major natal alignments and transits coincided with archetypal tendencies and potentialities that could create conceptual walls and barriers, shaping and confining the vision according to the individual biases reflecting a particular archetypal complex. But that same archetypal complex could also serve as a window or a door, permitting insights and openings into dimensions of experience and reality that might not otherwise be accessible. The difference between a wall and a window seemed to be not a matter of the planetary alignment or the archetypal complex per se, but rather of our own agency and capacity for discernment: to be reflectively aware of our own subjectivity while also relying on and developing it to enter the world we are exploring. For from this perspective, our subjectivity is itself our mode of access to the cosmic subjectivity of which ours is an outgrowth or flowering. The depth and clarity of our discernment in this respect, our capacity for self-lucidity, is the measure of our access to the more profound dimensions of the interiority of the cosmos.

Final Notes

An unexpected feature of pursuing astrological research in the area of psychedelic exploration is that the latter brings forth profound encounters with the deep psyche that can often include direct experiences of the archetypes themselves in various forms—transcendent, immanent, mythic, metaphysical, somatic, metaphoric, hermeneutic, and so forth. Such encounters provide a more vivid experiential ground for understanding astrological factors and can give one a better grasp of the multidimensional and multivalent character of the archetypal principles than might otherwise be possible. Systematic psychedelic exploration and archetypal astrological analysis thus proved mutually enhancing in remarkably fruitful ways. The richness and intensity of archetypal phenomena in psychedelic experiences, disclosing a detailed kaleidoscope of diverse expressions of the same archetypal principle, were immensely instructive concerning the nature of archetypes in not only psychological but also metaphysical terms—Jungian and Platonic, to be sure, but in many other inflections as well (Aristotle's formal causes and universals, Kant's a priori forms, Whitehead's eternal objects, Sheldrake's morphic fields, Freud's instincts, cross-cultural mythological expressions). Such experiences brought into focus both the multidimensional nature of archetypes generally and the specific range of interconnected meanings encompassed by each particular archetype being constellated, the many ways that principle could prismatically express itself while always reflecting its essential core.

Conversely, knowledge of one's birth chart, personal transits, and world transits can provide a uniquely calibrated instrument for more deeply understanding the dominant themes of psychedelic and holotropic sessions—bringing greater intelligibility to past explorations, more skillful and informed responses to present ones, and more thoughtful timing and preparation for sessions in the future. In this last regard, for researchers like ourselves, each such session was quite literally an "acid test" for assessing the validity of provisional conclusions as our research progressed and understanding deepened over the decades.

Psychedelic and holotropic experiences also tend to bring about a deep change in epistemological outlook, what might be called a dissolution of the Cartesian-Kantian double-bind of modern consciousness that experiences itself as evolved from and contextualized by a universe that is unconscious, purposeless, and ultimately unknowable. This shift of vision can lead to a recognition of the universe as fundamentally ensouled, and help mediate a spiritual-moral awakening—a shift of heart, not just of mind—that is necessary for entering into such an astrological perspective: strengthening a mature hermeneutics of trust to balance and integrate with our already robust modern and postmodern hermeneutics of suspicion. Such experiences can result in a new openness to the possibility of a cosmic intelligence that is coherent with and responsive to our own.

Using archetypal astrological analysis in close examination of our own and others' psychedelic experiences, we were able to assess with greater experimental precision which astrological factors tended to be most significant in this area, and what were the orbs (the range of degrees before and after exact alignment) within which planetary alignments seemed to be archetypally operative. Of all the many factors considered in the astrological tradition, we found that by far the most important factors in understanding these experiences were the planetary archetypes and the major aspects of the planets in natal charts, personal transits, and world transits. So too the planetary positions relative to the four angles of the chart: the horizontal and vertical axes, the Ascendant-Descendant and Midheaven-Imum Coeli. The approach we found most helpful was similar to that of the Renaissance astronomer and Copernican revolutionary Johannes Kepler, with his Pythagorean emphasis on the planetary aspects as the dominant astrological indicators, within an unfolding cosmic geometry of archetypal meaning centered on the moving Earth.[11]

The evidence we encountered also suggested the importance of recognizing larger orbs than have generally been used in traditional astrology. We came to see aspects not as acting like isolated on-and-off light switches within a narrow orb, but rather as indicating archetypal waveforms that enter into the individual or collective psychic field and interact with the larger complex whole of archetypal dynamics cumulatively operative in the field. These are then shaped and inflected by the specific circumstances and creative responses of the individuals and communities in question, and expressed as concrete events and experiences.

I am acutely conscious of a number of important issues that remain to be discussed in this context, but that space does not permit at this time. One issue is certainly the potential dangers of using and misusing astrology in this area. In general, one must maintain a constant epistemological discipline and self-awareness to avoid the projection of fears or wishes, the drawing of definite or premature conclusions on the basis of limited data, and the urge to control life rather than participate in it. On the practical side, the setting out of strategies for timing psychedelic sessions will require a separate publication. So too, the different operative orbs for the different forms of correspondence (natal charts, personal transits, and world transits); the differences between the hard or dynamic aspects (conjunction, opposition, square) and the soft or confluent aspects (trine, sextile); and the differential importance to be attributed to each of the multiple planets involved in transits at any given time.

Over the decades, Stan and I have discussed an enormous number of individuals' psychedelic and holotropic experiences that have come our way and examined the relevant birth charts and transits. In all cases where we had adequate data, the correlations consistently proved of great interest, instructive, and even after these many years, deeply impressive in their combination of precise archetypal correlation and seemingly infinite creative diversity. While in some ways extraordinarily elegant in its simplicity, the archetypal astrological perspective revealed such an intricate orchestration of cosmic movements and psychological patterns as to leave both of us at times shaking our heads in sheer admiration and awe at the universe's unimaginably powerful intelligence and creative artistry.

As Stan has often remarked, the great irony of our quest for understanding the variability of psychedelic experience was that when we finally found a method that illuminated the character and timing of people's psychedelic experiences, it was as controversial as psychedelics themselves. It seems that the greatest treasures are sometimes hidden in the most scorned and humble places. Jung often spoke of the stone the builders rejected that turns out to be the keystone. Archetypal astrology does indeed seem to be, as Stan has suggested, a kind of Rosetta Stone, allowing us to connect the symbolic language of the psyche with the symbolic language of the cosmos. Just as the expanded cartography of the psyche that emerged from psychedelic research has been immensely clarifying and even liberating as a comprehensive map of the realms of consciousness, we have found that the archetypal correlations with planetary movements have provided us with both an orienting compass and a detailed weather report, at once psychological and cosmic, that can be invaluable aids for the explorer of deep realms.

Any conscientious empirical approach to knowledge obliges us to overcome a natural human tendency to wall off evidence as illegitimate or even impossible when it does not fit our current framework of assumptions. That there could be a correlation, yet no visible causal connection, between planetary alignments and states of human consciousness is, in the end, perhaps no more surprising than that a minuscule change in body chemistry could be correlated with the experience of profound metaphysical and spiritual realities. Both sets of data, astrological and psychedelic, suggest the possibility of an ongoing direct human participation in cosmic psychospiritual processes, a prospect that, though quite antithetical to modern mechanistic materialism or Cartesian forms of dualism, has long been familiar to ancient, indigenous, and esoteric traditions of thought.

Beyond its potential usefulness for individuals pursuing deep self-exploration, what is perhaps especially thought-provoking and timely about this body of synchronistic evidence is that just when our Earth community as a whole is facing a great perinatal crisis of its own, we are discovering that the archetypal symbolism of the outermost planets of the solar system—the "ambassadors of the galaxy," as Dane Rudhyar called them—points with such vivid precision toward the perinatal threshold and the death-rebirth mystery.

II
PSYCHOTHERAPY & CONSCIOUSNESS RESEARCH

Mapping the Courses of Heavenly Bodies

The Varieties of Transcendent Sexual Experience

JENNY WADE

Tribute to Stanislav Grof

I wrote the following article in 2000 when I was almost finished with the largest transpersonally-oriented research project I had ever conducted on altered states triggered by sex—a groundbreaking study, as it turned out, and one that many of my colleagues viewed with jaundiced eyes simply because it involved sex. Not Stanislav Grof.

For years, I had been intrigued by his work on pre- and perinatal psychology, also a focal interest of mine. To say I was a star-struck admirer would not be an exaggeration. The fact that Stan so boldly ventured into controversial territory—including sexual altered states—with such a sweeping vision, innovative strategies, and critical discernment made him my academic hero. After I moved to California and started researching ecstatic sexual states, I shyly approached him to see whether he would share some of his records since I was having difficulty locating first-hand accounts of transcendent sexual experiences. Stan was warm, gracious, and generous with his time—but his house had just burned down, and all his research papers and records had been destroyed in the fire. No matter, he shared what few documents he still had, and most of all shared his wisdom with me in person.

His taxonomy of altered states resulting from all kinds of experiences formed the scaffolding I used to make sense of the complex array of data I was uncovering. Although I later developed different definitions and categories to reflect the nature of my sample with increased precision, this work would not have been possible without the foundation supplied by Stanislav Grof. I might say the same for his encouragement and support over the years.

* * *

One of the most controversial pairings in transpersonal studies is the relationship between sex and spirituality. Sex is an inborn drive to unite with something outside ourselves, foreshadowing the notion of the transpersonal (identifying with something beyond the personal). In addition to the drive to merge physically with another, human beings seek a psychological unification. Metaphorically and literally, male-female sexual unions have parallels in virtually all spiritual systems, ranging from the interplay of the sexes in indigenous religions; creation stories from around the world, such as that of Shiva and Shakti; male and female principles, such as yin and yang, and Shekinah and Sophia as divine female counterparts to the masculine god in esoteric Judaism and Christianity; and in the love poems of mystics in the ecstatic traditions, particularly Sufism.[1]

Furthermore, orgasm is, at least for a moment, a transcendence of the usual confines of time and space when the kernel of the self bursts open. There is a grain of truth in saying, "It was a religious experience" when referring to a particularly intense climax. Sex represents one of the most accessible roads to ecstatic experience, so it is hardly surprising that it is a path many people have found, whether their spiritual tradition recognizes, regulates, or forbids it.[2] Perhaps its very accessibility and highly personal quality, which bypass the intervention of ideology, authorities, and mediators, account for some of its troubled history in the annals of religion.

The fact is, whether spiritual traditions sanction sex positively or negatively in relation to the sacred, the ubiquity of non-ordinary experiences during lovemaking cannot be denied—nor can the inclination of experiencers to connect such events with spirituality. Researchers of spiritual experiences invariably seem to uncover their link with sex in a majority of cases, often to their own surprise.[3] And sex researchers publish records of spiritual breakthroughs and conversions, even among atheists.[4] Thus, events that people consider to be sacred happen fairly regularly through an act that 1) many have been taught by secular as well as religious authorities to be profane, or even shameful; and 2) under circumstances that may be forbidden, or even "sinful" (e.g., in extramarital or homosexual relations). Outside of Taoism, Tantra, Judaism, and other religions that may teach techniques deliberately designed to bring about altered states during sex (but that also regulate acceptable practices and results), a fair number of people (perhaps one in twenty, as suggested by Scantling and Browder's study) seem to have spontaneous, involuntary non-ordinary experiences while making love, regardless of their own beliefs and the mores of their societies. What happens to them, and how do they understand their experiences?

Despite the institutionalization of sacred sex in some religions and its documented recrudescence among the uninitiated, no one has attempted to study these states systematically as they occur in the absence of direction or bias provided by a particular ideology or technical instruction. This article, which describes part of an ongoing research project, represents the first report on the phenomenology and variety of transcendent sexual experiences as they occur naturally in a "naïve" (i.e., untrained) population.

Research Method

For a phenomenological inquiry into the nature of non-ordinary sexual experience, the overall plan was to recruit a sufficient number of highly articulate participants who could describe extremely subtle, probably virtually ineffable, subjective events. Since the range of experiences was unknown, however, there was no preconception of how many individuals would be adequate for this kind of survey. Males and females of any sexual preference, marital status, and ethnicity over the age of 21 were eligible, provided they had not had training in Tantra, Taoist sexual techniques, or similar erotic arts designed to bring about altered states and/or create an ideology or expectation that might shape either the state itself or its interpretation.

Owing to the private and sensitive nature of the subject matter, recruiting was limited to personal and professional contacts and word of mouth. While this introduced considerable bias, it seemed infinitely preferable to advertising through print and electronic media for two reasons: 1) in the initial stages of the research, it might be difficult to spot hoaxes or fantasies; 2) to screen out responses from people whose sexual interests would be at variance with the intent of the study. Recruiting took the form of email solicitations through professional and personal networks, mentions of the study at lectures given in various institutions of higher learning, and referrals from these sources. Solicitations invited anyone who had had a "non-ordinary, mystical, or transcendent experience during sex with a partner," or who knew someone who had, to participate. Anyone who self-selected based on that vague description was initially accepted. Only those who were using psychotropic substances at the time of the events in question were eliminated in order to minimize the artifacts of other altered-state triggers.

Recruiting qualified people was not particularly difficult, despite the minimal wording. People who said they had never had "transcendent" sex, whatever they imagined it might be, were absolutely certain. The momentary loss of self, time, and space during orgasm is so familiar that people seem to identify this sensation as "ordinary." A powerful orgasm may jokingly be a "religious experience," but people seem to classify even the most exquisite climax differently from an experience of transcendence. In fact, one researcher who compared powerful altered states he had had ("cosmic consciousness," psychedelics, and "the best orgasm of my life")[5] said definitively that the orgasms to which he was referring were "rather ordinary and certainly not 'mystical.'"[6] Thus, these stories come from self-identified volunteers with their own understanding of "non-ordinary, transcendent, mystical experiences" during sex, discussed below.

The following report is based on the results of 86 interviews with heterosexual, homosexual, and bisexual men and women ranging from 26-70. A surprisingly large number for a qualitative study, especially a phenomenological one, this number was necessary once the full range of experiences began to emerge. Since the thrust of this article is a phenomenological cartography, no attempt will be made to provide a quantitative analysis of the results or detailed demographic data. However, some general qualification of the sample is offered to indicate its limitations. It is not representative of American adults demographically for the reasons stated, nor is statistical validity a goal in this project. Furthermore, no attempt has been made to include representatives of the full range of spiritual practices and beliefs or the full range of sexual proclivities and experience.

The sample consists of fifty (58%) females and thirty-six (42%) males. The majority (77%) are heterosexual in their preference for partners (35 [41%] females; thirty-one [36%] males), followed by homosexual (16%: ten [12%] females; four [4%] males), and bisexual (7%: five [6%] females; one [1%] male). Marital status was not a useful category for various reasons: 1) individuals were sometimes speaking of past events unrelated to their present marital status; 2) if married, they were sometimes speaking of extra-marital events; 3) those with long, rich histories had had experiences while single, married, divorced, or outside of marriage; 4) for same-sex connections, marital status was not necessarily a relevant descriptor.

The sample is predominantly Caucasian (88%). Other groups represent 12% of the total: four (5%) each African-American and Latino; and one (1%) each Asian and Native American. Owing to the solicitation venues—mostly through universities and contacts in higher education—the sample is highly skewed toward people with advanced degrees: twenty-four (28%) doctoral degrees (PhD, MD, JD); thirty-four (40%) masters degrees; five (6%) some graduate education; eighteen (21%) bachelors degrees; two (2%) associates degrees; and three (3%) high school diplomas.

Concerning spiritual beliefs that might have a bearing on the subjective interpretation of non-ordinary states, people were asked about the traditions in which they had been reared and their present religious orientations. Forty (47%) had been brought up in households identified with various Protestant denominations; twenty-one (24%) Roman Catholic; sixteen (18%) agnostic or atheist; seven (8%) Jewish; and one (1%) each Buddhist and "eclectic." As adults, the vast majority have shifted their spiritual orientation—some as a direct result of their sexual experiences—though the reverse did not seem to be true: they noted little, if any, crossover from a spiritual practice, such as meditation, and the types of experiences they had during sex (a full discussion of these findings is beyond the scope of this paper). At present, most follow less traditional paths: forty (46%) consider themselves "eclectic" In their orientation; thirteen (15%), atheist/agnostic; ten (12%) "generally Christian" or a specified Protestant denomination; eight (9%) Buddhist; four (5%) Roman Catholic; four (5%) Yogic/Hindu; three (4%) Pagan/Nature mysticism; and one (1%) each Islam, "Jewish mysticism," "New Age," and Transcendental Meditation. (Interestingly, of the thirteen who said they were now atheists/agnostics, six used theistic references throughout their records in describing the events and meaning of their sexual experience.)

Respondents were interviewed in a mutually agreed upon location, or, if inconvenient, by telephone. The interviews were semi-structured and took anywhere from thirty minutes to two and a half hours. They consisted of standard open-ended questions, followed by additional probes to develop promising lines of inquiry and to clarify. A few preferred to write about their experiences, and in those instances, the same questions and probes were used in a dialogical correspondence.

How adequate is this sample? As already noted, the intent was exploratory, and the selection, deliberately skewed. This is an ongoing study, and it is possible that there are experiences outside those represented here, but the degree of overlap among 86 cases suggests that this is at least an adequate first attempt to chart new territory. How trustworthy is the information? People may exaggerate or fantasize about their sex lives or avoid telling the entire story. A further difficulty exists when the events in question involve an altered state, which may be difficult to capture and convey in an ordinary state. This is complicated by the difficulty inherent in communicating phenomenological information verbally (e.g., describing the quality and nature of an orgasm). Furthermore, the accounts represent the self-reported version of a subjective experience uncorroborated by anyone else. Even in the few cases where both partners agreed to be interviewed, their own experiences frequently differed significantly from each other.

It would be difficult indeed to make any claims for the data if any single experience had to stand on its own. But, just as with near-death experiences, when person after person describes the same type of event, the same feelings, and the same insights, patterns begin to emerge. Each story is a data point—the more there are, the clearer the image. The patterns they produce may not form the sharp delineation of representational art, but they can give the softer, more layered images of impressionism. The emergent images are as meaningful and clear as those of Monet, Van Gogh, or Seurat. Although only eighty-six people were interviewed, their stories represent more than eighty-six events. For some, a transcendent experience was a rare happening in a lifetime of sexual activity. Like any other profound spiritual event, such as a near-death experience, these one- or two-time occurrences were reported with great vividness from times and partners often long past, sometimes as much as thirty years later. For many others, however, transcendent sex occurred with some regularity in a particular relationship, often covering two or three different types of experience over time. A fortunate few commonly have transcendent experiences whenever they make love and wonder why anyone would "bother" to have ordinary sex. They sometimes had a range of experiences to report. The net result is that no category mentioned below represents a single case.

Defining a Transcendent Sexual Experience

Owing to the variety, no simple definition of transcendent sex is possible. However, two factors consistently emerge from the narratives as characteristic of transcendent sex. The first, not surprisingly, is participation in an altered state that could not be ascribed to the use of chemicals or deliberate techniques. In these states, as illustrated below, the ordinary sense of time, space, and/or agency (in Cartesian-Newtonian or Formal Operations terms) is transcended. Furthermore, the altered state includes an awareness of the lover, if only as a conduit, and is rooted in the union of the two during sex. These altered states appear to be more or less independent of orgasm, which is considered a discrete state of its own. That was invariably true for the men: they entered an altered state that had no relationship to the timing or duration of their climax. It was true for the majority of the women, as well. For a few, however, being non- or mono-orgasmic during sex is either infrequent or impossible. These women were having an unending chain of orgasms that could last indefinitely, usually, until their lovers withdrew the contact, although the events they described as transcendent had little or no (subjectively) discernable relationship to orgasm. (For most men and women in this sample, orgasm was either a non-event or a problem relative to the transcendent sexual state. Some could not recall whether they had had one as they were much more drawn into other events; some said it became "irrelevant;" others found it a somewhat irritating distraction, and for a couple, orgasm shattered or ended the state.)

The second factor that differentiates transcendent from ordinary sex is the felt experience of a cosmic force engaging one or both lovers in the context of their relationship, especially their lovemaking. This cosmic force is most often described in the terms reserved for Spirit: God, the Divine, the Oversoul, the Void, etc. Some described it spatially as a place they can enter, a world of cosmic power, intelligence, and love. For some, the force is implied: it is the invisible source of their journeys into other realms, or it makes the other realms possible. For others, it is a living entity that somehow arises from, but cannot be reduced to, the union of the lovers. The tendency to associate such experiences with the supernatural was marked, frequently resulting in deep personal transformation, spiritual insights, and even conversion.[7] Indeed, the interplay of both these factors—altered states and the felt sense of the numinous—creates a variety of sexual experiences that appear to have the same characteristics as those recognized in some of the major spiritual traditions.

The Varieties of Transcendent Sexual Experience

Classification of complex human experiences is fraught with difficulties as any taxonomy is somewhat arbitrary and can never reflect the dynamics or richness of an individual's experience per se. Rather than create a taxonomy specifically for this project, it seemed wise to determine whether an extant model would serve. The one chosen was developed by Stanislav Grof, and it was selected for several reasons.[8] In the first place, he originally created a comprehensive cartography of transpersonal states arising from a variety of sources: his psychedelic research, holotropic breathwork, and experience working with individuals having spontaneous episodes of non-ordinary states. It generally is congruent with perennial philosophy, although it contains more categories and differs in some details. Moreover, Grof has also frequently acknowledged the relationship between sex and various non-ordinary states in his writings.[9] His schema was also chosen for reasons of brevity (Rhea White's Exceptional Human Experiences classification is magnificent, but with over 200 categories, it would not be easily reproducible in this venue. As it is, Grof's 1988 addition of a third grouping labeled "Intentional Psychokinesis" is omitted because it involves the "ability to influence the material environment...by simply wishing events to happen," a category at variance with the involuntary, spontaneous nature of experiences sought for this study.)[10]

His model does indeed cover the kinds of episodes arising during sex. What follows, then, is a brief recapitulation of Grof's complete taxonomy, indicating where these sexual experiences fall and giving examples but also showing what types of experience have not (yet) been found to arise from sexual activity under the conditions noted above.[11]

Regrettably, in the brief space of this article, it is not possible to develop any of these cases fully outside the condensed content presented here for classification, which is a shame, because virtually all are ecstatic and transformative. The full records are luminous with joy, peace, discovery, insight, and inspiration. They have typically been life-changing events, but only such quotes as illustrating the phenomena for classification can be given here. Nevertheless, it is hoped that at least a tiny bit of the radiant bliss will shine through.[12]

Transpersonal Experiences

Experiential Extension within Consensus Reality and Space-Time

1. *Transcendence of Spatial Boundaries*

- **a.** *Experience of Dual Unity*—Loosening and melting of the boundaries of the body ego and a sense of merging with another person into a state of unity and oneness.[13]

  Perhaps the most common of transpersonal sexual experiences is the sense of the two lovers merging into one being. Respondents report not only a strong sense of psychological and spiritual union, but physical, as well, including, as seen in the second example, the sharing of kundalini. These are accompanied by profound feelings of sacredness and love, also noted by Grof with other subjects and relationships.[14]

  > Then any sense of separateness between us dissolved. I couldn't even tell whether I was making love to her or being made love to. I can hardly even tell you what our physical bodies were doing because it was like our bodies were part of the flow and ebb of all this energy and Spirit body. We were all mixed together in this mysterious, melting dance. Body awareness merged with all the other levels....We were one moving, touching mass of energy and awareness, not two separate poles of consciousness....I sort of felt like a woman *and* a man...where all we were was one being, one love, kind of a melting together.—Kyle

> Often...my partner will start to feel the energy in her pelvis, and it goes in shock waves up her spine. She'll begin to have *kriyas*... and my body will react. I'll feel it almost like my spine is a radar dish or something. Her spine is sending out this wave, and my spine catches it and responds....And sometimes vice versa.—Nanette

b. *Identification with Other Persons*—None

c. *Group Identification and Group Consciousness*—Expansion of consciousness to a global awareness of an entire group of people by race, nationality, cultural heritage, religions, profession, shared ideology, or destiny.

Some women identify with all women's experience, and in so doing, find their femininity affirmed, making them expand more into that part of themselves and feel more empowered.

> They feel like they're very, very close, all women. Not an identity, but *all women*. My identity falls away, and I'm identified with all women now and back in time, and their state of mind. There's not a separation....My way of being a woman is so much greater from having had this experience. I grew up Catholic...with the body being sinful and touching it being sinful....A sense of rapture that came from so many different places, and how wonderful it was to have that...and that sense of unity [with all women].—Kristin

d. *Identification with Animals*—Complete and quite realistic identification with members of various species.

Animal identification in this group may not be well-matched with Grof's because his examples display an abstruse knowledge such as what might be ordinarily be known only to biologists or animal specialists. Sex respondents report more feeling states than an unusual knowledge of actual animal behavior or attributes. The potential degree for cultural determination appears to be high in this sample, but perhaps also in this category of experience, given the reports of various indigenous groups to identify with culturally significant organisms, such as grubs, salamanders, vultures, spiders, scorpions, ants, snakes, frogs, weasels, baboons, rats, coyotes, and the like, while this North American sample mention only predators—usually mammals (wolves, bears, and the big cats) but also birds of prey. These most puissant animals in modern-day North America, however, are often the ones associated with shamans, who are among the most powerful individuals in their societies.[15] Metamorphosis into animals during lovemaking typically involves aggressive, rough sex play, which, along with the above, may suggest fantasy. Participant perception, however, is that of being overtaken (possession), confusing the causal link between sex play and the shift. Increased strength, bestial mannerisms—growling, snarling, scratching, biting, fighting or wrestling—and the altered-state sense of actually being and thinking like an animal manifest in ways quite contrary to subjects' everyday personalities and sex play.

> With that [electrical charge] comes a power and a strength that makes me feel invincible, like [I] could run through the woods and jump over trees any time now....It's that kind of strength. If I'm the wolf, my hips and butt and legs, especially the tops of my legs where the quadriceps are, feel different. Definitely leaner, sinewy—God, I wish I were sinewy!—kind of like a haunch sort of a musculature so you could spring. When I'm a wolf, it seems easier to move and jump about, and when I'm a bear, it doesn't. As the bear, I feel the weight of the head, really, really big, and very, very heavy.—Kim

e. *Identification with Plants and Botanical Processes*—None

f. *Oneness with Life and All Creation*—Identifying with the totality of life on this planet, including all humankind, flora, and fauna (extending to viruses and single-celled organisms).

> A number of individuals report an opening of self to identify with all creatures.

> There's a connection with the universe that happens. I sense a connection with the flora and fauna, all the other animals [and people] in the world. The experience has more of a horizontal plane to it, the immanent reality. But with the sense of oneness and connection certainly. It's a different plane, closer to the ground, about union with all living

> creatures....We're really connected, all of us with each other for all time and all space. And there's a sense of how vast the universe!—Terry

g. *Experience of Inanimate Matter and Inorganic Processes*—Identification with macroscopic and microscopic phenomena of inorganic nature (e.g., consciousness of fire, the ocean, iron, quartz).

Rarely do individuals go into the microcosm during sex, but when they do they see and become part of the interplay of energy and matter.

> It's almost like a biologist or physicist looking at subatomic particles or things under the electron microscope, the basic patterns of energy that the universe is made of: forms of waves and rays and points and the interplay of expansion and contraction, progressions, and the experience of densities. I'm having all kinds of visions and insights, the kind I imagine physicists have, where there's the confluence of energy and matter, the regeneration of life and death.—Donna

h. *Planetary Consciousness*—None

i. *Extraterrestrial Experiences*—Travel within this solar system, and to other stars, and galaxies, witnessing explosions of supernovas, contraction of stars, or identifying with these entities and events.

A number of people (mostly women) journey blissfully through outer space, flying along in the darkness surrounded by stars and planets.

> And then again, I felt myself in the cosmos, but this time it felt like in the solar system with solid planets and gaseous planets being present and around me. I seemed to be in outer space. There was a lot of space between what was out there, blackness with points of lights, stars....There seemed to be a presence of other planets too, and I was floating.—Vivian

> We were among the stars. It was black, and there were all these stars, and it was like we were flying, both of us flying. It was mostly a visual impression, but it was accompanied by that feeling of incredible openness and joy. I was just incredibly happy. It was very much a feeling of being my true self, being who I really am, authentic.—Jill

j. *Identification with the Entire Physical Universe*—None

k. *Psychic Phenomena Involving Transcendence of Space (out-of-body experiences, traveling clairvoyance and clairaudience, and telepathy)*—Leaving the body; moving to another place in the physical world; occasionally moving to experiential realms and subjective realities that appear to be entirely independent of material reality; reading the minds of other persons.

Out-of-body experiences are common, usually moving to a place suspended above the lovers and looking down at them in the here-and-now (very similar to those reported under other conditions, though usually accompanied by feelings of great peace and wonder.)[16]

> I suddenly found myself having for the first time what I would call a transcendent kind of experience. I was out of my body observing me and my lover lying in bed from above, perhaps near the ceiling. I'd never heard of such a thing and didn't know how to explain it, but it was a very strange phenomenon suddenly looking down at the two of us.—Dick

Along the other dimensions of spatial plasticity Grof includes in this category, some (mostly women) report being transported to different urban and pastoral locations, but most often under the sea. Transports are typically accompanied by ecstatic feelings.

> There was a time when we were like dolphins in the water. I saw other fish in the water, but I was more aware of us.... The water was very blue, and it was extraordinarily peaceful, joyful.—Jill

In a striking case of telepathy, a man making love with his date at a party became aware that his best friend, who was attending the same celebration, needed help....

> I went into an emptiness, and in that emptiness an awareness came to me that Tom was in danger.... And it was such *truth* that I *knew* he was in danger....So I immediately came back into my body. I said, "Oh my God, oh my God. Something's wrong with Tom."

> She said, "No, there's not. He's still at the reception. He's having a great time...."
>
> I started running down the street throwing on my clothes.... As soon as I turn the corner, there's this mob of people. In the center of it was Tom and this huge Hell's Angels type of guy brawling with him. Without hesitation—and I'm just a little guy—I lunged. I just jumped on this guy's back, trying to get him in a headlock. He was a huge, huge fellow.... This guy could have taken me out with his thumb, if he had wanted to, but I just had this tremendous energy.... That evening, Tom said, "You should really be hurting." And I should have been because that was a brick wall [the attacker] threw me against, several times. I was not bruised or in any kind of physical pain out of the experience, during it or afterward.—Esteban

2. *Transcendence of the Boundaries of Linear Time*

a. *Embryonal and Fetal Experiences*—None

b. *Ancestral Experiences*—Historical regression along biological lines to periods preceding the subject's conception and an identification with one's own ancestors.

Some respondents feel transferred back in time into the bodies and personalities of their forebears, which usually is something of a shock. For example, one woman found herself as a Celtic priestess during the Roman conquest of Europe.

> I was there during the time when the Roman empire was invading Gaul....I have this sense that I died in a Roman prison. Some of the most excruciating pain I've ever experienced was a vision that I had right after we'd made love, this feeling that they were burning my trees, burning my groves.—Ardrigh

At the time the incident occurred, she knew nothing consciously about ancient Celtic culture, nor about her own ancestry, but her identification with these events was so strong, she decided to check into her origins.

> I didn't even know the names of my great-grandparents....I did the research later. I found out that the groves were sacred in the Druid tradition. They actually worshipped in groves of trees, and that's why I was just so upset that my trees were burning. I've tried to verify some of the other stuff as actual historical occurrences, and of course, the Roman empire eventually overtook Gaul up almost into Ireland....But when I get in that guise, I can speak Gaelic fairly well, and I don't even know Gaelic. It's the damnedest thing. [My lover] has written down some of the stuff I said...and it's old, old Gaelic. We've actually looked it up.—Ardrigh

No claims are made here for the veridicality of such events, but their felt validity is sufficiently strong that most other participants spoke of conducting similar research or experiencing uncanny manifestations and synchronicities in conjunction with these past-life episodes.

In another example, a white man whose mother had always been troubled and fascinated by fair-skinned African-Americans who could almost "pass" for Caucasians found himself as the Creole mistress of a white plantation owner who later married a black man.

> I had six children in that life. Three of them passed for white because their father was a planter, a white planter. One of them, one of the children who passed for white, is my mother in this lifetime. She was my son in that lifetime. The person who is now my mother left New Orleans and came to Minneapolis and passed as white in that lifetime. In this life, she also went from New Orleans and Tennessee to Minneapolis.—Richard

c. *Racial and Collective Experiences*—None

d. *Past Incarnation Experiences*—Participation in episodes occurring at another time and place in history. The person maintains ego-identity, while experiencing the self in another form at another place and time.

Persons report past-life experiences without any sense of their biological lineage. These events frequently mimic those of past-life regression therapy in revealing (sometimes foretelling) fundamental conflicts that are not apparent to the lovers at the present time. In one, a couple on their second date relived a history of several years' duration in which they met, fell in love, had an illicit affair, created an unintended pregnancy, lost the child, and split up in Victorian times. The story was particularly disturbing, as they were just falling in love and had no belief whatsoever in reincarnation or other "nonscientific" phenomena. They married and were together for a long time, but by the time of the interview, they were separating.

> The story was dictating to us, like watching a movie, and you don't get to change it if you don't like what's going on. It was playing out, and we were being shown something....I didn't see any similarity in it with us at that time, though it did say something about why we felt so compelled to be together even though we didn't know each other that well, why we felt so familiar, like we'd always been together. It was overriding my logical mind, which was a big override for me. The rational part of me dismissed it. I'd say to myself, "Oh, you were just daydreaming." I believe we were given an opportunity to be with each other in a different context, but we haven't been able to do it. We made the wrong choices again in this lifetime. It's too painful to continue.—Carolyn

Unique among the sexual experiences, only those involving past-lives in some way seem to place both partners in a shared altered state. Respondents usually describe this as "finishing each other's sentences" or "like we both had the same movie going in our heads, but we weren't making it up."

e. *Phylogenetic Experiences*—None

f. *Experiences of Planetary Evolution*—None

g. *Cosmogenetic Experiences*—None

h. *Psychic Phenomena Involving Transcendence of Time*—Psi phenomena including precognition, clairvoyance, and clairaudience of past and future events; psychometry (obtaining history or information from touching an object); time travels.

Sex respondents report instances of precognition. Like the past-life experiences, future insights seem to involve negative outcomes, although the feelings associated with them tend to be peace and calm acceptance. For example, one woman who was resuming with great hope a relationship that had been suspended for a number of years, suddenly saw the future.

> I felt that we were lifted into levels of possibility, of unfettering connection, though perhaps not yet realized, which existed at some level still to be reached. I was thoroughly confident that the path was there even though I was perhaps on it alone. The feeling was of liberation...being freed...a sense of opening to another unknown dimension of experience....[In this one transcendent moment] it was as though I was lifted into a dimension of truth beyond ordinary rational understanding, wherein I knew it to be the realm of false hope. The disjunction of this overwhelming feeling and the uncertainty of what I knew to be the more likely outcome of our association was not at all upsetting. There was a refreshing calm that settled over and stayed with me for days.—Leah

i. *Physical Introversion and Narrowing of Consciousness*—Organ, tissue and cellular Consciousness

Cellular awareness tends to be reported in two ways, one of which is not necessarily congruent with Grof's definition in that it involves a sense of the "aliveness" of individual cells owing to an unusually charged sense of the entire body during sex.

> It was an experience of being both completely and totally aware of every cell in my body—and of every cell in my body vibrating at the same frequency—and yet not even *inhabiting* my body at exactly the same time.—Betty

The second type of experience seems to draw the individual down into the cellular or molecular microcosm.

> It's as if I dove [*sic!*] into your eyeballs, and I got into the interstitial spaces and into the vitreous humor and started to push my arms and legs out to make everything wider. And all of a sudden you feel that you operate from a very contracted place, relatively. Muscles relax where you normally don't even know that they're there because, at the time, you

thought they were part of an immovable structure. *Now* I can lean back, *now* I can stretch out and feel limber and really unfurl.—Sabu

Experiential Extension Beyond Consensus Reality and Space-Time[17]

a. *Spiritistic and Mediumistic Experiences*—Mediumistic and channeling experiences in which the individual's countenance, gestures and voice are dramatically changed. Sometimes the individuals perceive a discarnate entity but also may be taken over by an alien entity or energy form.

Perceiving other entities happens more often during sex than the sense of being changed into one (except for animals, noted above). Called "shapeshifting" or "morphing" by participants, these occurrences normally involve the rapid succession of a number of different faces superimposed on the partner's actual face (*trespasso*).

> At least half a dozen times, I've had an experience with my eyes open where I'd see my lover's face morph into all these different faces. [There] might be 20 different faces. I don't remember them all, but some are more memorable than others, so I'd recognize them when they came up again.—Vivian

Other experiences include the sense of physical manifestation and could involve not only physiological changes, but movements such as gestures and glossolalia.

> When she steps into that place, she gets rougher sexually. There's a rawness. She even smells different. And she said my eyes seemed a lot darker when I looked up. I have light hazel eyes, and she said they were so dark it was almost like black eyes. Her sense was that they weren't my eyes at all.—Lynn

> We would spontaneously start speaking in tongues, and at first I felt kind of embarrassed about that, like "What *is* this?" ....Either he would start speaking in tongues, or I would, and when that started happening, it was another signal that this wave of energy is coming over us, through the different movements of my tongue or different hand gestures, perhaps *mudras*.—Leona

b. *Energetic Phenomena of the Subtle Body*—Experience of various energy fields described in mystical traditions of ancient and non-Western origin (e.g., kundalini).

As might be expected, kundalini effects are well represented. They typically involve unusual percepts of light, heat, energy, and liquification, and they seem to behave in the same ways described in the Tantric literature, with which very few of these subjects had even a passing familiarity (one couple was upset by an episode of *amrita*).

> I'm aware of energy, patterns and electric colors....When it moves past certain areas of the body, the body will tend to get kind of tingly. It's like there's pure crackling, surging, grinding, burning energy. Boundaries tend to expand, and...the further the energy goes up the spine, it seems like the more intense it gets, and I start getting into an altered state.—Reginald.

> The light was just going through me.... and shooting out of the top of my head. I had feelings of white lights shooting out of the top of my head....You think of sunlight or lightning, but nothing like *this!*...I started crying because it was overwhelming, an emotional effect like "Oh, my God, I'm coming *home*. I got home." A feeling that I'd been separated for a long time, like coming together, and finally being *home.*—Suzette

c. *Experiences of Animal Spirits*—None

d. *Encounters with Spirit Guides and Suprahuman Beings*—Interactions with spirit guides, teachers, protectors, beings of light.

During sex, people may have visions of supranatural beings (e.g., angels, predeceased loved ones). In some cases they feel themselves or their partners manifesting divine forms (e.g., their Ideal Form in a platonic sense). More commonly, they sense an ephemeral Third presence or divine entity that is co-created with and by the lovers, yet also independent of them.

> I had a little sister who died when I was ten and she was six. And I felt her presence in the room right then....When I was growing up, she was really close to God because she was dying the whole time she was growing up. I felt her close to

the divine Presence, as if she came from God. So as an adult, whenever I had an altered state—well, not whenever but sometimes when I've had them—I would see her....I didn't have a sense of myself as being just me, but of a divine Presence being around.—Armand

Sometimes when we make love, I see my lover's face grow young, and she looks 20 years old to me. She has that radiant, lit-up-from-within beauty, almost as if she is divine herself in a way, as if she becomes a perfect version of herself. It's like looking at images in a church, or icons that can be suggestive of something beyond themselves. There's something transcendent even in the vision of her, like when great art gives rise to religious feeling. When she transforms before my eyes, I've been displaced to a different environment where everything has become radiant because *she's* my environment. I'm going into a holy place.—Blake

I've always had a sense of movement beyond myself. I have no illusions that my partner and I are creating that...I haven't come to a good term for that....Some of the native peoples call it the Great Mysterious....By our coming together... maybe there's something that we're doing that is the kindling in the midst of that fire, but...that fire comes from somewhere else, not just from the two of us....Making love... is a way to go to church, talk to God, connect with God, however you might understand that.—Roland

e. Visits to Other Universes and Meetings with their Inhabitants—None

f. Experiences of Mythological and Fairy-Tale Sequences—None

g. Experiences of Specific Blissful and Wrathful Deities—Experiences with deities associated with the forces of light and good or darkness and evil.

Somewhat rare, these experiences concern deities associated with archetypal forms (e.g., Sappho, Kali). The majority are ambiguous in their presentation, such as Kali, whose fierceness is as much relished as feared.

I beheld this dark man, fairly shadowy, I really couldn't come up with any features other than a rack of antlers... . It was Carnanuss, ...the [Celtic] lord of the forest and

> the hunt [who symbolizes to me] the seduction into the wilderness, the betrayal of that trust and murder. I think that's the degree of anxiety I have around this figure.... My Catholic background said it was the devil, but it was Carnanuss.—Zebediah

> All of a sudden, I heard this horrible roar...and a serpent came out of the water, the rainbow serpent, and he dived back in and I could never quite see its head but I could see its body where it came in and out and in and out. It was frightful and beautiful, shimmery, very bright.—Natasha

h. *Experiences of Universal Archetypes*—None

i. *Intuitive Understanding of Universal Symbols*—None

j. *Creative Inspiration and the Promethean Impulse*—None

k. *Experience of the Demiurg* [sic] *and Insights into Cosmic Creation*—None

l. *Experience of Cosmic Consciousness*— Identifying with the supreme and ultimate principle that represents all Being characterized by formless, dimensionless, infinite awareness knowledge and bliss.

Participants describe unitive experiences in the same language of the *unio mystica* of Western contemplative traditions, especially omniscience and overwhelming joy.

> It's a very boundariless time, and then comes the Light, mostly a flood of bright, whitish-yellowish light....It's as though the Light were all the universe that's not me, the part of the universe that I don't think I'm usually connected to. That Light definitely represents the life force, the universal force, the god force without the deity....I'm content in the Light because I'm also there observing it. I love it when this happens, but there is that moment when it begins, and I step in it without that awareness. It is maybe nondual for a few seconds, and then I'm there observing it, so it's not there.—Esther

> It's *real!*....At the same time I felt it coming from the outside, it was also coming from the inside. It was evident that it was also in me. I participate in divinity in a direct, inclusive way. There's no separation at all. All those mystical incantations about the entire universe being embedded in the person, or

that the soul of the human being participates in that, it was all suddenly manifest....Was there a face to God? No. Or a presence? *The presence was us.* It wasn't a call, it wasn't a sign. All the things I had interpreted as it weren't true. We were just there. It just *was.* Oh, yes, it's real. I become all of me but with my ego or subpersonalities all gone. Just God is left, and I am that.—Zebediah

m. *The Supracosmic and Metacosmic Void*—Experience of primordial emptiness, nothingness and silence; the Void that is supraordinated to and underlying the phenomenal world.

Several describe nontheistic nonduality in terms very like those of Buddhism, though it is hard to do them justice in a short excerpt.

> I sensed that I had gone some place, and that once I came out of it, I knew I had been there, but not when it happened. You know how there's the being in something, and then the knowing that you were there. So when you come out of just being and you get to the knowing part, you're an observer apart from what comes in that moment. Everything was dropping away, no sensory perception and there is to way to describe it. There was nothing but union in that moment, but I couldn't tell you union with what. And immediately afterward were the tears, tears of joy, incredible joy that I felt the privilege of having this incredible gratitude and awe.—Marta

> There was nothing there. The boundaries of the body *went,* and yet there was an awareness of doing, so there was an awareness of action, but the sense of the body was not. There was this dissolvingness, and this losing of boundaries. And then there was this incredible nothingness and everythingness. Out of this feeling of nothingness and no-self, there was yet all possibility and all potentiality. It was probably for only an instant, but in that instant there was no time and no space....Yet there was somehow an awareness of what was happening, an awareness of something and nothing all at the same time....There was just a more complete sense of really pure awareness, just no sense of self. The completeness of the nothingness was enormous.—Ann

Transpersonal Experiences of Psychoid Nature

1. *Synchronistic Links between Consciousness and Matter*—acausal connections linking various types of transpersonal experiences to physical events in the phenomenal world, especially the meaningful coincidences of events separated in time and/or space.

   Grof recognizes a number of phenomena in this general category that include experiences people report during sex, especially unusual connections between their actions and natural forces. As Grof notes, these are clearly subjective intrapsychic events but they are also meaningfully connected with specific physical changes in consensus reality.[18] Subjects in this study speak of both meaningful and magical synchronicities. A few think, for example, that their lovemaking creates a "bubble" in time.

> We must have made love in our customers' houses hundreds and hundreds of times. We always did it at our own pace, never hurried....It was the right thing to do in that moment, and in that moment, our sex ran the world. Once we started, we were completely safe. It was as though, somehow, through our strong desire to make love and our freedom from embarrassment, we were never caught. With this particular man, our love was so confident and sure it seemed that the universe honored the magic in us...but feeling as though the whole world is in me, and I'm the whole world. And it worked. We were never caught, ever....It amazed me that it was only just as we were putting ourselves back together that the customers came back. Never before.—Francine

Others have a sense of participating in the forces of nature.

> Our caressing and touching was slow and languorous, and we were moving very slowly together. There was a real sense of connectedness to each other. We were looking into each other's eyes a lot, and I started feeling as if our boundaries were melting....We had an orgasm together which was lovely, really lovely. We were both kind of trembling in that moment. And then the bed started to shake. And then the whole room started to vibrate. So right at the moment that we were having this intense climactic, orgasmic experience, suddenly the whole world was shaking. There was an earthquake!—Roland

2. *Spontaneous Psychoid Events*

a. *Supernormal Physical Feats*—Spectacular physiological changes in the body or seemingly impossible achievements (e.g., stigmata, luminosity of the body of various saints, etc.)

See the above description of unusual physical engagement in an unequal fight without physical injury (Esteban's narrative in 1.k. Psychic Phenomena Involving Transcendence of Space). According to Esteban, he threw himself on the back of a much larger man who was attacking his friend, attempting a headlock. Although Esteban notes above that he suffered no injuries, he was not, in this instance, supplied with superhuman strength:

> It was kind of comical in that [the attacker] just picked me up, plucked me off his neck and threw me against the [brick] wall. That should have hurt me, probably should have broken some bones. But I immediately got up and lunged at him again. In fact, I lunged at him about four or five times. I was obviously not in my right mind. All I knew was that I had to protect Tom, and yet Tom is bigger than I am. He's got a lot better chance than I do of taking care of himself.—Esteban

b. *Spiritistic Phenomena and Physical Mediumship*—None

c. *Recurrent Spontaneous Psychokinesis (Poltergeist)*—None

d. *Unidentified Flying Objects (UFO Phenomena)*—None

Conclusion

The numerous vague allusions in the literature to non-ordinary experiences occurring during sex thus appear to represent an astonishing array of altered and transpersonal states, many of them consonant with those described in the spiritual literature of various traditions, including the far-traveling and otherworldly episodes of indigenous religions and those associated with various stages of contemplative practice in the esoteric and perennial systems. Although another classification scheme would yield different results, this one allows for the extensive range of phenomena and tends to point to clusterings around higher-order entities and organisms, connectedness, terrestrial and organic (as opposed to inorganic) processes, and archetypal imagery.

Probably owing to the recruiting method, the majority of these experiences were positive—so positive, in fact, as to be ecstatic. These results can hardly be viewed independently from their context, even though not all of the sexual relationships were happy ones. Where negative emotions were either present or foreshadowed, they still seemed to be productive of valid realization and insights and thus were positively valued by the subjects. (Destabilization of ego boundaries as a consequence of transpersonal experiences is reported in this sample, but the experiences themselves were still quite positive.)

At this point, it seems clear that naturally-occurring sex can be just as effective as other processes in bringing about altered and transpersonal states, including those vaunted by traditions that highly regulate and repress bodily desires. Since these naïve subjects were not deliberately attempting to bring about the events, and since their spiritual beliefs were usually at variance with any expectation of a "religious experience" during sex, they were not especially helpful informants concerning causal factors. Several reasons suggest themselves, though.

To the extent that altered states result from changes in brain electrochemistry (an important qualification), they can be generated by many factors, such as lack of nourishment or sleep, sensory deprivation or overload, repetitive sounds or motions, drugs, contemplation, etc.[19] Sex often involves a number of these—especially repetitive motion, sensory overload, and a concentrated focus similar to meditation—not to mention the changes in body and brain chemistries as hormones are released into the bloodstream. Of course, some individuals are more inclined to be somatically attuned or swept into states of absorption. Other factors suggest that relaxation, letting go of all physical effort, and becoming highly concentrated in the present contribute as well. (A full treatment of these factors is outside the scope of this paper.)

The fact that sex can take people to the same realms as trance, meditation, drugs, etc., may well support the notion that the phenomenal worlds of human experience fall into certain universal patterns, as has often been suggested by materialists and transpersonalists alike. No matter what the trigger, there are only so many "destinations." Realizing that sex can transport uninitiated and disbelieving people to so many of these points, though, is powerful, new knowledge that can potentially benefit many.

For instance, just making this information public would open a discourse with those who are now shamed and silent. One of the saddest findings has been learning that with rare exceptions, subjects had never told another person—not even their partners—about their experiences prior to the interview for fear of being considered crazy. Just as in the early days of near-death research when people were afraid to talk about what happened to them, they expected ridicule—and often had to struggle alone trying to make sense of events that severely undermined their belief systems. Many also knew their partners were unsympathetic to "spiritual stuff." So, it was easier to suffer profound experiences alone than bring them out into the open, effectively creating more distance in the relationship regardless of the positive nature of the experience itself.

Finding a spiritual outlet for discourse was also difficult. As the results indicate, few people's traditions value non-ordinary experiences, and many—perhaps most—overtly discourage any thought of the convergence of sex and the sacred. Even for those who left mainstream practices, few found their experiences honored in recognized contemplative traditions, and instead have developed their own idiosyncratic, personal practice (as have others, such as near-dear experiencers who have had profound spiritual experiences at variance with the dogma of recognized paths). Only a handful of people said they were willing to discuss their sexual experiences in spiritual venues.

Acceptance of this type of embodied spirituality seems rare. Sex continues to have a troubled relationship with most spiritual traditions, but as an accessible medium for transpersonal experiences, greater awareness could foster discourse around its potential for transpersonal openings, personal transformation, and spiritual knowing. To the extent that sex is treated as a "lower" form, an unworthy desire, or a hazard to "true" spirituality, its potential will be repressed and its discoverers marginalized.

Certainly, sex is not the path for everyone (and the dangers of creating yet another hurdle for performance, however implicit, are awful). But it can be recognized as a powerful method of transpersonal journeying. Knowing that it can lead to the same places and realizations that recognized spiritual paths do can open new areas of exploration. With a map of the territory, those who have been there can come out into the open, enjoying the affirmation of knowing their experience is neither unique nor shameful and sharing their gifts with the larger community. Others can at least learn there is a new world out there, a vision of sex that goes way beyond more and better orgasms, whether they wish to attempt it or not. Just knowing how uplifting and inspiring sex can be may be helpful to those who have suffered sexual wounding and abuse in the past, including those who have been spiritually marginalized for their preference for partners. Indeed, the fact there seems to be no difference between the ability of homo- and hetero-sexuals to enjoy transpersonal experiences during sex could be extremely beneficial, especially for adolescents and others coming to terms with their sexual preference. The potential for working with such embodied forms of spirituality can open new horizons in everyday living and loving.

From 'Bad' Ritual to 'Good' Ritual

Transmutations of Child-Bearing Trauma in Holotropic Ritual

Gregg Lahood

For women, situations associated with motherhood can become another significant source of unitive experiences. By conceiving, carrying, and delivering a child, women directly participate in the process of cosmic creation. Under favourable circumstances, the sacred nature of these situations becomes apparent and is consciously experienced.
—Stanislav Grof

Introduction

In this chapter, a tentative and provisional theory is advanced on the holotropic treatment of birth-giving trauma. "Birth-giving-trauma" here refers to women (and men) who have been psychologically, physically, or emotionally traumatized during birth-giving. I will first outline anthropologist Robbie Davis-Floyd's argument that Western medicalized birthing can be constructed as a "modern" rite of passage which can negatively imprint disempowering images into women's minds, reinforce messages of inferiority, and traumatize the birth-giving mothers. I will then outline how the trauma catalyzed by the "bad" ritual of technocratic birth may need to be therapeutically treated or rather "ritually combated" with an equally powerful and reparative "good" holistic ritual. Stanislav Grof's holotropic breathwork is outlined as a preeminent ritual in which "good" transpersonal medicine is ritually made.

Feminist anthropologists have shown that birth in most cultures has been a ritualistic event[1] enveloped in protective rites of passage and spiritual procedures that lend emotional, psychological, and charismatic support to birth-giving women. They argue that fertility and birth are in all cultures embedded in social, psychological, cosmological, and spiritual systems.[2] Furthermore, the basic pattern of biological birth serves as a "model for structuring other rites of passage"[3] and ceremonial healing rituals.[4] Traditional helpers at birth, midwives and shamans, operated as "technicians of the sacred,"[5] and it has also been noted that transpersonal visions may be part of a contemporary birth-giving woman's reality[6] and the father's reality.[7] Grof writes, for example: "Delivering women and people participating in the delivery as assistants or observers can experience a powerful spiritual opening. This is particularly true if birth does not occur in the dehumanized context of a hospital, but under circumstances where it is possible to experience its full psychological and spiritual impact."[8]

Unfortunately, the Western biomedical approach to birth-giving does not value emotional or spiritual support, nor does it value visionary states,[9] and many women are traumatized by the dehumanized nature of "technocratic" childbirth rituals. Moreover, in the Western world most births do not occur in domestic environments but in hospitals.

While there is an increasing literature on 'birth trauma' relating to the fetal person, less attention has been paid to the trauma of the birth-givers. "Birth trauma" is a blanket term confusingly applied to the psychological and physical damage experienced by both women and neonates during the process of labour and childbirth. However, some fathers can also suffer from 'birth-trauma' in the form of post-traumatic stress disorder (PTSD). I think it would be useful to delineate between "birth-giving trauma," "neonatal birth trauma," and "witness trauma," and the treatments for these divergent phenomena.

This paper will contrast two important contemporary "rituals": modern childbirth and trans-modern holotropic breathwork.[10] I will suggest that while the former ritual system begets and amplifies birth-giving trauma, the later holotropic ritual process can be used to heal the trauma associated with birth-giving. Our aim here is to grasp the following nettle: *if it is in "bad" ritual where harm is caused, it may well be that it is in "good" ritual where trauma could be negated and healing found*. The purpose of this paper, then, is to offer those who suffer from birth-giving trauma—PTSD catalysed in childbirth, post-natal depression, grief and loss around miscarriage and abortion, those who feel emotionally, psychologically, and spiritually impinged upon by the medical system, and those who work with traumatized persons—a further treatment option, that of our species' oldest healing system: ritual.[11]

Background

I came to study birth-giving through a long-time interest in contemporary transpersonal rituals.[12] My post-graduate and doctoral studies were focused on the ritual dynamics of birthing in New Zealand and the transpersonal events experienced around birth-giving for contemporary women and men. I have described some of these research findings in several articles dealing with, for example, the encounter with death at birth,[13] the transpersonal dimensions of Indigenous midwifery,[14] fathers' near-death experiences around childbirth,[15] and women's transpersonal experience at birth-giving.[16]

Another complementary strand to my research life is that of a holotropic breathwork facilitator—a "ritual specialist," so to speak. I have been involved in broad holotropic breathing practice for almost two decades and this has given me an opportunity to gather data from a unique viewpoint, that of a participant/observer in the holotropic ritual itself (I also had a role as an antenatal educator in New Zealand).[17]

However, there is another link between holotropic ritual and transpersonal states of consciousness among contemporary women I should outline. Some of the women and men I spoke to during my doctoral research had experienced "non-ordinary states of consciousness" that bear a striking resemblance to what Stanislav Grof calls "holotropic consciousness."[18] These are profound "healing states of consciousness" having to do with the experience of death and rebirth.[19] Let me give an example of this, not from one of my informants, but from Jungian analyst Jean Shinoda Bolen, who, in the documentary, *The Goddess Remembered*, said this:

> My experience of a woman giving birth to a child put me in touch with the women's movement. Up until that time, I was a real medical student, intern, resident, kind of a person, who

felt quite different from other women because my path was different from most women's. But once I was in labour and delivery and was experiencing at the deepest ritual level and at the deepest life level, what it is to be a woman and how it hurt...and how it was also a miracle and how none of my training prepared me for this and what I was doing at that moment was what every woman who had ever given birth to a child has been doing through all time. I felt linked horizontally and through time with every woman that ever was.[20]

Note the strong link between birth and "the deepest level of ritual" and then "ritual" with the transpersonal domain. Her sense of becoming continuous with all women through time and space (as a healing and empowering event) is a becoming beyond the Cartesian box of time and space, which means that she has stepped outside of Western medicine's "body-as-a-machine"[21] image and the foundations of Western science in general. One of my informants said this at the birth of her daughter: "She was not breathing...not energetically so I breathed into her energetically... you are going to live! It's like I'm getting a vision...a sense of this line of women back through eons almost. It's like connecting with a line of all women. It had to do with the family of women through time like a line." Such experiences are also a recurrent theme in holotropic research.[22]

Rites of Passage

Rites of passage and their tripartite morphology were made famous by folklorist Arnold van Gennep. In his schema a rite of passage process has three basic patterns (although a pre-initiatory phase must also be assumed, e.g., the womb of childhood). They are 1) separation: the neophyte is removed from a previous social or cosmic world.[23] 2) Transition: a magico-religious space in which the initiate "wavers between two worlds;"[24] this *liminal* space was often a place of ordeal, chaos, and symbolic dismemberment. 3) Incorporation: a phase where the initiated is being absorbed or reintegrated into a new world.

Van Gennep also wrote that such rituals had a strong association with pregnancy and childbirth,[25] and it is interesting to contemplate the relationship between van Gennep's *rite de passage* template and the basic morphology of the fetal person's journey through the chaotic "gauntlet" of the perinatal passage. The child is separated from the "good womb," passing into a state of constriction, followed by an ordeal-like and laborious transition, and finally emerges from the dangers associated with the birth passage into the world and a new social or cosmic status.[26] In Grof's schema, this perinatal process structures the psychological experience of death and rebirth, and the holotropic therapeutic ritual is geared to support this transformational process.

A 'Bad' Rite of Passage

Feminist anthropologist and birth activist Robbie Davis-Floyd's *Birth as an American Rite of Passage*[27] is perhaps the most comprehensive study to date concerning ritual, cognition, and contemporary Western birth. She argues that contemporary hospital birthing can be constructed as a rite of passage operating tacitly within the medical birthing regimen. According to Davis-Floyd, in this context, the ritual process is deeply problematic because it is geared to indoctrinate women to its biomedical mythology by enacting its "body as machine" system of authoritative knowledge in a ritualized technological apotheosis: birth *as* medical operation.

Renowned British anthropologist Sheila Kitzinger has also suggested that modern birthing "rites of passage" do not function to provide emotional support (as traditional rites of passage would have) but rather they *reinforce* the established social system.[28] In the modern scenario, women are routinely stripped of bodily knowing, authoritative knowledge, and the status and charisma associated with birth-givers. Birth-giving is treated as a routine medical crisis indexed into a structure of hierarchical power running on an "assembly line" system bent to capitalist clock-time.[29] Its rhythms do not sway easily to the rhythms of a female birthing body. Kitzinger writes: "In achieving the depersonalization of childbirth and at the same time solving the problem of pain, our society may have lost more than it has gained. We are left with the physical husk; the *transcending significance* has been drained away. In doing so, we have reached the goal which is perhaps implicit in all highly developed technological cultures, mechanized control of the human body and the complete obliteration of all disturbing sensations."[30]

Using elements of biogenetic structuralism as a model (a model bound to transpersonalism),[31] Davis-Floyd's analysis suggests that women birthing can engage in the same neuro-cognitive processes that produce similar states to those found among ritual participants.[32] She argues that the climaxes and peaks found in ritual and meditation (after neuro-theologist d'Aquili)[33] when neuro-physiological subsystems fire simultaneously in the autonomic nervous system are also found in birthing women.[34] Once these ritual dynamics are catalyzed, the human cognitive system can be rendered open to gestalt perception,[35] and what is called "symbolic penetration"—that is, the ingression of symbols in the environment and their meaning into the opened mind of the ritual participant,[36] moving toward a peak, with the climactic experience resulting in the long term memory storage of symbolic messages.[37]

Davis-Floyd argues convincingly that it is the symbols of the Western technocratic medical system in all its hegemonic and patriarchal glory that are impressed into women's minds at childbirth, serving to reinforce its power and status over women. In other words, contemporary medicalized birthing rituals oppress women at a societal level through the use of a series of rituals that can be thought of as a dynamic rite of passage—a conversion process. It is a compelling argument. The price Western women pay for the belief in the Western hospital system's ability to control childbirth outcomes, its routine technological wizardry, its body as a machine mythology, and its efficiency in saving lives and reducing pain, is a reduction in participation, a reduction in emotional and spiritual life, the loss of personal autonomy and authoritative knowledge, and at worst, psychological, physical, emotional, and spiritual traumatization.

The Problem with Birth

I will not attempt an exhaustive account of the trauma of birth-giving here but will touch on a few key points, beginning with a definition from Cheryl Beck: "Birth trauma is an event during the labour and delivery process that involves actual or threatened serious injury or death to the mother or her infant. The birthing woman experiences intense fear, helplessness, loss of control and horror."[38]

Simkin and Klaus list the following: "a sudden emergency caesarean perhaps with inadequate anaesthesia; shoulder dystocia; severe perineal damage; fetal asphyxia; vacuum extractor or forceps injuries; severe haemorrhage; newborn disabilities or death."[39] We could add: prolonged labor, decelerations of heartbeat, anoxia and hypoxia (diminishing oxygen supply), meconium in the amniotic fluid, severe constriction, miscarriage, spontaneous abortion, and eclampsia. Birth trauma, then, is physically damaging, psychologically damaging, and may result in, or at least threaten, neonates and birth-givers with death. Research shows that Post Traumatic Stress Disorder (PTSD), as outlined by the DSM-IV[40] can also be catalyzed through birthing for mothers[41] and fathers as witnesses to traumatic events.[42]

Explicit in Davis-Floyd's account is a strong correspondence between modern technocratic ritual childbirth, patriarchal oppression, and women's birth-giving trauma. She refers to a process of "compartmentalization" by "the ones who were most totally effaced during their hospital births."[43] Often those anesthetized had placed their birth completely outside of their lives. Others awake and aware, but in "extreme terror and pain," also used this compartmentalizing process to divorce themselves from the traumatic experience. Some of Davis-Floyd's participants refused to even talk about the subject because it had been so traumatic.[44] Other women drive miles out of their way to avoid going near the hospital where the trauma occurred;[45] these behaviors are symptomatic of PTSD. However, these psychological defense mechanisms have psychotherapeutic problems since, as Davis-Floyd rightly points out (and as people-workers well understand), "unresolved traumas tend to resurface in various ways."[46]

So how to resolve the trauma caused by perinatal complications or technocratic rituals, or a combination of both? Meaning-making verbal therapies, such as narrative therapy,[47] might well be the first level of ameliorating this trauma, as Davis-Floyd suggests.[48] However, cognitive anthropologist Douglas Hollan says that many of our engagements in the world remain "unconceptualized, unverbalized, and outside of conscious awareness until they gain conscious representation through complex symbolic processes"[49] and that certain experiences can be so overwhelming and shattering of our normal everyday expectations that they "never become cognitively and linguistically processed and represented at all."[50] Furthermore, as Grof writes: "In the case of major traumas, particularly situations threatening the survival and body integrity of the individual....It is very likely that in situations of this kind, the original traumatic event was not really fully experienced at the time it was happening. [This] can lead to a situation where the experience is shut out partially rather than completely. As a result of it, the event cannot be psychologically 'digested' and integrated and remains in the psyche as a dissociated foreign element."[51]

Therefore, just as the birth-giving trauma began with a whole person, with bodily, social, political, psychological, sexual, existential, and transpersonal wounding—in a ritual context—healing may need to be attended to in a counter-ritual and counter symbolic/therapeutic social milieu.[52] This should be a ritual in which the creative healing potential of the woman's own psyche and soma are brought strongly to bear on the healing process, a process that restores her personal authoritative knowledge by engaging her intentional, volitional, bodily, emotional, intellectual, and transpersonal knowing. A "good" ritual should have the ability to bring the unprocessed material back into consciousness in a creative and emotionally intelligent environment (a set and setting) that is symbolically optimal for such a recovery.

In many rituals, the human body and nervous systems are "heated" through dancing, exertion, breathing, or emotional catharsis. An example would be the prolonged dancing and the !Kia trance of the Ju/'oansi (the !Kung Bushmen) of the Kalahari Desert: "!Kia can be considered a state of transcendence because during !Kia, a !Kung experiences himself as existing beyond his ordinary level of existence. !Kia itself is a very intense physical state. The body is straining against fatigue and struggling with convulsion-like tremors and heavy breathing. The emotions are aroused to an extraordinary level, whether they be fear, or exhilaration, or fervour."[53]

Heightened motor activity, bodily metaphors, gestures, vocalizations, and enactments, including a full embodied re-performance of birth-giving, might need to be engaged with. Here the same neuro-cognitive subsystems of the central nervous system can be harnessed in the healing ritual process. In a reparative ritual process, as the autonomic subsystems fire and open to gestalt perception, a new set of symbolic impressions could ameliorate or replace the traumatic imprint. This can be conceptualized as *ritual inversion*: replacing the hierarchy and hegemony of the traumatic governing system in the psyche to a position in which it is experienced as ultimately transitory and partial in an unfolding dialectical process.[54] A dynamic shift in the "governing system"[55] can move the participant through the negative system and attune her to more positive, nurturing, and healthy constellations.

The Problem with Sexuality

It is a strange situation where birth activists have to argue that birth-giving is an extension of normal female sexuality—nevertheless, due to the over-medicalization of birth, they must.[56] Davis-Floyd sees "obstetrical rituals" as having developed in tandem with a medicalized program to "desexualize" and render the mother's sexuality around birth-giving "tabu" and "defective."[57] She writes: "So effective are hospital routines at masking the intense sexuality of birth that most women today are not aware of birth's sexual nature."[58] We should also mention the father's recent presence at birth in Western cultures as symbolic of the couple's sexual power. Brian Bates and Allison Turner say that many "childbirth rituals found throughout the world appear to be of a sexual nature."[59] The stimuli used in such practices are symbolic of the man who fathered the child and, in particular, of his sexuality. They may thus inculcate some form of sexual imagery in the woman, albeit at the preconscious level, which then stimulates the physiological responses normally elicited by sexual stimuli—the release of hormones and contractions of the uterus which serve to aid the birth process.[60]

One of my informants spoke of her homebirth as intensely sexual, although it was a sexuality that incorporated a cosmic dimension. Frieda put it like this: "carrying a baby is such a deeply spiritual experience and giving birth is the ultimate spiritual orgasm…you tap into that greater energy, that greater consciousness," clearly a statement reflecting transpersonal dimensions of sexuality. Nevertheless, the vast amount of Western births occur in the hospital system—a system which, according to Davis-Floyd and others, has robbed women of vital and integral birthing energy.

However, there is another serious problem to take into account: the complications of sexual abuse. It was matter-of-fact among some of my midwife informants that childhood sexual abuse can seriously impact birth-giving women, protracting their labor and making it more exhausting and dangerous. In their book, *When Survivors Give Birth*, Penny Simkin and Phyllis Klaus write that some women can experience prolonged non-progressing labor and "extreme pelvic tension."[61] They speak of a woman whose "greatest fear was that something in labor would trigger 'body memories' or feelings of victimization."[62] Fear releases hormones (catecholamines) that are known to slow labor.[63] Thus, sexual abuse becomes a vicious and problem-laden cycle at the level of birth-giving.

Neva Walden's exploration of the relationship between holotropic ritual and healing, *Contributors of Transpersonal Perspectives to Understanding Sexual Abuse*, gives several examples of the relationship between sexual abuse, birth-giving, and holotropic consciousness: "I felt a strong build-up of tension in my genitals and bladder area. As it built I got extremely angry and sexually frustrated. My body was filled with it. First it was my rage and frustration. Then my body was filled with my mother's as well. Then, my grandmother's, and finally that of all womanhood throughout time."[64] "Opening to feel the pain and suffering of that little girl within, I cry. As I cry, I fully feel the cry of a wounded animal, and also the cry of all children being raped and abused. I feel the cry of all women in childbirth."[65]

Walden suggests that abuse victims (like victims of birth-giving trauma) are locked into the second phase of something like the ritual dialectic and that holotropic ritual can move her through the process, much like completing a gestalt: "Unfortunately, many sexual abuse survivors are still living the experiences of stage two of the initiation process. They are left in the turmoil of the separation, humiliation, shame, and death portions of the passage. With healing, however, the survivor can move into the integration phase. Experiencing the full range of emotions and physical feelings of sexual energy in [holotropic] states...brings the integration that is a result of any successful initiation. It produces a profound shift in their sense of themselves."[66]

In the same way, it could be argued that women traumatized by technocratic birth rituals or birth trauma and the encounter with death are in a similar phase in the ritual process; they are "betwixt and between" the traumatic situation and the successful integration. The trauma, largely unrecognized, is compartmentalized and isolated but at the same time unconsciously (psycho-dynamically) structuring her relationship with the world.

Let me give an example: Anna (25 years old) came to a holotropic session in New Zealand in which she replayed her birth-giving. Anna had been expecting a water birth with her partner and their friends. Unfortunately, this did not happen, and her dreams were dashed when her birthing became a highly medicalized event. She had also lost meaningful contact with her male partner during her birth crisis. While grateful for the intervention, she said that ever since she had experienced high levels of distress, anxiety, and nightmares that she strongly related to the birth of her child. It was as if, she said, there was an energetic, emotional, and spiritual aspect of her experience that was not brought into consciousness and this charge had been lying dormant ever since "just below the surface." Anna went through a very cathartic session involving a wide range of emotions, bodily movements (very obviously linked to labor pains and her own birth) oscillating with states of deep blissful relaxation and bouts of ecstasy. What was most remarkable was her conviction afterward that she had moved into a state of transpersonal consciousness where she somehow not only birthed all the babies in the world but all the creatures in the world and even all the *forms* in the world.

The Problem with Death

Some women's birth-giving narratives also point to a frightening encounter with death during parturition. Certainly, among the women (and men) I interviewed in New Zealand, this was a common factor.[67] The two following narratives disclose just how potent but also how unrecognized this feature of birthing is: "I was terrified when my daughter was born. I just knew I was going to split open and *bleed to death* right there on the table, but she was coming so fast, they didn't have any time to do anything to me."[68] "During the delivery process," writes Beck, "some women were shaken to the core by feeling abandoned and alone, as illustrated by the following quote: 'I had a major bleed and started shaking involuntarily all over. Even my jaw shook and I couldn't stop. I heard the specialist say he was having trouble stopping the bleeding. I was very frightened, and then it hit me. *I might not make it!* I can still recall the sick dread of real fear. I needed urgent reassurance, but none was offered.'"[69]

There is a serious knowledge gap surrounding birth-giving women and the impact of the potential psychological encounter with death during parturition. Davis-Floyd, for example, does not acknowledge the encounter with death as a central aspect of the ritual process for contemporary women in her study. Furthermore, I believe Beck is, in fact, naming two traumas here. First, the "primary": an acute and dreadful encounter with death, and then a "secondary" trauma occurs when this experience is not offered any social support, empathy, or understanding and is left isolated. Was she shaken to the core because she was "abandoned" as Beck suggests, or was it because she feels she is really dying?

According to some anthropologists, the fear of death (like sexual abuse) plays a crucial role in reproductive crises. Carol Laderman writes that "The prolongation of labor because of fear is associated with much higher than normal perinatal mortality rates."[70] James McClenon notes that "Fear results in muscle tension, which inhibits the normal dilation of the cervix."[71] He also suggests that fear and stress can increase the likelihood of childbirth complications, psychosomatic infertility, spontaneous abortion and miscarriage, postpartum hemorrhage, and obstructed labor,[72] all of which call forth further medical interventions. It is important to note, however, that the human encounter with death is not always necessarily a negative experience (perhaps in the same way that losing control is not always a negative experience and one mandatory for birth-giving). It can also become a part of the ritual process and a doorway into transpersonal consciousness.[73] Kitzinger, following Levi-Strauss, writes that "birth and death are rich with meanings which have penetrated the whole of social life. But in the West, as part of a process of "scientific praxis" we have emptied birth and death of everything not corresponding to mere physiological processes."[74] For example, anthropologist Megan Biesele in her study of birth and trance dancing among the Ju/'oansi of Africa's Kalahari Desert writes: "Daring death seems to be part of cultural maturation for the Ju/'hoansi, as it is in fact for many other groups of people. Both the men's and the women's daring—in trance and in giving birth—seem to function as transformational rites of passage in Ju/'hoan society."[75]

Indeed, Grof and anthropologist Joan Halifax wrote that an encounter with death is at the very core of rites of passage: "profound experiences of symbolic death result not only in an overwhelming realisation of the impermanence of biological existence but also in an illuminating insight into the transcendent and eternal spiritual nature intrinsic to human consciousness."[76] Women in the Kalahari have access to ritual status through the processes of transformation and self-actualization by giving birth and encountering death.[77] This ritual or charismatic status is denied in the Western birthing system. Good ritual, then, must be potent enough to rework the encounter with death at the somatic and symbolic level and restore ritual status and charisma to women.

Ritual Amplification of Alienation

Reproduction creates an intersection between the three biological powers of birth, sex, and death (outlined above) with Western culture where it is *isolation* in its various guises (compartmentalization, separation, solitude, alienation, lack of empathy, and obstructed relations) that can become critical. As we saw, each of these biological categories and their emotional and psychological responses can be highly constricted and hegemonically controlled by Western biomedicine: women until very recently were routinely *separated* from their families, their husbands and lovers, even the newborn, the psychological encounter with death largely ignored or denied, sexual relationships obliterated, sexual-abuse isolated or even amplified.

According to Jeanne Achterberg, this sense of "alienation" from "family, community, the environment, the self, and the spirit world"[78] is axiomatic with illness in many tribal societies and requires transpersonal rituals for its amelioration. Yet, as anthropologist Jurgen Kremer points out, these are the very relational fields severed by the march of Western progress (including Western medicine). The Eurocentric ego is "constructed dissociatively from nature, community, ancestors."[79] Indeed, the categories equating with *alienation* are the very ones associated with the *demonic* in many traditional societies. For example, anthropologist Bruce Kapferer says of Buddhist exorcism: "In Sinhalese cultural understandings a demonic victim approximates what I refer to as an existential state of solitude in the world. The demonic as conceptualized by the Sinhalese is similar to that which Goethe recognized from within the worldview of European culture as ultimately everything that is individual and separates one from others. Demons attack individuals who are understood to be in a state of physical and mental aloneness. Solitude and its correlate, fear, are among the key essences of the demonic."[80]

We have seen already how fear plays an important part in obstructing a woman's labor. Kapferer writes, "At the paradigmatic level and in accordance with Buddhist cosmological view and worldview, demons are at the base of a hierarchy dominated by the Buddha along with a host of major and lesser deities" [similar perhaps to Christian hierarchies of angels].[81] Kapferer, arrestingly, links Buddhist thought to Goethe's Romantic, participatory thought—both of which are *seminal* ancestors of the transpersonal movement.[82]

Here is the crux of the matter: the modern European worldview, as spelled out by Richard Tarnas, is very much an ego-centric one and therefore, according to Kapferer's Buddhist/Goethe formulation, categorically "demonic." The picture Tarnas paints of the Western ego is one of absolute solitude—solipsistic, alone, and isolated. Our "cosmological estrangement...ontological estrangement [and] epistemological estrangement [results in] a threefold mutually enforced prison of modern alienation."[83] Seen from the participatory standpoint, the European mind is cathected to a flawed image of the universe. The mystery of nature is demystified through "objectivity" and we are severed from participation in the sacred worlds of our ancestors. But perhaps more importantly for this article, the Western birthing system can be seen as a product of this worldview. Thus, birthing women and their partners are participants in a ritual process that can amplify the demons of isolation, separation, alienation, and fear.

By way of an example of the spirit of alienation promoted in technocratic rituals, I would like to use the following statement from Robbie Davis-Floyd:

> The Cesarean itself felt like somebody stepping on my stomach with a boot, and pulling up the skin for laces. It was cold in the room, and the table was cold, and that cold penetrated my opened insides till I felt cold throughout my entire being—lonely-cold, as if I were floating naked on an iceberg. And my mouth was dry as sand, and I asked for ice chips, but the anesthesiologist just shook his head. So this is how I felt during the Cesarean—stepped on like the floor, laced up like a boot, cold as the Arctic, dry as the desert, and just as alienated from my experience as if I had been on another planet.[84]

Combating Trauma with 'Good' Ritual

From a participatory worldview, the antidote to the demon of isolation would be what Tarnas calls "radical kinship with the universe"[85] catalysed by "good" ritual. Or, to follow Kapferer, "the languages of ritual contain varying potential for bringing together the Particular and the Universal."[86] If ritual is the "foundation for transpersonal medicine,"[87] as Jeanne Achterberg claims, then holotropic breathwork is transpersonal ritual medicine-making *par-excellence* and one geared for our times.[88] While the method is used for healing psycho-trauma, an approach in transpersonal research, and self-exploration,[89] it can also be conceived as a ritual and rite of passage and, for the purpose of this article, I will conceive it as such.

In holotropic ritual, while in a holotropic state of consciousness, participants can organically retrieve and "relive" traumatic events, abuses, accidents, birth and birth-giving wounds that compound a person's sense of alienation. These are non-ordinary states of consciousness moving toward a greater sense of wholeness. This sense of wholeness is often accentuated in transpersonal states of consciousness accompanied by a shift in meaning of the traumatic experience.

During holotropic breathing sessions some women seem to "relive" their birth-giving experience. The unconscious material, coupled with the dynamic urge to re-enact birth-giving, seemed to arise naturally from the women's psyche when placed in the therapeutic holotropic environment. As Grof wrote 30 years ago: "It was frequently observed that female subjects reliving their own birth re-experienced the delivery of their own children. Both experiences were usually relived simultaneously, so that these women often could not tell whether they were giving birth or being born themselves."[90]

The following statement is from a woman, Leone, who participated in a holotropic setting in England: "Then I carried on sharing around my own birthing experience and the only way I could describe it, was as if my spirit had been born again and again and again. It was like I was giving birth, I was birthing my own children and I was my children in the birthing process, and I was aborting and I was being born…it was like I was coming down through the birth canal."

As we know, language fails to convey the essence of the post-conceptual nature of the transpersonal condition and the lived experience of healing. No less difficult to convey are the exact healing mechanisms of the holotropic breathing ritual. Something that I hear time and time again is that a shift in consciousness happens somewhere during the process and people feel themselves cradled by a deeper wisdom, or higher power, a Gaian holism, or a Great Mother and as they entrust themselves to that wisdom a profound emotional, somatic, and transpersonal unfolding can begin which seems to have an intelligence all of its own. Once held by this intelligence, the process is often likened to a purification, which is to say that anything felt by this intelligence to be inorganic or not healthy begins to emerge into consciousness and moves toward "a climax of expression,"[91] which then allows for the unconscious material to be integrated in a therapeutic/ symbolic social milieu.

Healing a Traumatic Birth

Let me give a typical example: Beth, a woman of about 55 years, came to the breathing ritual. Although this occurred some years ago, I remember her well because I was so struck by her story. During her breathing session she became extremely primitive and (as she told us later) gave birth to all her five children again. Beth told us that she had been brought up a Catholic and that "down there," meaning her reproductive organs and genitals, were never talked about in any positive way. Beth said she felt strong injunctions about living "in her body" and in particular "down there." Thus, with the onset of her first labor she had been extremely overwhelmed and shocked by the depth of her biological power and process but had struggled to keep herself from occupying her lower body because of the shameful associations from her upbringing. In her words, she had felt "split off" from herself. Beth had not come to the workshop with any agenda about replaying her birth-giving, but this is where her process took her. She had also had to deal with admonitions from hospital staff not to make any sound when giving birth. Indeed, she was told to "shut up" when she swore with pain. She described her anguish to be giving birth from a body that was held to be shameful, surrounded by strangers, who were reinforcing the denial of her physiology and her need to express pain and outrage.

During the breathwork ritual she actually relived the birth of all her five children and made a point of bringing her awareness into her birthing body as a sacred vehicle and with each birth she roared, swore, and labored and roared some more. At one stage I remember her powerfully discharging her anger, frustration, disbelief, and fear at the medical staff, her parents, and the Catholic Church, for the ways they had negatively contributed to her birth-giving. Thus, in the course of her breathing she revisited the archaeology of her traumatic or oppressive birth-giving history and re-enacted her births with deeper awareness, with vocal expressions fitting her needs, a greater sense of autonomy and power in the situation, and freed-up emotional and motor responses as a result. In my opinion, and most certainly in Beth's, she had transformed herself by re-birthing her children and claimed for herself some of the charisma and status that she believed were rightfully hers. But more importantly, she felt she had allowed her emotional body to finally go through the process of birth: an "act hunger" she had held back ever since then.

Jungian Psychoanalyst Edward Whitmont in *Return of the Goddess* said that, "differentiation from others, and hence self-definition occurs through struggle," and then this: "Grof has described the close association of birth and rebirth experiences with violence, upheaval, and death as they emerged in [holotropic] research. He describes the arousal of feelings and urges of violence during the passage through the birth canal. The subject experiences overcoming a state of deadlock and inertia, of feeling oppressed and hemmed in. Subsequently, urges of violence and aggression are likely to be aroused by any stagnating or deadlocked life situation which calls for the need for regeneration, a *new birth.* This is true collectively as well as individually."[92]

Another woman, Karen, 35 years of age, came to a group in Australia and relived her birth in a most extraordinary way. During her session she was lying very still on her mat and I motioned my co-facilitator over and said to her that I had a sense she was conceiving. This intuition seemed to be more or less correct, because during the course of her four-hour session we watched her become pregnant and then give birth. What I remember most about Karen's session is that she had turned her sitter (her assistant during her breathing) into her husband. She appeared to be deeply engaged in birth-giving and he with her process. She was sitting, sometimes squatting or standing, other times on her hands and knees—her "husband" was holding her, encouraging her and breathing with her. Sweat was pouring off both of them. It was a most remarkable thing, the magic of it tangible, and many of the other sitters in the room were drawn to their performance.

Later she told us that the birth of her son had hurt them both and that their relationship had suffered from the trauma they had caused each other as if there had always been a very primitive and intimate anger between them. She said that the breathing had enabled her to go right back to conception and replay somehow the whole reproductive cycle. But this time she said it was like doing it all with a deeper wisdom that she felt pervading the ritual space. She said she was not hampered by fears, embarrassment, and the directives of the hospital staff (or lying on her back in stirrups) but was able to return to this defining moment in her life with the wisdom of the group and a healing intention and that somehow she had found herself re-doing birth. It was her belief afterward that she had changed a major distress pattern and that it was her hope that this would have an effect on her relationship with her now teenage son.

Not Altogether Sure How It Works

I should say from the outset that since the healing is orchestrated by the breather and her integral wisdom, and is deeply idiosyncratic and unique, I can't ethically offer an authoritative meta-narrative about what it is that heals—other than this one. Secondly, anthropologists are well aware that ritual has an uncanny way of doing magical things, which is to say that there are question-marks about how ritual *really* works—and also how the human psyche *really* operates.[93] However, as a ritual facilitator and an anthropologist I have observed *some* things and I offer these thoughts, however modest.

I can say that if the container is co-created by the ritual participants, in an atmosphere of positive regard and the "inner healer" is evoked, then healing seems to happen, but again, not necessarily in quantifiable ways. Preliminary discussions (re-mythologizing the human body and psyche) with participants describes and negotiates a broad map of possible perinatal and transpersonal experiences. Thus, at the beginning of a breathwork group a contemporary "myth or map" of the universe is offered which embeds the participant in an ever-widening non-Cartesian worldview and paradigm. Ritual participants then move into a liminal stage as they enter into holotropic consciousness in what amounts to a healing ceremony. At the same time we see people often discharging very primitive levels of pain, anger, grief, and fear. This appears to be similar to Victor Turner's description of ritual: "Powerful drives and emotions associated with human physiology, especially the physiology of reproduction, are divested in the ritual process of their antisocial quality and attached to components of the normative order, energizing the latter with a borrowed vitality."[94]

This is to say that "negative" perinatal energies are at some level transmuted by the group structure itself. During the liminal phase, "breathers" can enter into a healing ritual crisis which can include re-connecting or bonding with the wider universe, nature, society/group and something like a sacred-mind.[95]

As participants emerge from the holotropic state, they pass into a post-liminal stage. Here they make artwork of their experience and present their knowledge to the co-ritual participants. This presentational knowledge,[96] I suggest, becomes part of the symbol system of the group and helps to canalize the psychodynamic, perinatal, and transpersonal energies into each person's idiosyncratic symbol system. This gives the energies unleashed in the healing crisis an artistic, embodied, and communal container around which meaning-making and self-reflection coalesce. During the sharing circle, or "reflection phase," participants are now seen, heard, and acknowledged as being at the crest of their own transpersonal being and becoming (to borrow a phrase from Heron),[97] and importantly, ritual charisma or "mana" (a Polynesian variant) is associated with the breathers and not appropriated by the ritual specialist (a restoration of charismatic self).

Holotropic breathers take it in turns to breathe while evocative music is played over several hours. Each "breather" has a "sitter," a personal guardian who behaves (a bit like a midwife) supporting and not interfering in the unfolding process. Below are two accounts from sitters. The first is from myself, whose breather was working with issues of infertility, followed by an account from Leone, speaking of her sitting experience:

> I am sitting for Donna and she is in some deep process and I am staying with her somehow. I do feel very 'with' her on some energetic level. She gets up to go to the toilet and I escort her to the door and wait outside. She is cradling something invisible in her arms and she turns to me and without words puts this invisible, weightless 'object' into my arms. I carefully cradle the object and while I am not sure but it feels like a baby. She comes out of the toilet and gently takes the baby in her arms and we go back to her mattress. Later she speaks to me of how healing this gesture for her was. She tells me and the group that she wants to have a child but thinks maybe she won't because she is in her mid-forties. Apparently by holding her 'spiritual' baby for her was powerful affirmation of her healing process.

> [Leone] The following morning we all came together to talk about our mandalas (symbolic drawings made after sessions) and to share, as I could not contain my excitement, I was the second person to share. I spoke about how awesome and amazing this was and that I felt like I was in a limbic state and things

were coming in on many levels and there was just so much information. I went back to the sitting experience and talked about, how, I never in my wildest dreams could I encapsulate getting something like that back, which from the workshop I had, which was to do with the birth of my son. I had to have an emergency section and was not able to hold him, and this is what I was re-connecting to—a sense of being able to hold him. It was real powerful stuff.

On occasion I have been present with women working through elective abortion, miscarriage, spontaneous abortion, unwanted caesarean section, loss of fertility, the frightening encounter with death at birth, and the traumatic sense of abandonment that can occur when a woman loses contact with her partner, or when her desire or expectation of a natural birth is obliterated during medical interventions. The following two stories are from women who have experienced various reproductive crises and then relived those experiences as part of their healing in the holotropic ritual. Holotropic sessions are generally serial, and, in terms of depth, cumulative. Usually after several sessions, when the traumatic material is well managed and integrated, the "gestalt" finds "closure" in a full-blown transpersonal experience and initiation. Each of the following sessions can be seen as "a work in process" with the final session (Jeni's) an example of the movement toward integration.

Imam's Story

My daughter's birth was very long. I had had a pause in the middle where I had been sent home and felt frustrated. I had felt embarrassed when a group of medical students had come to watch, I hadn't been asked if that was ok, they asked my husband who said yes. I also tore the ligaments at the front of my pelvis on delivery.

[Imam's 3rd holotropic session:] I started this session again with extreme heat in my body and lots of pain. There was intense pain in my womb, the front of my pelvis and lower back. I felt myself go rigid. Then I was visited by my mischievous man. He has appeared to me several times before. We dance together and he has dragged me out of my body before [in an NDE experience during childbirth]. Although he is mischievous and fun there is also a deep side to him and an immense feeling of power, so he deserves great respect. This time I was rigid and he danced around me rattling his red rattles, his blue eyes glinting with mischief. He gave me a song to sing. When I sung it, it came straight from my womb. I was under my blanket for this. I then had to leave the room for a toilet break. When I returned I still had the pain in my back and womb and couldn't get back into it. I felt extremely frustrated.

Then Gregg came over and asked to check in with me about what was happening. I realized that these pains related to my daughter's birth. It felt like I was going through a birthing process. I wanted to go under my blanket again but talking with Gregg I realized that this shame related to the feelings I had had at the birth with the medical students. Gregg suggested I chant powerfully as a way of deliberately releasing the distress. I chanted the song given to me by the mischievous man as I pushed and sang [and] the pain began to subside. I had another go and the pain from the womb went but some of the pain remained in my back. When I rested Gregg made a comment about having felt like rattling or drumming around me. [Gregg: I had heard from midwives that sounding during birth, especially powerful sounds were useful. They had told me that fearful sounds release hormones that can slow things up so I suggested this to her to discharge by chanting. I also had a very strong urge to grab my red rattle or a small drum and play for Imam, something I would not do during a session, and had suggested to her that I would be willing to rattle for her—but only if she thought it would support her in her birth-giving. I was not aware that in her transpersonal world she had been visited by a bearded figure who had given her a song to chant and was rattling and dancing around her. Her vision had occurred prior to my engagement with her].

The following morning I felt faint and nauseous again and went to my bed. I came round from what felt like an anaesthetic by a friend calling my name and it took a while for my body to regain feeling. We shared the mandalas from the session. I talked about my session and how I had felt strange this morning. I realized there were threads of all my experiences in my mother's, her mother's, my father's mother's, my husband's mother's, and even my ex-husband's mother's biographies. I also felt the abuse of the hospital system as well. It seemed to link all three sessions together, like a deepening and interconnecting. The birth experiences, rejections and abuse all interwoven together in some way.

Jeni's Story: A Spontaneous Abortion

The following is from Jeni, a Scottish woman who attended a workshop in North England:

> The peace of being pregnant settled within as I took every precaution to nourish my growing bundle. I sang my songs, made plans, paid close attention to the doctor's advice and attended the scans. I smiled, seeing the fetus develop into a recognisable form, my pleasure and expectation mirroring the growth. The magical nature of the following months heightened my sensitivity to the wonder of creation. I occupied a space which held my baby and I in an inviolable bond.
>
> Disaster struck just before my sixth month as I stood in the bathroom one morning. I blacked out. Coming round I crawled to the bedroom followed by my youngest daughter whose face was streaked with tears. Her distress was palpable and I was caught between trying to comfort her and the pain in my belly. My husband called the doctor who came sometime later. When he examined me he told me I was miscarrying. The bed was soaked with blood and the contractions strengthened doubling my body with pain, the weakened womb pacing my mounting distress joining with the stark realisation that my baby's life was transformed into death's dark and still hand. My son was stillborn, my grief overwhelming as I held his small, formed lifeless body, the doctor gently mouthing his sorry. Time stood still.
>
> [Jeni's 4th holotropic session:] I remember entering the coldest blackest space I have ever encountered. My whole self shivers to the bone and beyond, as chilly fingers flex their ice laden grip stilling all hope. I curve my body as tightly as it will allow; fear transplanting the warm blood in my veins. I descend into the agony of loss, seeing my broken baby, dead in my womb. My son, his lifeless body held gently in the weakening womb, is small, formed and silent. Holding him with my mother's

love, I struggle to rise for him, for me. I am caught in a density, which threatens to overwhelm me. Pinioned by strong hands, I smell the maleness of raging desire assaulting my nostrils and stare desolation in the face [Jeni is re-experiencing a sexually abusive episode]. I flee to that place of non-identity. I know not how long I wander desolate or how I find myself again. My broken baby is no more and my womb is no longer with me.

I hear my name being called softly, it is whispering still on the wind, warming my limbs, bidding me rise. I feel a powerful presence—primal in nature. A huge yellow and black cat softly pads across my path—sinuous, lithe and, familiar. Mouth open, enormous yellow sabre teeth displayed, ears sifting meaning, tail long and gently flicking side to side. Cat stops and gazes in my direction, slanted eyes focussing. Without warning we merge and, I become Cat—savage, ancient and flushed with natural instinct. My cub has been killed and I hunt his slayer. My humanness sits inside quiescent and accepting. Tears flow inside as I roar 'you killed my baby'. The force of my distress marked by a loss of control, warm urine rushes down my legs. I feel no shame. Our flattened tight body hugs the grass, eyes centred on my prey. A thrust of speed and claws and teeth fasten; rending, tearing, cracking muscle, sinew and bone. Life's blood spurts, spills, soaking fur teeth and tongue. My hunger sated, I curl my body into contentment's shape and rest. Fat Cat.

A huge deep orange sun hangs just above the horizon. Seven tall, tall dark men, twig thin, stand in front of this sun; their bodies glistening with effort. Startled by this image, I come to a standstill. They are moving in unison, to an internal rhythm, spears held in hands as brown as the soil they dance upon. Their heads are each decorated with four points floating just above. Fascinated by this and their primordial ways, it takes time to realise they are dancing for me, and for him—the broken woman and the broken baby. A sense of awe and a feeling of pure connection spirals within.

Summary and Conclusion

Holotropic ritual can break through the hegemony imposed on the psyche by a traumatic event, reworking the traumatic event until a new sense of self is born. The data these women gather in transpersonal states of consciousness generated in holotropic ritual suggests that the frozen energy bound up in blocked emotional or psychosomatic symptoms is converted "into a stream of experience"[98] coupled with a sense of "flow" after.[99] This ritual process has been likened to the death and rebirth mysteries of many cultures. It is a process that seems to be a universal one, naturally occurring when the psyche seeks to rebalance and re-tune to its integral healthiness. The experience is also educative: after the climax of the session and the breakthrough into transpersonal consciousness, "the remainder of the session will be spent in these spaces as one's education continues against an often ecstatic background."[100] There is a feeling of community and cooperative endeavour that pervades the ritual process, which when heightened to its zenith, bears fruit as *communitas*—a sense of deep psycho-spiritual bonding with the universe and its particulars beyond *all* hierarchies. It is in the state of communitas where further transpersonal potentials open and flower.

A final word on authoritative knowledge: in holotropic ritual, authoritative knowledge rests with the ritual participants. While the map and model (or myth) of the holotropic cosmology is given as authoritative and warranted—it is nevertheless a *provisional* map. Indeed, because Grof's transpersonal paradigm is perpetually open to revision,[101] ritual participants can contribute to this revision (as my participants have), therefore Grof's transpersonal cosmology is creatively open and is not only demonstrated or *legitimated* in holotropic ritual but extended. Therefore, holotropic breathers can participate in and share in the construction of authoritative knowledge which is empowering and restorative.

As anthropologist Richard Katz put it in relation to healing in non-ordinary states: "During the experience itself, cultural concepts and descriptions are not available. So, while there is conceptual clarity, there is experiential mystery."[102] Grof's model has conceptual clarity yet the ritual's "experiential mystery" leaves the door open for idiosyncratic healing events.[103] That is to say that ultimate authoritative knowledge rests in the hands of the protagonist and his or her "inner healer"—this is very important for persons who have been ritually robbed of their spiritual authority and embodied knowing.

I have presented here the tentative foundations of a theory suggesting that the traumas accrued in modern technocratic birthing rituals, and reproductive crises in general, could be healed in holotropic ritual. I have offered several examples pointing in this direction that further women's "epistemic exploration through narratives"[104] and enactment of their birth-giving and ritualized healing experiences.

Ideally, the original ritual where the birth-giving trauma was maximized would be changed; however, attempts to re-ritualize hospital birth beginning in the mid-1960s seem to have (debatably) failed. Nevertheless, as our exploration into childbirth and healing expands into the 21st century, new possibilities and paradigms open and older ones slowly disappear. The transpersonal movement, among other movements, will continue to offer alternatives to the dominant system in the hope of changing those structures for the better, or until the alternatives become mainstream.

The Use of Holotropic Breathwork in the Integrated Treatment of Mood Disorders

Paul Grof and Arlene Fox

Introduction

Recent epidemiological data indicate a gradual but disconcerting increase in the prevalence of mood disorders and demonstrate the need for more efficacious treatment.[1] The current therapeutic armamentarium for these conditions includes primarily pharmacotherapy, psychotherapy, and alternative approaches. Unfortunately, confidence in the results of therapeutic intervention studies has recently been waning,[2] and there is growing awareness that none of the available approaches offer a satisfactory outcome, particularly when used in isolation.[3]

In the past, the above-mentioned three approaches were strictly delineated, with psychotherapists avoiding any use of medications, alternative therapists practicing exclusively on the fringe, and psychiatrists paying lip service to the patients' psychological support. Times have changed. Increasing numbers of psychotherapists have been combining their therapy with alternative healing techniques and psychotropic drugs ranging from antidepressants and mood stabilizers to psychedelics.[4] Psychopharmacologists are more frequently recognizing the value of supportive psychotherapy. Biological techniques such as brain imaging are now employed to document the physiological changes induced in the brain by psychotherapy.[5] And not only do the professionals have a different attitude, but many of their clients have an expectation of treatment that is multi-faceted and comprehensive. Such combinations are intended to improve the situation and are particularly helpful if the client's predicament is difficult to manage.

Using a similar rationale to enhance our therapeutic potential, we incorporated holotropic breathwork into our approach. The holotropic strategy touches the deeper layers of human consciousness and was developed primarily for in-depth self-exploration, but because it has transformative potential, it can also be applied to psychotherapy. The following is a brief explanation of holotropic breathwork and several case vignettes that illustrate our experience with such an application.

Background and Principles of Holotropic Breathwork

In the history of psychotherapy, holotropic breathwork is a relatively new approach. This technique was developed in the mid-1970s by Stanislav and Christina Grof.[6] The simplest translation of holotropic is "striving for wholeness." Holotropic breathwork integrates relevant elements from various dimensions of depth psychotherapy and psychology. It has grown out of the groundbreaking work of the Freudian, Reichian, Rankian, and Jungian schools and is enriched by insights from modern consciousness research.

In practice, holotropic breathwork also shares certain techniques with the experiential therapies that have developed mainly in the context of humanistic psychology. The creation of this approach may not have been possible without the discovery of the potent psychoactive effects of LSD and without the experience the discoverers had with psychedelic therapy.

It was this exploration into the deeper layers of the unconscious that enabled Grof to chart a new, expanded cartography of the human unconscious.[7] These discoveries also contributed significantly to the appreciation of the remarkable transformative potential of non-ordinary states of consciousness.

A vastly extended map of the psyche, which is essential for the practice of holotropic breathwork, includes not only the postnatal biography and the individual unconscious described by Freud, but also the perinatal (the unconscious memory imprints of biological birth) and the transpersonal (the collective unconscious of archetypes and human history) dimensions. This approach also enlarges the therapeutic armamentarium by including powerful therapeutic mechanisms that in non-ordinary states of consciousness become available on the perinatal and transpersonal levels of the psyche.

The new insights concerning the strategy of self-exploration and therapy signify the most radical innovation of holotropic breathwork. Unlike the traditional psychotherapeutic schools that start with a conceptual speculation as to how the mind functions and then proceed to attempt to remove the obstacles by verbal techniques, the utilization of non-ordinary states of consciousness offers a simpler but a more radical experiential option. In non-ordinary states the mind/brain that operates as a self-regulating complex system activates an "inner detective" that identifies the intensely emotionally charged areas of the mind and transposes them into consciousness for resolution. During this process, the main role of the psychotherapist is to provide a well-informed, skillful partnership with the client, ensure his or her safety, and facilitate a productive outcome. Those who facilitate holotropic breathwork need to be particularly knowledgeable in transpersonal psychology, and must draw from their own direct personal experience which has been acquired during training.

Throughout human history, use of the breath, music, and nourishing physical contact have played an important role in various sacred and healing ceremonies and indigenous rituals,[8] and more recently in the Body Oriented Psychotherapies.[9] These are all essential elements of holotropic breathwork. Each session begins with a guided relaxation to encourage the thoughts of the day to ease and encourage a heightened awareness of the physicality of the body. Participants are then encouraged to begin to breathe very deeply and with more force. Carefully chosen music is integrated and, at the beginning, encourages lively involvement. For the next couple of hours, the music follows a pattern that best enhances an inner journey, and no one but the participants know where the music is taking them.

Obviously, it is essential to create a safe physical environment and a system of support. A preliminary explanation of the process is essential to prepare the participants, and some participants may feel reticent at this point and decline the opportunity. Principles have been developed to screen participants for emotional and physical contraindications prior to holotropic breathing, and create a safe physical setting and an interpersonal support system which will prepare the subjects, theoretically and practically, for the sessions.[10]

The final outcome of each holotropic breathwork session depends on good integration of the experience into the client's psyche. The painting of mandalas has been employed to assist the process of incorporation of experiences that took place during the session. Following the mandala drawing, there is group sharing (after a break for eating as everyone is usually in need of sustenance). During the share, group participants typically explain the relevance of the painting. There is frequently a highly charged atmosphere in the sharing as it is a time for the formation of words to begin the integration of the experience.

Our Approach

Over a period of nine years, we included 60 subjects in holotropic breathwork sessions. The first year we worked with a group of meditators who were intensely interested in self-exploration but did not request psychiatric help. Subsequently, we started working with patients suffering from mood disorders in groups ranging from eight to sixteen members. Each group was supported by two or three facilitators. The sessions took place once per month, on average, and the material that emerged in the sessions was used for psychotherapeutic work. Sessions were structured in the traditional format as taught by in Grof Training and as described in several manuals.[11]

A total of 51 patients participated, all suffering from bipolar disorder or recurrent unipolar depressions). Some of these participants were considered difficult to stabilize with the available treatments. Some felt that they needed more insight into the nature of the disorder and the psychodynamic mechanisms involved, and some were in search of a meaning for their illness. Each subject was in individual psychiatric care in addition to the holotropic breathwork. Notes were kept for each session, and the outcome of each session was reviewed at regular appointments thereafter. The sessions took place in a large basement of a private home that was adapted for this purpose.

We were interested in observing how much each individual might gain by adding a series of holotropic breathwork to our clinical care. Although our program for mood disorders in the Royal Ottawa Hospital has always integrated clinical practice with research into the nature and treatment of mood disorders, when working with holotropic breathwork, we limited ourselves to straightforward clinical observations without a research approach or quantifications. The environment in which we practiced proved not to be ready to investigate, with a research design, questions that included transpersonal issues.

As a consequence, the observations we offer here were, of necessity, generated not from a specifically designed quantitative research model but within a qualitative framework of individual, client-oriented therapeutic clinical practice.

Outcome Observations

For the reasons explained above, we will characterize the outcome simply by providing several typical, individual vignettes. The holotropic approach succeeded in improving symptoms and well-being of some patients suffering from mood disorders resistant to psycho- and pharmacotherapy. In some cases, the psychotherapeutic process was facilitated, and in others, helpful insights resulted. Illustrative case histories of such changes are described. The names and a few personal details of the persons mentioned in the case vignettes have been adjusted to protect their privacy.

Vignette One

Leah was 56 years old when she came to us for help. She had suffered from recurrent depressions for more than 30 years. Despite twelve years of psychoanalysis, three years of other psychotherapies, and fifteen years of acute, as well as maintenance, treatments with available antidepressants, she continued to struggle and was periodically incapacitated by depressive episodes. As a young mother, she had been given little hope that she would ever live a normal, happy, and balanced life. After completing a comprehensive psychiatric and psychological assessment and laboratory tests of physical health and hormonal and metabolic balance, we once again attempted to treat her the traditional way. This time we used the help of the newest antidepressant in a full dosage, combined with supportive psychotherapy, but again to no avail.

We concluded that repeating her treatment of many years was not sufficient and offered her a series of sessions of holotropic breathwork to which she consented. From the beginning of the interviews, she seemed unusually preoccupied with negative thoughts and anger directed at her deceased mother. Clearly, she and her mother had an uneasy relationship since the patient's childhood. Although she could offer some Oedipal interpretations acquired during her years of psychoanalysis, this did not reduce the intensity of her resentment. Instead, her preoccupation with these feelings seemed to play an important role in triggering more acute depressions.

Her first two sessions were relatively unremarkable although in subsequent sharing after and between the sessions, she talked about the problems in her marriage and her unhappy childhood. Then in her third session, she relived a situation that she dated to the time when she was still in her crib. She recognized it because of the images that she remembered: the parallel, colored beams of wood that surrounded her in the crib. She experienced intense terror as she became aware that her mother was entering the room. This scene of intense emotional discomfort lasted for some time during the session. She suspected that her mother had scared and punished her, but she never fully re-experienced what her mother actually did do after entering her room. Surprisingly, just reliving with full awareness the negative feeling complex, with the terror and anxiety that was associated with the retrieved memory, was enough for a very positive, lasting outcome. As the music finally shifted to a more emotionally calming segment, at least in her perception, she found herself in the midst of a large cathedral, and she was in touch with a deeper spiritual core that she had not been attentive to before.

Part of her insight following this session was that she had always hated horizontal blinds with a passion when they were down and open. Closed, they were fine. She had resisted any attempt by her husband to have blinds on the windows of their home. Leah realized that the open blinds resembled the bars of her crib when she was lying down.

She participated in four more sessions of holotropic breathwork, assuming that she must go back and fully relive the repressed encounter with her mother during her crib time in order to have a lasting benefit. However, that did not happen. The subsequent sessions were fairly peaceful. She came out of the sessions feeling free, relieved of her depression, and has remained that way for twelve years now.

After completing the series, she remained in regular follow-up. She experienced occasional mild ups and downs in her mood, depending on the usual life stresses, but has never become clinically depressed again.

The relationship with her husband had been void of any physical intimacy for many years, and instead, she had maintained an affair with a close friend. However, after the holotropic sessions she became able to make important changes in her marriage. She connected with the feelings that attracted her to her husband initially, and her affair of many years began to feel empty and meaningless. She terminated the affair and happily resumed emotional intimacy with her husband.

The description of her experiences from her third session and her subsequent well-being are captured on videotape.[12]

Vignette Two

James is a health care professional who, in his 30s, started experiencing episodes of depression and hypomania. His bipolar illness wreaked havoc with his life. His wife divorced him, he lost contact with his four children, his work was negatively affected, and he was released from his position.

With the help of systematic lithium treatment, he regained his overall emotional composure. In spite of this, he continued being troubled by dysphoric episodes and was unable to make sense of the disruptions that the abnormal moods caused in his life.

Holotropic breathwork and supportive psychotherapy were instrumental in helping him work through his personal past. During the initial sessions, he often revisited his depressive states which were tainted with stress, tension, despair, and hopelessness. For quite a while, his sessions vacillated. Some were deeper and more meaningful—affecting his dreams—while others were superficial. As the number of sessions increased, his preoccupation with his personal problems and past failings disappeared from the content, and the process focused more on the nature of human suffering and healing. He gradually gained deeper insights into these issues in general and into his own depression in particular. The preoccupation with his melancholic past vanished, and he started experiencing an increased level of overall adjustment.

He became able to accept the breakdown of his marriage and of his family connections. He now felt increasingly free to make changes in his personal life, which gave him satisfaction and greater peace of mind. He created a new life for himself and became involved in a relationship. He no longer identified so closely with his medical practice; his new wife and her children became the main focus of his life.

He decided to continue with his lithium treatment as he knew he needed good emotional stability for his professional life and did not want to risk the social destruction that often follows untreated hypomanic and manic episodes. The paintings of mandalas he made at the end of each session gradually became more numinous and finally became dominated by a beautiful flying white bird—a commonly used symbol of a free human spirit.

The positive outcome has been long-lasting. Over time he has even been able to reconnect with his own four children. I have followed James for the subsequent twelve years, seeing him two or three times a year. He recognizes that, since the completion of his sessions over a decade ago, he has been free of any black moods and attributes it to the gains made in the holotropic sessions.

Vignette Three

When Ray was referred for help by his family physician, he was in his early fifties. Employed by the government as an economist, Ray was in a senior position because of his professional accomplishments. Despite his ability to function at such a high level, he was chronically dysphoric, with his low mood often deepening into a full clinical depression with suicidal tendencies requiring intense pharmacotherapy.

Prior to his referral to us, he had undergone intensive psychotherapy and psychiatric treatment for a number of years. He was acutely aware of his difficult childhood mainly due to the physical and emotional abuse at the hands of his father, but however much he had tried to resolve his childhood conflicts in talk therapy, there seemed to be no beneficial impact on his painful moods.

When he started a series of holotropic sessions, he often relived the intense pain and acute distress of his early conflicts with his abusive father. Nevertheless, as he progressed, he gradually realized that, as his inner life had a powerful visual dimension that kept searching for artistic expression, he was deeply dissatisfied with his present life as a civil servant. He had become an economist to please his father at the cost of suppressing his creative aspirations. The dream he had as a child to become a visual artist was not being addressed in his present state.

He started seeing more clearly the deeper source of his discontent and the inner conflicts underlying it. He came to the realization that dwelling on painful memories from the past was preventing him from moving toward a positive resolution.

Despite the huge challenges one encounters by giving up a senior government career for the economic insecurity of a struggling artist, he switched to art photography. He experimented with expressing the unusual images from his fertile mind. He found an accomplished mentor and studied photography in order to learn art photography.

Major changes gradually took place: he transformed from an unhappy, frustrated economist who spent much of the time blaming his father for his unhappy childhood, to a productive, creative, successful artist who expressed his profound feelings through dramatic visuals. Through the discovery of his different orientation, he found peace of mind, contentment and has not been troubled by depressive episodes. His tolerance for the difficulties of everyday life markedly shifted, and he became able to see the positive and negative aspects of life as two linked inseparable components, both reflected in his art. His experiences are also captured on videotape.[13]

Discussion

We outlined how we integrated holotropic breathwork with traditional psychiatric care and illustrated how we witnessed some very positive changes in individuals.

Holotropic breathwork radically diverges from conventional psychotherapy as it utilizes the induction of non-ordinary states of consciousness, takes advantage of the effects on the mind of evocative, intense, carefully selected music, encourages the expression of strongly felt emotions and welcomes supportive physical contact with clients. Through this process, holotropic breathwork can have positive, healing effects not only on emotional disorders but also on an individual's personality, life strategy, and hierarchy of values.

We propose that the mechanisms involved in the observed changes include the reduction of psychological defense mechanisms, the retrieval of emotionally relevant important but repressed traumatic events, and the discovery of personally powerful insights. The process can also activate personal and transpersonal mechanisms so far considered beyond the conscious reach of the mind: reliving formative memories from the earliest years of life, reactivating imprints of biological birth and prenatal traumas, triggering of what clients may label as past life memories, and encountering archetypal figures and mystical experiences. The alleviation of symptoms is, at times, accompanied by spiritual experiences and a leap in the development of compassion, tolerance, and changes in values.

Holotropic breathwork is almost always practiced in a group, and this seems to play an important role. While the participants had no verbal contact during the sessions, observations suggest that there were meaningful emotional links. In the group sharing, an integral part of the sessions, participants would often talk about others in the group and how their own experience incorporated the experiences of other group members.

Group context obviously colored the experiences. Sometimes this would even be at a depth that would be difficult to explain in any other way than emerging from a group unconscious or a group mind. The deeper links among some of the participants were sometimes expressed by a striking complementarity of their shared experiences, particularly when the participants were describing their "past lives" and when they were independently drawing the same symbols in their mandalas.

Such observations suggest that group involvement may be very therapeutic, particularly for clients who rarely experience togetherness in their daily lives. It is worth noting that we did an investigation of the personal characteristics of those who stayed with the group and continued with the sessions and those who did not connect and dropped out.[14]

It was also very interesting to see that different members of each group had their most intense, peak experiences at different times, which often coincided with signs of elevated brain arousal (elevated cortisol in dexamethasone-suppression test). To us, this implies that there probably was an interplay of individual biological and transpersonal arousal rhythms. From the practical point of view this was advantageous as at any point in time there were only a few members of the group requiring the full attention of the facilitators.

Holotropic breathwork is ordinarily not recommended for people suffering from, or with a history of major psychiatric disorders. We were not, however, in an "ordinary" situation. Dissimilar to the holotropic workshops, which are open to the prescreened public,[15] all our patients had continuous care during the series of sessions along with prompt access to us in case of need. There was a team available, should there be the need for an extended session or close observation afterward, until the activated problem was resolved. We were also in the particularly fortunate situation of having access to a hospital bed should someone acutely decompensate and require it, although such a situation did not arise.

Still, we must stress not only the advantages but also the need for caution, for a protective environment, and for expertise when using this approach with subjects with major mood disorders. It is important to keep in mind that dealing with the various challenging situations that might emerge in patients with major psychiatric disorders and bringing the sessions to a productive closure requires careful preparation and an extensive familiarity with non-ordinary states of consciousness. Those who plan to conduct sessions with such patients should, in addition to adequate psychiatric training, first undergo thorough training in holotropic breathwork under proper supervision.

Conclusions

In order to improve the outcome in the long-term treatment of patients with recurrent mood disorders, we incorporated holotropic breathwork into our practice. While following the principles of the process described in the literature, we achieved success in some patients who previously had failed to respond to expertly applied long-term psychotherapy and pharmacotherapy. Overall, our patients were helped by new valuable insights into the psychodynamic sources of their suffering and their deeper conflicts.

Breathing New Life into Social Transformation

Holotropic Breathwork for Social, Cultural, and Political Leaders

WILLIAM KEEPIN

Introduction and Background

This chapter summarizes key insights and breakthroughs in experiences of holotropic breathwork that emerged in working with professional leaders across several disciplines, including environmental leadership, politics, science, social activism, and the health professions. Most of these leaders would not likely have engaged in breathwork were it not for their participation in a special leadership training program. Given their active lives and commitments, many of them would probably have dismissed the value of such focused "inner work," until they experienced breathwork.

By way of personal background, I completed one of the earliest two Grof Training programs conducted over three years in the late 1980s with Stanislav and Christina Grof. Since then, I have had the privilege of applying holotropic breathwork in many different professional contexts. For six years, I worked closely with two clinical psychologists, Dr. Sharyn Faro and Dr. Bonnie Morrison, conducting holotropic breathwork sessions for their clientele, many with histories of trauma, sexual violation, and other abuses. I co-facilitated numerous breathwork workshops with other certified facilitators, including Laurie Weaver and especially Diane Haug who had trained alongside me in the parallel three-year training.

In the early 1990s, I co-founded the Satyana Institute and began developing 'Leading with Spirit' trainings and "Gender Reconciliation" programs aimed at transforming gender relations in society. Breathwork was an integral component of both initiatives. My primary professional work at that time was environmental science and climate change mitigation, and I began to implement holotropic breathwork in contemplative retreats for fellow environmental scientists and social activists. I co-led a series of retreats for environmental leaders in several states and in the federal government.

In the early 2000s, I launched the first professional training in Gender Reconciliation with Diane Haug as a key co-trainer. This program evolved over time to become Gender Equity and Reconciliation International (GERI), which today conducts dozens of intensive trainings in multiple countries. Breathwork remains a key component of the facilitator training curriculum.[1]

Breathwork Experiences of Social Change Leaders

Several accounts of breathwork experiences are recounted below, mostly in the clients' own words, including the impacts on their professional work. The application of breathwork to Gender Equity and Reconciliation work is primarily documented elsewhere, and so is not repeated here.[2] The examples below are recounted with permission and have not been published previously. The first few examples are brief vignettes to give a sense of the breathwork territory, followed by further examples recounted in greater detail.

A brilliant, dynamic scientist in one of our retreats for environmentalists several years ago was initially resistant to doing the holotropic breathwork: "I can't believe we are devoting an entire day to some New Agey breathing practice," he exclaimed, "when we have an Earth to save!" He worked tirelessly for the Nature Conservancy and spent weeks at a time in the wilderness meticulously tracking herds of elk and their forced migration patterns as their habitats were being destroyed by encroaching suburban sprawl and industrial infrastructure. During his first breathwork session, to his complete surprise and terror at first, he heard a loud "whooshing" sound, and suddenly he *became* an elk. His hands became hooves, his limbs became massive elk legs, and he found himself galloping at full speed as one of the herd. Glancing left and right, his own bulging elk eyes met the eyes of the other elk on either side of him as they galloped together at breakneck speed across the high mountain plain. He came out of this session totally exhilarated, and still half-terrified. This consummate scientist had a full-blown shamanic shape-shifting experience. It radically transformed his understanding of his own work and the elk he so dearly loved. It also greatly transformed his previous scientific worldview to embrace shamanic epistemologies. We helped him to understand that, from the perspective of the worldview we inhabit, *he* didn't just become an elk; rather, the elk family *invited* him into their midst and intimate identification, precisely because he was that rare human being who deeply cared about the elk and their future destiny. Thereafter, he was a strong advocate of breathwork in our retreats.

A prominent colleague who was the director of a national environmental department developed an aggressive brain tumor and died within four months. Our community was shocked and devastated. In our subsequent annual retreat, we did breathwork, which catalyzed a remarkable collective grieving process. This precipitated a new, unanticipated level of deep bonding within the group. We all felt the palpable presence of our deceased colleague, who effectively brought us together in a profound new level of depth and relational intimacy. The breathwork facilitated an unexpected breakthrough of deep blessing and newfound trust and love within this prominent professional community.

After decades of no contact, we received an email communication from a participant in one of our earliest workshops on Gender Equity and Reconciliation 29 years ago. Ms. Kaia Svien wrote, out of the blue, to report about her experience:

> Dear Will,
>
> We spent time together in the mid-80s (I'm guessing) at a five-day Breathwork event in rural Osceola, Wisconsin. The experience was designed by you and others to explore and heal the abyss between the genders....
>
> I have been forever grateful for that experience because it taught me, profoundly, how to transform rage into compassion. I've since understood that there is an almost equal relationship between the amount of anger, disgust, grief, etc. I experience—and the amount of resulting compassion that is possible.
>
> The compassion I got from this understanding is immeasurable

> and one of my best gurus....I look to that workshop as having a great influence on me as well. That shift from rage to compassion has been a powerful, bright guide for me over the years in so many outer and inner situations.

Libby and Len Traubman founded a remarkable method of conflict resolution in their living room, which grew to become the Jewish-Palestinian Living Room Dialogue Group. Their Sustained Dialogue method entails one-to-one, face-to-face contact and conversation between adversaries. In the mid-1990s, when their work was still in its early stages, the Traubmans experienced holotropic breathwork in one of our spiritual leadership programs called *Leading with Spirit* conducted at the Foundation for Global Community in Palo Alto, California.

As Libby describes her breathwork session,

> It was one of the deepest and most profound experiences I have ever had. I remember seeing and being with people from all over the world. The diversity was so rich, and it was a parade of color, from faces to dress. It was sooooo beautiful and moving and the most powerful feeling of We Are One. That image and those feelings are still with me today, all these years later, and inspire everything I do. In my heart, I believe we are interconnected, interdependent, and interrelated, not only to each other but with all of Life. It is my deepest faith.

Len also recounts a poignant visionary experience.

> [There were] dead people—a shared experience of Jews and Native Americans. Libby [was there] bringing her music and light over the rim of the planet, and [I was] going up the mountain carrying the Torah and its principles to the people.

Len's breathwork session culminated in a remarkably prescient vision of his and Libby's future role in facilitating transformative healing dialogue:

> Deep into the experience, I remember weeping and calling out: "It's so beautiful, it's so beautiful" over and over. It was one of the deepest, most insightful, unforgettable, and undeniable subjective experiences of my life. We are forever indebted to [this breathwork experience] for deepening ourselves and helping us fasten down our life activity in Sustained Dialogue.

Over the subsequent twenty-five years, the Traubmans' Sustained Dialogue grew into a remarkable manifestation of their visionary insights in breathwork—bridging hearts of divided peoples. Sustained Dialogue has been applied widely around the globe by educators, researchers, journalists, and strategists, including the US Department of State, which distributes their group's instructive films in Africa.

Oneness of All Beings, and Blessings from the Ancestors

Betsy Taylor is the founder of New Dream, a pioneering service organization that strives to transform consumerist values and lifestyles to improve the quality of life for people and the Earth. During its start-up phase, Betsy attended one of our *Leading with Spirit* training programs, conducted at the Fetzer Institute in Kalamazoo, Michigan.

Betsy entered into a deep inner journey during her first breathwork session, and as the session drew to a close, I checked in with her. Betsy was deeply immersed in an intensive inner experience, so I encouraged her to remain in the process. She continued in her breathwork journey through lunch and all through the second session, for a total of more than seven hours. Although this is unusual, there are cases when this is warranted, and this was such a case.

Betsy recounts her breathwork experience in excerpts from her personal journal below. The first two entries were written in the days immediately following the breathwork session, and the last entry was written thirteen years later, summarizing the long-term impacts of this experience.

> *February 6, 1998.* I am at the Fetzer Institute for a retreat called Leading with Spirit, offered by the staff of the Satyana Institute. It has been a very rare event. We did a breathing exercise that was designed to help us transcend the immediate dimension of time and space. Wow, did I!!!
>
> I had two journeys. The first was indescribable. Neil came—my older brother who died of brain cancer in 1990—as a swan. He took me back in time. I was with my mother (who died in 1991 of leukemia) and a host of ancestors: a great grandfather I never knew, my Grandma, a great aunt, and I met my [other] grandmother as a girl. Then we went back further to Boston and the Mayflower (yes, my family came over on the ship), and to Scotland and my ancestors there, and then to a deep and

beautiful canyon somewhere in the Western US. My ancestors supported and loved me. I loved my mother and was with her, rubbing her back as a little girl when her own mother died of a brain tumor (she was only ten when her mother died).

Then suddenly, my brother Neil had me stand on the side of the Grand Canyon, and he told me to jump and that I would be safe. I was frightened. Neil was a huge swan. The swan had Neil's face but otherwise a powerful body and wingspan. Neil told me to jump off the side of the Grand Canyon, and I would be okay—he laughed. I was very frightened, but finally, I jumped!

Neil caught me in his wings and *whoosh*—we flew together. I rode on his back. I laughed out loud with joy. I was briefly my own swan, doing turns in the sky. While riding with Neil, I was happy. He was happy and healthy—no pain, no cancer, no death. He then told me he was going to let me go to fall but that I would be okay. I was scared, but he let me go and I fell—to my mother who was also a swan. She caught, cradled, and rocked me. I felt so united with her again after missing her so much. Then she let go, and I was caught by a large Eagle, and I hung from his talons. I was then gently and joyfully tossed from ancestor to ancestor until I fell at the bottom of the Canyon into a soft pile of hay. It was really blissful. I felt so totally safe.

My mother became a swan again, and I was on her back. We flew to the concentration camps of Germany and brought bones of the dead Jews back to the Canyon, where they awakened and joined us at the fire. Many people were drumming, dancing, and gathering around the light. The ancestors, including many from hundreds if not thousands of years ago, gathered around me and celebrated me, including my kind surrogate grandfather who had died a few years earlier.

I came away from this first journey with two strong feelings: (1) we are all deeply connected to each other across time and space, and we must just be present for each other; (2) in an eternal sense, we are absolutely safe. We will be cared for in perfect, communal ways when we die. Death will not hurt. It will be exhilarating.

My second journey was shorter. Neil was a swan. He asked me to get on his back because he had just one more thing to show me. We flew to Africa and to a refugee camp. There were some children with a soccer ball, but they were too weak to kick it. I felt myself pulling out of the journey, I did not want to look. Everyone was emaciated and weak. People were starving. I felt myself leaving Neil.

He spoke quietly and said, "Okay, just one more minute, Betsy. I just want you to see these two children, and then we will leave." We flew down for a closer look. Two small children had their backs to us. They were African, very thin, very weak. Neil asked them to turn around, and when they did, they were *my* children! They had the faces of my daughter and son. Again, the message was clear. We are one. There is no real separation.

> *February 10, 1998. further reflection in my journal.* My breathing experience was really incredible. Either I went deeply into my unconscious, or I crossed into a different dimension. It took four people to get me to the bathroom. I was out of my body. I reconnected deeply with relatives, friends, and ultimately with strangers and all creatures. We are one. We must care more, risk more, listen more to each other. We must surrender to God's dream. That's all.
>
> I felt so restored after this. [And] deep peace of knowing I'm on God's/Love's path.

Thirteen years later, Betsy Taylor reflected back on her breathwork session:

> *September 2011.* This breathwork journey has stayed with me over the years. I have shared the story with perhaps a dozen close friends and family members. Swans have remained central, and they have come to me at key moments relating to birth and death—both in my dreams but also in reality, actually flying near to me as a loved one departs. It is a mystery.
>
> The journey is a source of enormous peace for me. I turn to it, along with my regular turning to silence and meditation, as one source of connection to something much deeper than the transient, changing nature of things.
>
> When I lose my way, feel despair, or get overwhelmed by issues like climate change, injustice, species extinction, and more, I can turn back to the fire and the loving circle of the

ancestors – or to the field of animals – and simply rest into the silence of faith that there is something more we can lean on. We do not have to define it. We can't. It is accessible through surrender—not of our rational self, but of our clinging, dominating, individual self.

This is accessible to everyone who simply risks stopping, turning off all gadgets, and listens, breathes, and seeks love. We just have to stop, give thanks for what is good, make apologies for our wrongdoings or oversights, and then just open up to the potential of love.

We will discover a whole new realm of potential communication and connection, one that has always been there and that was probably accessed by our ancestors, who were more attuned to the natural world. I think we can connect across time and space. I, of course, can forget this at times, but when I return to it, this journey is always one of my anchors.

Reuniting with Departed Children's Souls

This example comes from a skilled professional facilitator in our Gender Equity and Reconciliation International (GERI) training program. This man and his wife have two young children, and once had three earlier children, all now deceased, which has caused them tremendous grief.

> At first, I had no idea what to expect. The first thirty minutes were a mixture of listening to the music—with full awareness and deliberate exaggerated activation of my breathing—and a big mental conversation inside me about: What am I doing this for? How am I going to know how to let go? My previous experience in facilitating experiential and wilderness-based therapeutic work helped me to surrender to whatever this process was going to be for me.
>
> At some point, my experience shifted to finding myself completely in a dark tunnel and going through it to finally reaching the end of it, and being welcomed by a beautiful, remarkable sight: Steve Foster cuddling my late son Mbuso, who had died of Sudden Infant Death Syndrome. Steven Foster is also deceased, and he was one of my teachers who trained me to be a wilderness guide.[3]
>
> I could not contain my joy and sadness, pouring out

simultaneously from the depth of my belly. I cried and laughed, rejoiced and became sad altogether! I spoke to them, and without answering me, Steve smiled and acknowledged what I was saying to him and then turned and walked away from me.

I followed him and could not reach them (Steve with Mbuso in his arms)—laughing and enjoying this profound moment and being reunited with my son. Then I saw where they were taking me. We came upon a beautiful lawn where Steve put Mbuso down, and he crawled towards two other children who were there playing and having a ball of a time. Then suddenly another bout of powerful and mixed feelings and emotions emerged for me, as I realized that these other two children were none other than Mbuso's sister and brother (Lilitha and Nkululo)—my two other late children [who had been stillborn].

My three deceased children were reunited with me! Their passing had created such excruciating suffering for my wife and me, and now here they were—all together—so happy, carefree, and joyous to be reunited! How I cried and laughed, rejoiced and became sad altogether again! Not far from where my children were playing were seated my wife's grandmother, her mother, and her aunt. They are also deceased in reality, but there I was seeing them smiling warmly and looking at me.

My body twitched and trembled, and I had a knot at the pit of my stomach. I arched my back, trying to release the pain and discomfort this knot was causing. One of the facilitators provided a gentle resistance on my stomach, which elicited a tingling that went from my stomach through my leg and then to my toes, as well as through my chest and my head. I found myself smiling warmly, and then became aware that I was forcing a smile and had been holding myself together in this special encounter all along.

Then, like a thunderbolt, it hit me that I did not have to be the Tiger, The Man, the one who holds himself together. At this realization, I broke out again into a raw, deep, and sincere crying bout—coupled with anger, grieving, and despair. I grieved the loss and passing of Makhulu, Sis'Thamie, Sis'Phakama, Nkululo, Lilitha, Steve Foster, and Mbuso. I let it rip and emptied myself of everything. Eventually, the pain and discomfort in my stomach disappeared, the tension on my upper and lower back disappeared, and the tingling left my body completely.

After some time, my loud, deep cry shifted and morphed into an uncontrollable laughter. I could feel my cheeks fill up with a broad smile, and I noticed that all along, the reaction and response from my family on the other side has been that of warmth, smiling, and positive nodding to all I was doing. I felt my heart sing, and the words "Arise and Rise" came to me. My heart continued to sing this beautiful song, which I had never heard before, and to this day, I cannot even recall.

I don't remember saying any farewell to any one of them, but eventually, I became aware again of the music in the room and how broad my smile was as I gently re-orientated myself. I wrote these words on the back of my mandala painting work: Release, Deep Grieving, Exceptional Comfort, Deep Healing! Great Gratitude! GOD! Spirit! Great Peace! Great Vision!

Birthing the Earth and Healing the Deep Wound of the Feminine

Lowell Brook is an artist, community activist, mother of five grown children, and an ordained Unitarian Universalist minister. Lowell's breathwork experiences recounted in her own words below, inspired her to create a remarkable series of magnificent sculptures and paintings.

> My mother taught me throughout my childhood, in word, and by her example, that men are stronger, more important, smarter, more interesting, more valuable, and more fun than women. She believed that my survival depended on understanding this fundamental truth. From my mother's perspective, this was the way it was supposed to be, and contradicting or struggling against it would only put me at risk of isolation and ultimately cause my downfall.
>
> My real struggle for survival was rooted in overcoming this primary wound—this powerful conditioning to negate my own essential feminine nature. This personal work has been reflected and held in the culture by the women's movement and the contemporary openness of emerging spirituality.
>
> During three sessions of Breathwork in the context of Gender Equity and Reconciliation work, I opened into

non-ordinary realms—beyond the usual boundaries of rational/intellectual/mental understanding and processing.

In the first Breathwork session, I slid easily beneath the radar of previously conditioned belief systems and found myself catapulted out of conventional structures into boundless space. Out there, beyond the blinding light of the sun, the mysteries of the universe were revealed to me in astounding new ways. In the "black void" appeared the shape of a woman giving birth. It was not a woman's body floating in space, it was space itself shaped as a large, naked woman. She was in a birthing position and "I" could move all around her, viewing her from every direction.

Then, unexpectedly, I merged with her, and found myself giving birth to the Earth. The Earth emerged exquisitely beautiful as the "blue ball" we've seen in photos from space, luminous and vital. I was not "Mother Earth." I was the Universe giving birth to the daughter Earth! My womb was the Cosmic Womb. I beheld what had emerged from my space-body, with the tender and adoring love of a mother, in awe of creation and with overwhelming desire to protect it from harm.

This was the healing gift of experiencing Birth, the amazing ability of the feminine of every species, as the original source of life. The significance and essential value of who I am as a woman was thus permanently secured in me by this extraordinary, embodied yet mystical experience.

In the second Breathwork session, a huge sun appeared, with glorious rays streaming out and all around me. I stayed enwrapped in this radiance until I was swept away into an undulating tube that seemed like an umbilical cord. Then a question appeared, and I responded, "Yes! I want to go all the way." I entered a narrow chute that was leading out of the radiant light world that previously had been dazzling me. Thus, I was given the opportunity to fearlessly trust the unknowable mystery that was approaching. Headfirst, I allowed, even welcomed, a back dive out of this world.

I felt the cacophony of stimulating vibration recede into the distance as my journey took me away. Completely surrendering my will, I dove through the original cervix, sliding into perfect stillness, caught by the midnight hands of absolute Now.

The silence was divine. The Great Silence. No light, no sound, no stimulation or vibration. The Peace was divine. I wanted nothing. I rested in this ultimate haven with a feeling of total yet simple gratitude for being relieved of the complicated burdens of living in the world. This was the gift of being birthed into dying and experiencing Bliss on the other side.

In the third Breathwork session, at one point, I saw in the distance a woman being raped and abused by a group of men. She was being beaten, and I felt her agony as she struggled in pain. Coming closer, my body merged with her body, and I found myself being buried in desert sand up to my neck... then the men all around were throwing stones at my exposed and defenseless head. It was unbearable. Absolutely unbearable!!!

I started sobbing and reaching out for my husband, who was my witness ("sitter") in the Breathwork session. I told him to hold me while I sobbed and sobbed, my body and soul wracked with the pain of embodying the cruelty, desecration, and complete helplessness I felt. As I cried away my pain in the safe arms of my husband, I came to feel these abusers' total ignorance and oblivious state—like animals violently destroying what they felt threatened by. It seemed they had to kill this woman because they were terrified of the power she had over them. But they didn't understand that; they just hated her and were unconscious of their motivation, or the consequences.

It was my great good fortune to be able to return to my usual life with abundant retention of all the teachings from these Breathwork journeys, complete with precise and acute visual memory. I turned these visions into a series of sculptures and paintings, comprising a body of work I considered a Feminine Cosmology and a chronicle of Becoming Free.

Although the transformation from masculinism to feminism to egalitarianism in our world is far from complete, there has been a significant shift in the past fifty years. May an era of cooperation between genders bring new healing ways and a possible future for this gorgeous planet and all her species.

A Rare Birth Trauma and Visitation in India

Gaus Sayyed was born into a poor family in a rural village in India. When he was still very young, his father became seriously ill, which meant a loss of income for the family. The father's condition deteriorated rapidly, to the point that he was paralyzed and unable to speak. The family financial situation soon became dire, so Gaus began to work—at age five—washing cars at a local garage, for which he was paid a pittance. One day, when Gaus was seven years old, Sr. Lucy Kurien of the well-known Maher project in India happened to drive into the garage where Gaus was working. Seeing his circumstances, she offered to take Gaus into her shelter for destitute women and children. With his mother's blessing, Gaus moved to Maher, and his life was utterly transformed. He entered school for the first time, and before long, he began to excel at everything he did. Gaus was a gifted child and a natural leader, and ten years later, he completed high school with honors. Then he entered university and earned his bachelor's degree, and received an award from the President of India for being the top student in his university. Gaus then entered graduate school and completed his MBA. Beyond his academic achievements, Gaus is also an extremely talented dancer, performer, public speaker, group leader, and facilitator.

Gaus and another gifted young man named Mongesh from Maher joined our Gender Equity and Reconciliation training program in 2014. Holotropic breathwork is part of the curriculum, and when it came time for their first breathwork experience, we trainers anticipated that both Gaus and Mongesh would likely take to the breathwork very well, given their innate exuberance and positive attitude toward new experiences in life. During the breathwork session, Mongesh responded as expected, but early in the session, Gaus entered into what was clearly a deeply challenging experience. As facilitators, we didn't know exactly what was transpiring for him, but he was writhing in pain much of the time, and his body language indicated that he was experiencing something relating to the birth process. As facilitators, we provided extensive bodywork support for Gaus during the session, with great compassion for whatever he was going through. After the session, Gaus took quite some time to come back to normal waking consciousness, and he then recounted a gripping story, the likes of which we had never heard before.

In his breathwork session, Gaus was taken back to the time surrounding his birth, which entailed a remarkably painful drama. He recounted the circumstances to us, many details of which he said that he had not known before.

Gaus's mother had earlier given birth to four girls, two of whom died at birth, and two survived. Gaus' father now dearly wanted a son. In India, tremendous value is placed on having sons, to an extreme degree that produces social aberrations such as female infanticide, which although illegal, is a significant affliction in many parts of India. In Gaus' case, because there were already two daughters in the family, there was tremendous pressure to have a son.

In his earnest desperation for a son, Gaus' father had entered into a bizarre "contract" with God—a superstitious Faustian bargain not uncommon in rural villages. The "contract" was this: God must give him a son, and the father's side of the bargain was that when the child was born, the father must throw his newborn son from the top of the village mosque—a height of more than three stories—to a small group of men on the ground below holding a green mat, onto which they would try to catch the falling infant. If the infant survived, then the newborn son belonged to the father; and if not, then he belonged to God, and in dying was thereby "returned" to God. This contract was deemed an impartial test of God's will, unfettered by the father's ardent desire to have a son. Such superstitious rituals are all too common in rural villages in India.

Gaus' mother was beside herself in the weeks leading up to his birth. Many male infants in their village had already died as a result of this ritual—dashed to the ground after missing the mat, or careening off the roof structures of the mosque, or else slamming onto the mat in an unfavorable position. She secretly prayed that her child would be another girl, so as to avoid this entire ordeal. When Gaus was born, she pleaded with her husband not to go through with the ritual. The father was also distraught, yet determined, because he held the strong conviction that he was "bound by God" to follow through. The couple entered into a profound conflict, and each day the wife begged her husband, "Not today! Not today!" So they would wait another day, and this excruciating drama continued for a full eighteen days, during which time they hid the fact of Gaus' birth from the community.

Gaus relived this entire drama during his breathwork session. As he reported afterward, he had known only the most basic facts about this history, but in the breathwork he re-experienced the terrifying details of the entire episode, including his mother's agony while still in utero prior to his birth and the intense stress of keeping his birth a secret for more than two weeks, the abject terror of his own likely impending death from the fall, and the conflict between his parents.

"I felt like I was a baby, literally," Gaus said. "I totally forgot that I am twenty-two years old. I was back in my Mom's womb, and then I was hidden in the home for those eighteen days. But I was feeling so sad. There was lots of noise and chaos, and my headache started. I witnessed this whole situation, as if in some kind of time machine. I was thinking maybe this was all just my imagination, but I was there! It was real, and I was like a witness."

Finally, on the eighteenth day, Gaus' father insisted that the ritual must be fulfilled, because the secret of Gaus' birth was about to be found out anyway, and then the family would become the target of the wrath of the villagers for attempting to circumvent "God's will."

Gaus' uncle proposed that he be the one to throw the infant, but the father insisted that he must be the one to throw his son. Gaus' mother fainted and was taken to the village clinic, where she remained in utter distress, fully expecting to receive news of her dead son. Before throwing Gaus, prayers were spoken in the mosque. "Before my father threw me, I was really scared because I had the feeling I would die. Then my father took me with one hand under my head, and one hand on my leg. There was this tremendous pressure on my head, and my father threw me."

The journey down was only a few moments in real-time, but in the breathwork session, "it was like an hour-long process—to fall from my father's hand down to the carpet." In the breathwork session Gaus repeatedly relived the harrowing trajectory of being released from his father's hands, the free fall, and hitting the mat with a massive blow to his head, which knocked him unconscious. "During that time, I was witnessing this whole situation, and at the same time I was fully connected to the experience. My head was blasting [in pain], and I think the music helped me to get through it. I was literally back in time 22 years, fully in that experience of being that falling infant."

"Some of my uncles and other people from the mosque caught me in the green carpet, and I lived. It was like my rebirth." Gaus was unconscious after his little body slammed into the mat, and he was taken to the clinic. He eventually woke up and began crying, and joined his mother. "And then the celebrations started, because I was alive."

Reliving of pre- and perinatal experiences, as in this case, are relatively frequent occurrences in breathwork. Gaus came out of the breathwork session with a major headache that did not go away as we continued to work with him. Later, he told us it took seven days for his head to return to normal. "It was hell, and I took lots of medicines, but only after seven days was I fully recovered." But the experience was profound. "I saw my birth. The breathwork is amazing. It was a journey, it was just two hours, but it felt like twenty-five years in that two hours."

Gaus subsequently visited his mother, who confirmed the details he had relived during his breathwork session.

> We spoke for three hours. I told her my story, and she was very surprised, and she demanded to know, 'Who told you all this!?' She was shocked because she had never told me most of these details, so she naturally assumed that someone else must have told all this to me. But I was simply relating what had happened in my breathwork session, and she confirmed that my account was indeed accurate in the various details and had actually happened the way I experienced it in breathwork.
>
> She was very moved by all this, and she was literally crying. She had wanted to bury that whole memory forever, and she was afraid that I might think she is mad, or something. But when I told her my experience, then she realized, 'I need to tell him.' So she told me the whole story again, from her perspective. And then *I* was surprised, because it was like a mirror, like magic. It was such an amazing talk with my Mom. When I was little, people had used to say that I was God's child, and this was the reason. Because I had survived this ordeal, people in the village had speculated I was a child from God, or something like that. My sister was there when they threw me, and she was crying too. My mother thought that somebody told me all this, like maybe my uncles or someone, but they hadn't. As I narrate this story, my headache returns; the same headache I had during the breathwork session.

Before the breathwork session, Gaus had never seen the mosque because his family had moved away from that village soon after he was born: "In the breathwork, I saw the mosque, in detail. Later, when I visited my Mom, some relatives had photographs of the mosque, which they brought out and showed me. It was virtually identical to the mosque I had seen during the breathwork."

A few months later, during another training module, Gaus did a second breathwork session. Given his first experience, it took real courage on his part to enter into another breathwork experience. During this session his headache returned, and initially Gaus experienced further details relating to his birth trauma. But before long, the entire breathwork experience shifted dramatically.

"I disappeared from the music and the room, and I was somewhere else altogether," Gaus said. "There was nobody there, and I was all alone. And then there came a light, very bright. It was too much light. And then this light came close to me, and there was a person there who I couldn't recognize through the fog. But then the fog lifted, and I saw my father's face. I couldn't believe it. I asked him, 'Appa, are you here?' And he replied yes, and he started talking to me."

Gaus felt somewhat preoccupied at first but then opened himself to this most unusual experience, as he realized this really was his father, who had passed away when Gaus was still a small boy.

> It was really him! And we sat there and talked, and my father told him many things about himself, including the ordeal of my birth. It was very difficult time for him, because as a father, it's very difficult to throw your child. But he had made a promise, and so he threw me from the mosque.
>
> He asked me how I am doing, and he knew all kinds of things about me, where I had been, what I was doing. And he asked me how my mother was doing, and Sr. Lucy (at Maher), and he told me that I should take care of them. Then he talked very beautifully about the world, about what was happening in various places, including Syria, Israel, and Palestine, and he told me that I should work for the people and spread happiness in the world.
>
> Then my father started crying, and it was very touching. My head was still hurting, and he apologized to me. And I said,

> "No, for what? And he apologized for not being able to give me what I needed as a child, and for my having to work as a child, because he was paralyzed, and so I had to take his place. And he started crying, like a baby, and I started crying too. I was very moved, and I told him it was not his fault, he couldn't help it. But I had never known that he felt sorry or guilty about this."
>
> Then my father took my hand and held it. His hand was very soft, and then he put his hand on my head and began to stroke me very tenderly. He was not like this when I was a child, at that time, he was a very tough guy. But now he was so tender, and he took my hand and placed it on top of my other hand. "You are your own support," he told me. "Never worry that you don't have a father in the world. You have yourself, you are your own support." And he kept my hands there, together.
>
> At this point, the light became brighter, and then I realized that I was alone. And I thought, okay, I have to go back now, and it was like magic, and then I was back in the room, and the music was going on. Before that, the music had been completely gone, I hadn't heard it at all.
>
> After this breathwork session, I tried to contact my father again, in meditation, hoping to see him again. But this hasn't happened; I think it may not be possible. But I know he is there. He watches over me, and his presence is there, and he knows what I am doing, and what is going on. I never knew this before.

Many times throughout his life, Gaus had inquired of his mother to tell him about his father, what he was like, and Gaus had sorely missed his father growing up. Now he had this direct experience of meeting his father face to face. Most of his breathwork session was spent in this father-son encounter. Thus, Gaus' life-long dream to know his father was answered in this remarkable breathwork experience.

Breathwork Contributes to Reversal of Destructive AIDS Policy in South Africa

Another remarkable story about breathwork comes from South Africa in relation to the AIDS crisis. By way of background, as the AIDS epidemic mounted in South Africa, then-President Thabo Mbeki denied the scientific consensus and declared that HIV was not the cause of AIDS. His government policy effectively banned antiretroviral (ARV) drugs for most HIV and AIDS patients, despite the fact that ARVs were freely offered to South Africa and had proven highly effective in slowing the growth of the HIV virus.

Mbeki's policy was an unmitigated disaster. AIDS deaths skyrocketed, and by 2005 an estimated nine hundred people were dying daily. Archbishop Desmond Tutu compared it to a mid-air collision between two Boeing 747 jumbo jets every day. Both Tutu and Nelson Mandela ardently implored President Mbeki to abandon his AIDS policy, but Mbeki would not budge. A study from the Harvard University School of Public Health estimated that in the period from 2000 to 2005, Mbeki's policy was directly responsible for more than 330,000 unnecessary deaths from AIDS in South Africa.[4]

Deputy Minister of Health Nozizwe Madlala-Routledge conducted her own research into the AIDS crisis, challenged Mbeki's policies, and advocated introducing ARV drugs into South Africa. Mbeki's swift response was to silence her for more than two years. She was prohibited from speaking or doing further research on the HIV/ AIDS issue. Her superior, the Minister of Health "Manto" Tshabalala-Msimang, was vehemently supportive of Mbeki's policies.

Gender Reconciliation Program for Members of Parliament

While all this was going on, Deputy Minister of Health Nozizwe Madlala-Routledge hosted an invitational Gender Equity and Reconciliation program for Members of Parliament and other prominent leaders inside and outside of government. The program was conducted by myself and Rev. Cynthia Brix with our team of colleagues from Gender Equity and Reconciliation International.

I had first met Nozizwe three years earlier when she was Deputy Minister of Defense, and together we conceived a plan to conduct Gender Equity and Reconciliation programs for Members of Parliament. This was the long-awaited fruition of that dream. A Quaker and pacifist, Nozizwe had earlier been an activist against Apartheid. She was imprisoned three times and spent a year in solitary confinement without a trial.

During the program, we conducted two sessions of holotropic breathwork. When Nozizwe heard our introduction to the breathwork process, she confided with me privately that this process would not likely work or be of much value for her, but she agreed to give it a try.

Nozizwe's breathwork session was relatively "quiet" outwardly. Afterward, when I checked in with her, she told me she felt fine, and the only other thing she said was, "I had an experience with the mother of a colleague."

"African Minister Ends Decade of Denial on AIDS"

This was the headline two weeks later in *The Daily Telegraph* when Nozizwe Madlala-Routledge suddenly announced a bold and sweeping reversal of South Africa's AIDS policy.[5] On World AIDS Day (December 1, 2006), Nozizwe officially overturned President Mbeki's AIDS policy, declaring it a "serious violation of human rights." She announced a major new government treatment campaign and committed billions of Rand to make ARV drugs widely available in South Africa. For this, Nozizwe was hailed as a hero in South Africa and the wider international community. The treatment campaign she launched remains intact today, and she is credited with saving untold thousands of lives.

Mbeki was enraged, and eight months later, he fired Nozizwe on a specious pretext.[6]

This caused a huge public outcry, and South Africa's leading political cartoonist captured the ensuing political drama by depicting Nozizwe crucified upon a large iconic AIDS red ribbon "cross." Public support for Nozizwe increased dramatically and declined for Mbeki, contributing to his forced resignation from the presidency on September 21, 2008.

Nozizwe's Breathwork Experience

Four days after Mbeki's resignation, Nozizwe became Deputy Speaker of the National Assembly (equivalent to Parliament). She gave a keynote speech a year later at a conference on Patriarchy in South Africa,[7] where she spoke to the promise of Gender Equity and Reconciliation. At the end of her formal speech, Nozizwe put her notes aside, and proceeded to share on a more personal and intimate level. For the first time, she shared the story of her breathwork experience.

Nozizwe recounted that during the breathwork, she had had a remarkable and unbidden visionary experience, in which the mother of President Thabo Mbeki appeared to her in a vision. This was not a dream; Nozizwe was fully alert and awake, and the sense of the presence of Epainette Mbeki (mother of Thabo) was vivid and palpable, even though Nozizwe had never actually met her in person. In this vision, Nozizwe saw a reversal of roles between mother and grown son, where the President shrank in stature to a boy, and his mother expanded greatly, towering above him.

A remarkable forthright exchange ensued between Nozizwe and Mbeki's mother, who expressed great love for her son and said she was proud of him as President in many ways. But on the AIDS issue, she exclaimed that she and Nozizwe both knew he was woefully mistaken. She gave Nozizwe a firm message of empowerment and admonition to challenge her son. As his mother, she explained that she knew Thabo inside and out, and although he was well-intentioned, she also knew very well his stubbornness and manipulative tactics. She admonished Nozizwe not to be intimidated by him or his antics, and not to be afraid to challenge Thabo directly and repudiate his disastrous AIDS policy.

Nozizwe completed her keynote speech by explaining that this visitation from President Mbeki's mother was the much-needed impetus that gave her the courage and final determination to openly defy the President and overturn his AIDS policy.

The Zulu name "Nozizwe" means "mother of nations." In Nozizwe's breathwork session, the mother of the President called upon the 'mother of nations' to defy the President of the nation—thereby saving countless lives among the people of the nation. It was an exquisite example of feminine wisdom speaking truth to patriarchal power.

Conclusion: Holotropic Breathwork Offers Social and Cultural Leaders Multiple Benefits Which Remain Largely Untapped Today

Holotropic breathwork is a practical methodology that facilitates a remarkable awakening of consciousness by unveiling deeper levels of reality and truth on multiple levels. Breathwork is utterly natural, "fueled" by the breath alone, and it reveals and applies the hidden powers and secrets of the breath, which are widely known in esoteric traditions yet equally widely obscured in contemporary society.

Holotropic breathwork offers unique and valuable benefits to innovative social change, cultural, and political leaders because it helps to awaken, liberate, and expand their consciousness beyond the narrow and often rigid confines of traditional institutions and cultural biases.

As one example, in mainstream science today, physical matter (*physis*) is still regarded to be the sole cause and foundation of all existence. Yet physis is only one pole of a vast continuum, the other pole being *psyche* (and spirit). Breathwork reveals the essence and nature of the psyche to be profound and stupendously magnificent beyond measure, as aptly reflected in the title of this anthology, *Psyche Unbound*. To reduce all existence and human consciousness to what materialist science currently understands is like reducing the vast ocean to the mere waves on its surface—utterly missing its life-giving depths, while blithely insisting with complete confidence that one has fully grasped them.

From what I have observed in more than thirty years of facilitating breathwork with social change innovators and leaders, a few key benefits of breathwork stand out clearly.

Holotropic Breathwork:

- *Reconnects people to their higher purpose, vision, creativity, and goals,* which is valuable for busy leaders who have precious little time for "inner work," yet urgently need the awareness and benefits it brings.
- *Awakens deeper awareness, helps to "clear out" emotional baggage, and purifies mixed motivations, and thereby liberates people from the grip of their personal attachments and biographical history,* all of which can otherwise interfere with their professional and personal lives.
- *Expands horizons, shifts personal values and worldviews toward more compassionate, humanitarian perspectives.*
- *Inspires new insights and visions and opens refined awareness of inner dimensions*—often in surprising or unexpected ways, yet frequently providing direct guidance or inspiration for practical challenges and issues in professional or personal life.

- *Opens people to subtle dimensions of reality and other ways of knowing*—which in turn usher in a more expansive view of life and humanity's place in the cosmos.
- *Strengthens connection to one's own psyche, soul, and innate wellsprings of inspiration and wisdom* which contributes to a sense of well-being and oneness with all of life.
- *Restores deep appreciation for life as a grand adventure of discovery* and helps leaders not take themselves too seriously and maintain non-attachment to the outcome of their work.

Finally, when done in a group context, breathwork serves to "convene the group soul" of a professional community or social group by fostering palpable bonds of intimacy, trust, and interpersonal connection and support—going well beyond the usual collegial relationships in the workplace.

In conclusion, holotropic breathwork offers profound and far-reaching benefits to professionals and leaders in virtually any field. Yet this rich potential remains largely untapped today.

Personal Reflections on the Mystery of Death and Rebirth in LSD Therapy

CHRISTOPHER M. BACHE

It is with deep pleasure that I join my colleagues in celebrating Stanislav Grof's extraordinary accomplishments as a scholar, healer, and psychonaut. Stan changed the intellectual landscape of his generation and triggered a revolution in self-exploration. His courage engaging the deep psyche opened uncharted territory, and he modeled how one could critically and systematically explore this territory, bringing back new clinical and cosmological insights. I hope this essay will convey a small measure of my deep gratitude for his enormous influence on my life and on all of our lives.

The experience of dying and being reborn is one of the central dynamics of deep psychedelic work. Death is simply the price one is asked to pay to gain access to the myriad worlds that lie beyond space-time, death not as a metaphor but the agonizing loss of everything we know to be real and true. Death comes in many shapes and sizes. It may steal in softly, melting our resistance slowly, or break through the door violently with drums pounding. Either way, if we want to experience the deeper currents of the cosmos, sooner or later, death calls to us.

In this essay, I would like to share some personal reflections on the dynamics of death and rebirth as they surfaced in my psychedelic practice. I know this account reflects only one person's journey following one protocol and using only one sacred medicine, but I hope it will resonate with other psychonauts as we together explore the infinite being of our cosmos. I want to ask three basic questions of death and dying here:

1. Why does death become as large as it sometimes does in our psychedelic practice?
2. Why do death and rebirth repeat themselves so many times?
3. What is actually dying and being reborn in this extended transformational process?

Framing the Inquiry

I was 29 years old and just starting my academic career when I read *Realms of the Human Unconscious* and my life pivoted. As a result of this encounter, I began an intense psychedelic practice that ended up lasting twenty years. Between 1979 and 1999, I did seventy-three therapeutically structured LSD sessions. I worked for four years, stopped the work for six years, and then resumed it for ten years, averaging about five sessions a year. After getting my bearings in three medium-dose sessions, I worked consistently at 500-600 mcg for the remaining sessions, following Grof's protocol for psychedelic therapy.[1] All the sessions were conducted in a private residence with a sitter, eyeshades, headphones, and music. Set and setting were standardized and spiritually focused. They started in the morning after a period of yoga and meditation and lasted all day. An account of each session was written up within 24-48 hours.

In a sustained psychedelic inquiry like this, one is repeatedly forced to rethink one's assumptions and recalculate what is possible. This was certainly true for my understanding of death and rebirth. This paper comes from years of reflecting on my experiences in the context of reports published by Grof and other researchers.[2] A more complete account can be found in my book *LSD and the Mind of the Universe: Diamonds from Heaven.*

For the purposes of this discussion, I'm going to divide this journey into three broad phases: a first phase leading up to ego-death, a second phase centering on the collective death-rebirth experiences reported in *Dark Night, Early Dawn*, and a third phase marked by an ever-deepening spiral into the Divine. Psychedelic states are inherently multidimensional, of course. Experiences of great depth may open at any point along the way, and there is always a doubling back and reprocessing of material at deeper levels. With that said, it was my experience that when the variables were standardized as much as possible, a gradual and systematic unfolding took place into the psychic, subtle, and causal levels of consciousness.[3]

When I began this work, I was primarily interested in spiritual enlightenment. Like many in my generation, I wanted to cleanse my system of its habitual constrictions and realize spiritual freedom. Because the spiritual traditions I had studied emphasized death of self as the gateway to liberation, this was my initial focus—challenging my ego, facing my shadow, and reconnecting with my Essential Nature. I initially chose to work with high doses because time for inner journeying was hard to arrange in a dual-career marriage, and I simply wanted to make the most of each session. The spiritual literature I had read described one's karmic conditioning as being finite, and I thought that I could work through mine faster by using this accelerated method of transformation, in effect biting off larger pieces of karma in each session. I knew from Grof's early books that the sessions would be more challenging, but I thought that if I confronted my shadow conscientiously and could endure the intensity of the work, it would get me to my goal of liberation faster. It turned out that I was completely wrong about this, or rather that *all the assumptions I was making were wrong.*

I began this work thinking in terms of a clinical model focused on personal transformation. I found, however, that working in a sustained manner with doses of LSD this high activates consciousness so powerfully that it expands the work beyond the individual and beyond enlightenment. It wasn't simply a matter of eating the same karmic meal in fewer bites as I had naively thought. Because the fabric of reality is an integrated whole from the very start, working at these levels changes both who the patient is and what the goal of the project can be. In my work, personal transformation expanded into humanity's collective transformation, and spiritual awakening expanded into cosmological exploration.

I should mention that with the wisdom of hindsight, I don't recommend taking such an aggressive approach to the deep psyche. Though I deeply cherish the many blessings I received on this journey, this extreme protocol came with costs. Were I to begin this journey over again today, I would take a gentler approach, balancing low and high doses with both synthetic and organic psychedelics. And if one's focus is personal enlightenment, this work is better done closer to where the ego lives in the world, and this means working with lower doses.[4]

The first phase of my journey lasted two and a half years and ten sessions. These sessions were classically perinatal in nature, involving intense existential confrontations, convulsive seizures, and fetal experiences combined with many forms of surrender and dying. This series eventually culminated in a poignant ego-death in which my identity was turned inside out and shattered. Starting the day as a middle-class, white, male philosopher obsessed with the meaning of life, I became trapped within lives that were the complete opposite of "me." I became countless women of all shapes and sizes, women of color at the laundromat with no prospects, women trained in the art of living by television with no horizons beyond the here and now. It was the perfect hell for a male professor with layers of metaphysical and existential anguish folded into it. It wasn't women that were the problem, of course, or race or poverty; it was the tight grip that my physical and social identity had on me, telling me that "I" was not any of these. When I let go of my life as I had known it, I died...and was reborn into the extraordinarily beautiful world of women explored under the arm of the Great Mother.

1) Why does death become as large as it sometimes does in our psychedelic practice?

After this sequence had completed itself, a second phase began that brought with it a new set of death-rebirth experiences. After a transition in which my compassion for humanity was deeply aroused, I entered a domain of collective anguish that was more challenging than anything I had previously faced and completely different from the personal meltdowns that had preceded it. In session after session, I was brought back to the same landscape and systematically taken deeper into its mayhem. I came to call this domain the "ocean of suffering," for it was a vast ocean of fury and pain, enormous in scope and intensity. This phase lasted fourteen sessions spread over two years—one year before the six-year hiatus and one year after it.[5] One excerpt will give the flavor of these episodes.

> I don't know how to describe what I've been through today, the places I was in, the destruction I was part of, the searing pain and torment of thousands and thousands of beings, myself with them, tortured to their breaking points and then

> beyond, and beyond. Not individuals but waves of people. The tortures not specific but legion....Driving sitar and drums tearing me apart, plunging me into more and more primitive levels of anguish. Passing through previous levels, I eventually reached a level I can only liken to hell itself. Excruciating pain. Unspeakable horror beyond any imaginings. I was lost in a rampaging savagery that was without bounds....The world of the damned. The worst pictures of the world's religions showing the tortures of hell only touch the surface. And yet, the torment cleanses one's being.[6]

Grof's early interpretation of collective ordeals like these was to see them as clusters of memories housed in the collective unconscious that get pulled into one's personal death-rebirth process at the perinatal level of consciousness. Essentially, he saw them as bleed-through from the collective unconscious, drawn in by their resonance to some aspect of personal ego-death. If this were an opera, ego-death would be singing lead, and these collective experiences would be the chorus. Accordingly, my early understanding of these experiences was to see them as a deepening of my ego-death, as leading to a more complete ego-death. My assumption was that if any form of death and rebirth were taking place in my sessions, even these very collective deaths, there had to be pieces of ego dying somewhere in the mix. What did the sensation of dying attach to if not to an ego?

Eventually, however, this assumption was overwhelmed by the sheer scale and intensity of the collective suffering involved. These episodes went on for too many years and were too extreme in their scope and content for me to see them as collective experiences drawn in through resonance to a core of unfinished ego-death. This eventually forced me to reassess the boundaries of the entire therapeutic enterprise. The conclusion I came to, both experientially and intellectually, was that these collective episodes were *not* aimed primarily at the transformation of my personal consciousness. I became convinced that they were aimed at nothing less than the transformation of the species-mind as a whole.

Dark Night, Early Dawn was written in large part to answer the question: Why did death become as large as it did in my psychedelic journey? What is driving the healing process when it opens to such collective tracts? In that book, I abandoned the person-centered narrative I had been assuming and adopted an expanded narrative. By integrating Rupert Sheldrake's concept of morphic fields into Grof's paradigm,[7] the way opened to viewing these collective ordeals as part of a larger cathartic process aimed at healing the scars of the collective unconscious, scars accumulated over thousands of years and still carried in our collective memory. I argued that in highly energized psychedelic states, the collective unconscious is sometimes activated to such a degree or in such a manner that it triggers a collective healing process. The "dying" in these episodes lies in surrendering completely and irrevocably to these vast fields of suffering and violence, letting them redefine the boundaries of one's being. Through some fractal flip or quantum entanglement I had not anticipated or even thought possible at the time, the "patient" in my sessions shifted from being me to being some portion of humanity itself.

Moving to a model of collective transformation represented an enormous transition for me because with this pivot we are no longer speaking of an individual "having transpersonal experiences." Here the individual dissolves into the preexisting fields of the species-mind. It is these collective fields that become the "working unit" of experience in these sessions. This required a new way of thinking about what is taking place in our sessions and a new therapeutic calculus.

In *Dark Night, Early Dawn,* I kept the term "perinatal" for these collective ordeals and tried to stretch it by proposing that one's experience of the perinatal domain could "slant" toward the personal side or the transpersonal side of the perinatal interface. I further suggested that there were two tiers of death and rebirth intertwined in these complex episodes, one aimed at personal ego-death and a second aimed at the death and rebirth of the species-mind.[8]

At the present time, however, I have shifted to what I think is a simpler and more elegant way of understanding these experiences. My current thinking emphasizes that death and rebirth is a *cycle* that repeats itself many times as one moves into deeper layers of consciousness. Being an archetypal cycle, any single death-rebirth experience may incorporate material from multiple levels of consciousness. Looking back at the trajectory of all my sessions, however, I now recognize that this cycle was repeating itself in different forms as I moved into progressively deeper levels of consciousness.

Rather than seeing these episodes of collective suffering as a protracted first turning of the wheel of death and rebirth at the perinatal level of consciousness, I now see them as a second turning of this wheel, one coming after the personal ego-death that had already taken place and before the cycles of death and rebirth that followed in later years. They are the second movement of a larger symphony, a movement taking place at the subtle level of consciousness, a movement whose dynamics are inherently collective, focused on a collective patient, and aimed at a collective transformation. The memory clusters being engaged here are not Basic Perinatal Matrices but META-COEX systems operating deep within the collective psyche.[9]

In offering this assessment, I am not suggesting that the bleed-through interpretation that Grof originally proposed is wrong, only that it applies in some circumstances and not others. It applies best, I think when these collective elements show up in sessions where the individual is clearly engaged in a perinatal process with prominent fetal features. But in circumstances where one has already moved through the perinatal process and undergone a solid ego-death, I think we must expand our frame of reference to comprehend these collective episodes. This is a revision that Stan supports. In a personal exchange on this topic, he wrote: "I feel that for clarity, we should change the terminology and make it clear that the term perinatal should be used only for fetal experiences...and not for experiences of death and rebirth on higher levels of the transpersonal spectrum with no relation to biological birth."[10]

The transpersonal vistas that opened during this second phase of the journey compensated me generously for the time spent in these hell realms of collective suffering. They took two forms, focused on the psychic and subtle levels of consciousness. First, there was a systematic opening into what I came to call "Deep Time"—a transtemporal consciousness in which the rules of linear time are suspended. In seven consecutive sessions, I experienced my entire life from birth to death as a completed whole, with deep insights into its core themes and relationships. After the six-year hiatus, the platform of discovery shifted far beyond my personal reality. Now, each time I came through the ocean of suffering, I was taken on a series of ecstatic initiations deep into the universe. It felt like an infinite intelligence was educating me, reminding me of things long ago forgotten but now in need of being remembered.

2) Why does dying repeat itself so many times in psychedelic therapy?

When I began this journey, I thought there would come a point where the dying would eventually stop, that I would reach some final destination or ultimate condition, be it oneness with God or absorption into the Primal Void. This is not what happened, however. No matter how complete the deaths I underwent were or how ecstatic the homecomings that followed, death always came back into my sessions somewhere down the road. I did not understand then what I have come to understand now—that dying is part of an endless cycle of discovery and that it will keep returning as long as one keeps challenging the limits of one's experience. I had expected closure. What I received instead was the infinite depth of the Beloved.

When the ocean of suffering reached its explosive crescendo in the 24th session, it ended and did not return in future sessions. At this point, I was catapulted into the greater real of archetypal reality for two years. The energy of this domain was extremely powerful—rivers of liquid fire, white-hot lava flows, and exploding sun flares. It took multiple sessions for me to stabilize my awareness at this level while receiving carefully orchestrated initiations into how reality functions at the subtle level of consciousness.[11] Then I was taken through another round of deaths and rebirths that lifted me into the ecstatic bliss of causal consciousness for two years, as the Uranus-Neptune conjunction in the sky squared my nodal axis exactly. The dynamics of dying and being reborn were different at this level than I had experienced at previous levels. There was a distinctly effortless quality to some of these transitions, as though effort would contradict the ever-present, all-inclusive wholeness I was entering.[12]

At this point in my journey, I had been given so many blessings that I felt completely and utterly satisfied. I had explored the universe, been taken deep into Oneness, been dissolved into the Fertile Void, and drenched in Cosmic Love. What more could one ask for? So I was surprised when death returned yet again, this time dissolving me into a new level of reality that was so infinitely clear and ecstatically radiant that it immediately eclipsed everything that had gone before. It extinguished any interest I had in exploring dimensions of existence that had previously fascinated me. I came to call this reality the domain of Diamond Luminosity. It is, I think, what Buddhism calls *dharmakaya,* the Clear Light of Absolute Reality. Over the next five years and twenty-six sessions, I was taken into this reality four times. These four sessions were the most precious gifts of my journey, the true diamonds from heaven.

Because this is where the spiral of death and rebirth eventually brought me, let me share the first of these four initiations before continuing with these reflections. The session prior to this had been spent entirely in death and dying, so this session moved immediately into the harvest of rebirth.

> After a long period of opening, I found myself repeatedly saying, "I have earned the right to die." Far from fearing death, I was seeking it out, demanding that it come to me. I was deflecting half-measures and insisting on my right to a complete and final death. I had done my work; I had earned the right to die, and I was calling on this right. My litany focused me and carried me deeper and deeper to a point of complete concentration.
>
> From this position of absolute focus, I began to die. Oh, what sweet death! I began to savor what was happening! What I had previously feared now opened to me as incredible sweetness. How wonderful to experience death! What a surprising reversal! Thank you, thank you.
>
> Upon dying, I moved into an ecstatic, extended mode of experience I cannot describe adequately. It was a different mode of experience from anything I had known in previous sessions; the entire flow of the experience was different. Light-filled, yes; a universe composed of nothing but light. What stood out for me, however, was something I cannot articulate well.
>
> It was as if I had moved inside the inner flow of God's being, as if my life was now bending and flowing through a being of infinite dimensions. There was nothing amorphous or fuzzy about the experience; on the contrary, it was extraordinarily clear and precise. The boundaries of this clarity exceeded anything I had previously known.
>
> Apparently one death was not enough to get the job done in my case. I found myself standing in the middle of a circle, surrounded by a spinning torus of colorful bands of light that comprised my entire life. All the time-moments of my life were present in these bands. I fell into this circle and touched some part of my life, but as soon as I did, it suddenly "died out from under me" and I instantly found myself in the luminous

death-state beyond individual identity. Then I was returned to the center of the circle, and the process repeated itself, now falling in a different direction. Over and over again, I went through this process of "dying in all directions," driving home the point that there was nothing left unfinished here. Wherever I turned, there was no resistance, only effortless death, and incredible sweetness.

The repetition kept expanding the scope of the transition I was making, taking me deeper and deeper into ecstasy until eventually there was no center to return to, only the pure, seamless condition of the death-state. What strange language to describe our true nature.

The death-state.
Incredibly clear.
Luminous beyond measure.
Incredible age.
Incredible extension.
A seamless intelligence running not above but inside existence.
Pieces of reaching out and moving into large "wholes" of experience.
Blocks of experience encompassing thousands, perhaps millions of beings.
Human-experience folded into Earth-experience.
Just touches, tastes.
Ecstatic reverence for the integrated movement of life throughout the universe.

For the next several hours, I was carried along the currents of this condition. About this state, one says either too little or too much. The price of saying nothing is to risk forgetting the subtler textures of the experience, yet to speak creates the illusion that words are adequate to the task and they are not. Even after so many years, today was so unlike any previous mode of experience that language truly fails. Silent appreciation seems the best recourse, combined with ceaseless prayers of thanksgiving.

How can something so crystal clear,
so devoid of earthly form evoke tears of homecoming?
What are we that such imprisoned splendor,
once released, floods us with rivers of gratitude?
Whom shall we thank for what we are?
Where do I direct my deep appreciation?
There is no one place,
so I send my prayer into the seamless fabric of existence
left and right, high and low,
in infinite dimensions all around.
My attempts to describe the experience keep breaking down,
and I end up repeating the same words over and over.
I was home
...and free
...and Light.
There is nothing more I can say.[13]

As complete as this rebirth was, it was in time followed by more experiences of dying as the spiral of initiation continued. My understanding of this pattern is that death repeats itself not because it has failed to hit its mark and something of ego survives, but because the Divine is an infinite landscape with countless levels to explore. As our transpersonal experience deepens and refines itself, one undergoes many deaths, for each death is but a gateway to what lies beyond. Levels of reality that we are born *into* at one stage we are challenged to die *out of* later—dying out of humanity, out of space-time, out of the *bardo* echoes of space-time, out of the archetypal expanse.

While death took many forms on this journey, the *core experience* was always the same—a complete surrender to whatever is happening, a loss of control and collapse of all reference points, and a disorientation so deep that it dissolves reality as we have known it. Similarly, the essence of rebirth was consistent across levels—the experience of awakening within new and unanticipated dimensions of existence, the birth of a new identity with new capacities, and the experience of absolute grace, of having been given infinitely more than one has given up. Within this basic structure, however, the details of dying and being reborn reflected the level of consciousness at which I was working. The experiential texture of each death, its flavor, focus, and function changed as the stages of initiation deepened. I think we can even say that *what* was dying changed, as we will see below.

After one has died and been reborn many times, eventually the very concept of death begins to lose its meaning. One learns through repetition that at the deepest level of one's being, it is impossible to die. The form that we are can be shattered, our reality can be repeatedly destroyed, but our innermost essence always reemerges. The phoenix always rises. This forced me to look closely at what exactly is taking place inside this repeating cycle.

Death and the Cycle of Purification

Over the years, I came to recognize that each step deeper into our multi-dimensional universe is a step into a more intense field of energy. *Deeper* states of consciousness are *higher* states of energy. This is an unmistakable sensation and a widely recognized principle in spiritual traditions. One may have glancing contact with these deeper levels without this becoming apparent, but to have stabilized experience of a given level of reality, one must acclimate to its level of energy. Otherwise, our experience there will be fragmented and chaotic. Just as when climbing a mountain we must acclimate to lower levels of oxygen, here we must acclimate to higher levels of energy, and this activates intense purification processes.

As I have experienced it, the *cycle of purification* is the combustion cycle of growth in sustained psychedelic work. The essence of the cycle is this. Being propelled into a deeper level of reality shifts one into a higher energetic state, and this higher energy "shakes loose" impurities from one's mental, emotional, and physical being.[14] In subsequent sessions, one's system works to empty itself of these toxins as it continues to absorb the purity and intensity of this higher energy. By sweeping out the old to make way for the new, eventually a clearer and stronger energetic platform is established on which future sessions will build.

This pattern of breakthrough followed by detoxification repeated itself like clockwork in my sessions. I found that after each major breakthrough to a deeper level of consciousness, there was often a turgid "carrying out the garbage" quality to the sessions that followed. This was so consistent that I came to dread the sessions that followed a breakthrough. An analogy from mining comes to mind. After an explosion opens a new vein of ore deep in the mountain, you still have to carry away the rocks to get complete access to its riches.

Let me take this one step further. When the process of purification reaches particularly deep, it becomes *purification unto death*. When the cycle of purification reaches so deep that it begins to dissolve the structure of our life as we have known it, it becomes a *cycle of death and rebirth*. When it empties us of all that we have known and all that we have been, a crisis is reached in which what we have been collapses and we are carried forward into a new reality. In the rebirth that follows, we have become a different *kind* of being with new capacities and access to new categories of experience. If, after stabilizing at this new level, we continue to drive forward, the cycle of purification will begin again, and we will eventually enter a new cycle of death and rebirth as new levels of reality continue to open. *Death, therefore, is an intense form of purification.*

Understanding death as purification helped me make sense of another pattern that emerged at this stage. I found myself in situations where I could choose how much or how little I would die in a session. I discovered that there are many *degrees of dying* in this work and that the deepest breakthroughs tend to follow the deepest deaths. With this discovery, death became my closest ally in my sessions, something I actively sought out, not because I was a glutton for pain but because I was a glutton for what lay on the other side of this pain.

It was during this time that I had an experience that showed me that the cycle of death and rebirth is never-ending. It happened during my second initiation into the Diamond Luminosity. I was as deep as I would ever go on my journey, resting in a state of hyper-clarity far beyond spacetime, completely at peace and one with the Light. Suddenly my visual field rotated 90 degrees and a gap opened to reveal entire worlds beyond the world I was in. Shining through them was the most sublime, exquisite Absolute Light, an exponential increase in clarity beyond even the Diamond Luminosity. This glimpse completely transfixed me and left me stunned in rapture. In only seconds it completely redefined my vision of reality. After so many years and so many breakthroughs, I saw that the progressive realization is endless because the Divine Cosmos is infinite. The dying stops when one's capacity for discovery is exhausted, and we can simply take in no more.

3) What is dying and being reborn in this transformational process? This brings me to the third and final question: What exactly is dying and being reborn in this repeating spiral of initiation? There is often an acute sensation of dying at the deeper levels, but what does this sensation attach to? Is it simply ego dying over and over again, or is there something more going on? I want to briefly suggest four overlapping answers to this question.

In what follows, I want to affirm a delicate both/and balance. On the one hand, I want to affirm the position of the individual. The individual registers and absorbs these successive deaths. They become part of his or her life story, and so they "belong" to the individual in ways that I do not want to negate or deny. On the other hand, I think these deaths also "belong" to the universe in ways that transcend the individual. They are something *it* is doing. The universe appears to use these opportunities to heal itself and commune with itself in ways that reach beyond the individual. I think both these perspectives are true and important.

1. The ego

In the early stages of the journey, what is dying is our physical identity, the body-mind ego, the bundle of habits, beliefs, and aptitudes generated by our earthly history. In order to enter what lies beyond physical existence, we must let go of everything our physical existence has told us that we are. This death is well mapped in the psychedelic literature. But in later sessions, after ego has surrendered its grip on our consciousness, what exactly is dying then?

2. The species ego

When the wheel of death and rebirth is turning at the subtle level of consciousness where the deaths are largely collective, what is dying, I think, is some portion of the *species ego*. If the patient in these sessions expands beyond the personal psyche to the collective psyche, then what is being engaged and healed is some matrix of traumatic memories held in the collective unconscious. The pulse of the life one is living in these hours is the pulse of human history. The COEX systems that are resolving themselves are not personal complexes but META-COEX systems within the collective psyche that contribute to our self-identity as a species. Though we participate in these deaths and feel them acutely, what "we" are at this stage has shifted. In these sessions, we are no longer our private selves but have expanded to encompass some aspect of the species-mind.

But after our work at the collective level has run its course and reached its conclusion, what is dying when the spiral of death and rebirth continues to turn? As broad a reality as the species ego is, in the end it is a human-centric phenomenon. In the context of the vast cosmos, our species consciousness is a relatively small thing. Sooner or later, transpersonal experience outgrows these human proportions, and we must look for larger explanations of what is dying in these transitions.

3. The shamanic persona

A third answer to the question of what is dying in these later sessions is what I call the *shamanic persona*.[15] I think all psychedelic journeyers have had the experience that after a session ends, we sometimes cannot remember all the experiences and insights that we had during our session, and yet when we reenter psychedelic space in our next session, this "missing knowledge" is present once again, intact and waiting for us. Something in our psyche has remembered our experience at a level deeper than our egoic awareness. Similarly, when our sessions get underway, we often have a sense of resuming our "psychedelic identity," a deeper and larger identity that is familiar to us and different than our egoic identity. When we enter the psychedelic state, we "walk with the Old Ones" once again.

I want to suggest that in the repeated opening and closing of awareness in well-structured psychedelic sessions, a semi-autonomous, state-specific consciousness is formed that retains and integrates all our psychedelic experiences, including those that the egoic self cannot retain. This consciousness not only remembers our experiences, it preserves the knowledge and capacities we acquired in them. It represents a higher order of memory and capacity within the psyche. I call this psychedelically-generated self-awareness the shamanic persona. The shamanic persona reflects the natural aggregating function of life. As Rupert Sheldrake and Ervin Laszlo have argued, life remembers and holds on to its experience at all levels,[16] and now even at psychedelic levels.

The shamanic persona can be thought of as a *state-specific alter ego*. It is a psychological construct of the deep psyche, a mechanism of assimilation and integration that allows the psyche to manage the extreme swings of consciousness generated in a long psychedelic regimen. In calling this entity the shamanic *persona,* I am not suggesting that it has a masking function but am simply drawing attention to the fact that it is a living identity that changes as our psychedelic experience deepens. It could also be called the *shamanic self.*[17]

The more transpersonal experience we have accumulated in our sessions, the stronger our shamanic self will be. If our psychedelic experiences have been chaotic and fragmented, our shamanic self will be weaker. If our experiences have been well focused and clear, it will be stronger and more stable. The more successfully we have integrated our psychedelic experiences into our conscious awareness, the "closer," and more familiar our shamanic self will feel to our ordinary sense of self. It will feel more like "us," a natural extension of our earthly identity. Conversely, the less well-integrated our experiences have been—either because of poor session management or because the content of a session was particularly deep—the more "distant" and "other" our shamanic self will feel.

Because the shamanic persona is a synthesis of a specific set of experiences, it is a specific entity with a specific identity. It is not an archetype. It is the living memory of our psychedelic history, and as such, it has built into it the limits and specifics of that history. A shamanic persona that embodies stabilized *psychic level* experience, for example, is a very different entity than a shamanic persona that embodies stabilized *causal level* experience. By "stabilized experience," I mean that we have entered a certain level of consciousness often enough that we have acclimated to the territory and learned the terrain. Our psycho-physical system has undergone the necessary purifications and adaptations for us to maintain coherent awareness at this level.

The concept of the shamanic persona adds a third answer to the question of what is dying and being reborn in the deepening spiral of psychedelic initiation. When our consciousness begins to open to levels of awareness that are deeper than those levels we have previously experienced, our earlier psychedelic history must yield to this new territory. Just as our Earth identity dies in ego-death, the psychedelic identity that holds our session experiences must surrender before a still deeper mode of transpersonal experience can fully emerge in our sessions. In essence, then, what I think is dying in these later openings is our shamanic persona, the living memory of our psychedelic history. Though the death of the shamanic persona may feel like a personal death, it is not the ego that is dying here but a deeper identity that has been birthed *inside* our sessions. It is a hybrid identity that is both personal and transpersonal.

In a sustained psychedelic regimen, our shamanic persona may die and be reborn multiple times as new levels of reality continue to open. After ego-death, a shamanic persona emerges as the living integration of our early transpersonal experiences. As our experience continues to deepen in subsequent sessions, this first shamanic persona will eventually have to surrender and die, giving birth to a second shamanic persona that will integrate our new transpersonal experiences. This second persona will retain the memories and knowledge of the first persona and add to it the new knowledge and capacities gained at this second level. If we continue to push the boundaries of experience, sooner or later this second shamanic persona will also have to surrender in order for a still deeper dimension of consciousness to open and become fully operational.

This pattern of successive deaths and rebirths is the natural rhythm of all operational identities that emerge on the psychedelic path. The *essence* of these identities, like all identities in our world, is the Absolute, but the *form* these identities take reflects the changing depth and breadth of our psychedelic experience.[18]

4. A dimension of the cosmos

Let me now suggest a fourth answer to the question of what is dying and being reborn by returning to the concept of the psychic, subtle, and causal levels of consciousness.

The threads that make up the shamanic persona at the psychic level of experience are largely personal or soul-centered in nature. At the low subtle level where collective patterns begin to predominate, the threads become increasingly collective or species-wide. At higher subtle levels where the currents of experience become archetypal, the threads of the shamanic persona become correspondingly archetypal. If we continue this progression into causal levels of reality, eventually the threads of experience become so vast in scope that I think the category of shamanic persona becomes less useful to describe what is actually taking place in these meltdowns.

For me, the category of shamanic persona is always tinged with personhood; it is an extension and deepening of my personal history. As such, it is too small an entity to adequately describe the experiential texture of these later deaths, at least as I have experienced them. It may still apply, but by itself it is an incomplete description. Clearly, we're marking stages on a continuum here, but eventually it feels more accurate to let the category of the shamanic persona go and try to conceptualize these transitions from within a still larger frame of reference. But here, the clear road runs out, and things become more uncertain. At this point I can only share intuitions I've formed over the years, and these intuitions are tentative and incomplete.

How does one describe the larger arc of life that sparks in these deep meltdowns? How much can we truly know about what function these meta-deaths may serve in the larger web of life? And what need do such deep levels of reality have of "rebirth" at all? I do not know the final balancing of these accounts, but I believe we should begin by viewing everything that takes place in our sessions from the perspective of the Great Chain of Being as a whole.

Let me begin with an observation I made in the closing pages of *Dark Night, Early Dawn*:

> To use Ken Wilber's vocabulary, if we are a holon functioning as a part within a series of ever-enlarging wholes, then the death-rebirth dynamic may have different functions for different levels of reality, all of which are being realized *simultaneously*. From the perspective of the smaller holon, for example, the effect of death-rebirth may be liberation into that which is larger, while the effect of the *same* transition from the perspective of the larger holon may be to allow it greater access to and integration with the smaller field. An event that functions as spiritual "ascent" from below may simultaneously function as "descent" from above.[19]

This observation invites us to look at the dynamics of death and rebirth multi-dimensionally. It is an insight that generalizes across multiple levels of reality. At the subtle level of consciousness, for example, death and rebirth may open a portal that not only serves to drain destructive energies *out of* the species mind but also to infuse healing energies *into* the species mind from a higher source.[20] At still deeper levels, such portals may allow any number of transcendental blessings to be infused directly into "lower" orders of existence. The question then becomes: How does the addition of this principle of *infusion-from-above* influence the question of how we might conceptualize what is dying and being reborn at these deep levels of psychedelic experience?

Let me draw upon Sri Aurobindo's involutionary/evolutionary cosmology here because it resonates strongly with the cosmology that emerges in psychedelic work, as Grof has demonstrated in his beautiful book, *The Cosmic Game.* According to Sri Aurobindo, in the cascading involution of the Divine, many levels of existence are manifested. While these levels may be porous from "above," they are less porous from "below." Like looking through a series of one-way mirrors, the Divine looking "down" sees everything It has become, but looking "up" from lower levels, the Divine sees less.[21] When we who are below manage either by hard labor or by grace to access some of these higher levels, one sometimes has a sense of participating in a far-reaching cosmic communion.

Assume for the moment that through the fiery exercises of psychedelic initiation we have managed to stabilize experience at the high subtle level of consciousness. In order to reach this level, what "we" are has changed. In deep psychedelic work, *one learns by becoming.* To know the universe at this level, we must temporarily *become* some aspect of subtle level reality. Inside our sessions, we live as a life-form that breathes this rarified air. When through further exercises a doorway opens to still deeper levels of consciousness at the causal level, it allows a *cosmic communion* to take place between the causal and subtle realms. Deep communes with deep.

It has been my experience that bringing different levels of spiritual reality into conscious communion with each other, even if for only a few hours, nourishes and brings joy to the weave of existence as the "below" remembers the "above" and the blessings of "above" pour more freely into the "below." What is taking place is a cosmic dance between deep levels of the fabric of existence, far beyond our personal and collective being. It is God communing with God, nourishing Its self-manifesting, self-emergent being in ways we may touch but perhaps never fully comprehend.

What is dying and being reborn, then, in these advanced cycles of death and rebirth? Beyond the ego, beyond the species ego, beyond even the shamanic persona, what "dies," I think, is something of truly cosmic proportions. Something deep in the fabric of the universe surrenders, and in that surrender is nourished from above. Some dimension of existence extraordinarily vast awakens more completely to Itself, and much to our surprise, the Divine appears to genuinely appreciate our collaboration in facilitating this communion.

Lessons for a Career in Psychedelic and Holotropic Therapies

MICHAEL MITHOEFER

Stanislav Grof's contributions to the understanding of the psyche and the healing potential of holotropic states of consciousness are only beginning to be realized in current Western psychiatry and psychology. His work has been conveyed largely through his scientific papers, books, and lectures. I'm grateful for this opportunity to write about another avenue through which Stan taught and influenced many people who have carried his teachings into their own work: the Grof Training. For many of us, our time in Grof Training was life-changing; for some of us, it also changed the direction of our careers.

In 1990, after ten years of practicing emergency medicine, I was ready for a change. In the emergency department, the suffering I was treating so often had its roots in psychological trauma and longstanding societal forces. I was longing for a different, more collaborative approach to supporting people in their healing, and curious about possibilities for addressing the underlying causes of suffering. Casting about for a new direction, I came across a book in the library at the Medical School in Charleston, South Carolina: *Beyond the Brain* by Stanislav Grof. I was impressed that they had that book on their shelves at all, and not surprised that I was the first to check it out. The "cartography of the psyche" Stan described in the book was a startlingly powerful map for understanding human consciousness, including some of my own psychedelic experiences from twenty years earlier.

The impact of the book on my career is evidenced by the result; I applied to Psychiatry Residency and the Grof training, and started both in 1991.

Over three years of psychiatry training I made periodic trips to California for a total of nine six-day Grof training modules. This made for an interesting contrast. The psychopharmacology taught in psychiatry training was aimed at suppressing symptoms, and Stan was demonstrating that "a symptom is something that's halfway out," so it may become more pronounced in the process of releasing. Most people at the Institute of Psychiatry in Charleston thought that lying on the floor with patients and doing bodywork during hours-long breathwork sessions was unethical and dangerous. Many people in the Grof training thought psychiatric drugs and psychiatry, in general, were dangerous. While I was grateful to learn from each of these very different approaches, without question I found the Grof model far more compelling in terms of understanding and bringing healing to human suffering. A few years later, my wife, Annie, also completed the Grof training. For the next ten years, we worked with patients in our outpatient psychotherapy practice and in monthly holotropic breathwork groups that we facilitated together. From the beginning, our therapy practice was oriented toward the principles we learned from Stan, that our role is to trust and support each individual's inner healing intelligence.

Over those ten years facilitating holotropic breathwork groups, we saw many examples of profound healing and growth, often in people who had not responded to years of therapy and medications. We also recognized that breathwork wasn't effective for everyone. As in most areas of medicine and psychology, we need an array of different tools for different people, and there was already a body of work exploring other tools for catalyzing access to holotropic states. Stan Grof and others had published clinical research as well as case reports using psychedelics in the setting of psychotherapy, and some psychedelic plant medicines had been used for healing and exploration in other cultures for centuries. Knowing this, Annie and I felt a duty to our patients to research psychedelic treatments aimed at making them available for clinical use.

In early 2000, we approached Rick Doblin, who had formed MAPS for this purpose in 1986. It is no coincidence that Rick had also trained with Stan and Christina Grof in one of their earlier training groups, so we shared that foundational orientation. The inspiration for the MAPS research, and the therapeutic approach we have been using in clinical trials of MDMA-assisted psychotherapy for posttraumatic stress disorder (PTSD), stems directly from the training that Rick and Annie, and I had with Stan, Christina, and the other wonderful teachers at Grof Training. This influence includes the opportunity to engage in our own deep personal work as an essential element of learning to support others in deep healing. Now, after twenty years of successful MAPS-sponsored studies, MDMA-assisted psychotherapy is likely on the verge of FDA approval for use in conjunction with psychotherapy. This will be a unique event: a drug brought to market through a non-profit and public benefit model, and used not just as a pharmacological agent but as a catalyst to access a deep healing. To use Stan's language, it will be an FDA indication for a drug aimed at accessing "the healing potential of non-ordinary states of consciousness" guided by the "inner healing intelligence."

I will describe some of the seminal moments and teaching points I recall from Grof Training, which Stan conveyed with theories, stories, and by his own example. Here is some of what I learned from Stan:

Inviting someone to do inner work and agreeing to support them in non-ordinary states of consciousness is a commitment that takes compassionate presence, energy, flexibility, a sense of humor, and stamina. The most powerful lessons for me came from spending hours watching Stan and other teachers during holotropic breathwork sessions, as they stayed with the last people in the room at the end of long, intense breathwork sessions. We saw again and again that taking extra time at the end of a session often allowed a level of healing that would not have been reached in therapy with a rigid schedule. Stan was almost always there until the end of the session, and later in the evening, often the last person awake in the hot tub at midnight ready to play another round of Higgledy-piggledy, a very complex word game I think he and Christina had invented.

Supporting someone who has the willingness and courage to do deep inner work is a great privilege that can be as nourishing as it is demanding, and as energizing as it is tiring. One of the most inspiring things I've heard Stan say was in Atlanta at what was billed at the time as his last public holotropic breathwork group before retiring at 70 (none of us really thought he would stay retired). We were standing around and someone asked him how it was to turn 70. His reply was something like,

"Oh, it's wonderful! The older you get, the more interesting life should become because you've had so many experiences to draw from." At a time when I was already hearing friends in their late 50s or early 60s bemoaning the way things weren't "what they used to be," I knew which example I would try to follow.

Curiosity, respect, and trust in the inner healing intelligence are usually more helpful than theories about pathology. This is expressed in another of the most striking things I heard Stan say during a training module (I haven't seen this written down anywhere, so this is from my memory). One of the trainees had asked him what he would do if he thought someone is being manipulative during a session rather than genuinely in a deep process. Since Stan had many years of psychoanalytic training in the past, I'm sure he could have come up with explanations about character pathology based on psychodynamic theory, but my memory is that he said this, "In that kind of situation, I choose to believe that there is something going on that I do not understand." The unspoken extension of that choice is that he would then continue to stay present with an attitude of curiosity, love, and trust in that individual's own ability to discover their own path to healing. This is true to what Stan has often quoted his friend Joseph Campbell as saying: "If you're going to have a story, have a big story, or none at all."

Preparation to support others in deep process includes the therapist's own inner work. The Grof training teaches a framework for understanding the nature of consciousness and an approach to healing based on facilitating each individual's own innate healing intelligence. Stan's extensive knowledge about the therapeutic and cultural use of psychedelics and other technologies for shifting consciousness, combined with his "cartography of the psyche," brings wonderful depth and color to his teaching. However, maintaining balanced presence in the face of suffering and emotional intensity requires more than a cognitive framework. Essential elements of Stan's teaching are conveyed by his example, his modeling of compassionate, courageous, and seemingly tireless presence with people during holotropic breathwork sessions. And at every Grof training module, trainees participate in their own breathwork sessions as well as acting as sitter and facilitator for others. For me, this direct experience was the heart of the training. Although I arrived for my first training module after having had a number of years of psychotherapy, as well as past psychedelic experiences, it was quickly obvious to me how much more I had to do. Happily, it also soon became obvious what an extraordinary opportunity was presented for us all to learn through the direct experience of facing our own challenges while supported with the skill and love of facilitators who trusted our inner healing intelligence even when we were not yet able to trust it ourselves. In line with this, one of the requirements in the Grof training is that trainees can take as long as they want to complete the training modules, but are not allowed to do it in less than two years because it takes at least that long for trainees to integrate the major shifts in perspective that may come over the course of their own holotropic breathwork sessions.

Some specific skills are important to have, such as how to offer nurturing touch or do bodywork safely and effectively to support release in the body. More important is knowing when not to use any skills other than compassionate presence. This requires that the facilitators have sufficient self-awareness and capacity to refrain from acting on their own fears in the face of intensity, or acting from a need to feel skillful or useful. Any direct work with the body should come only after time and encouragement to bring attention to the body, and to allow any movement, shaking, or other spontaneous process to unfold. This is often sufficient, without intervention by the therapists, which is needed only if the process becomes stuck. Premature or unnecessary intervention may be a distraction from an unfolding process that might have resolved without intervention, and may have led in unexpected and helpful directions. Stan has used the example of a sitter in a breathwork session who decided her "breather" needed to be comforted, so she started stroking his head and coddling him like a baby. She found out later that the breather had been having an experience as a Viking, so comforting was not what he needed!

On the other hand, support offered skillfully at the appropriate time can be a great service. I watched Stan during a breathwork session at a training module; a woman had been in an apparently very emotionally upsetting process with lots of writhing and crying on the mat for a long time. Her sitter was next to her, very attentive but not engaging. Stan walked up and stood watching for minutes. He then slowly lay down beside her and watched for several minutes more without her being aware of his presence. His pace and manner conveyed deep caring and deep respect for her process. Finally, he said gently, "Do you really want to do this alone?" The dam broke. When she felt the loving connection and Stan's offer of non-intrusive support her process shifted dramatically to one of welcome release.

The experiences illustrated by these vignettes have profoundly influenced my personal life and my work. They are the underpinnings of our approach to MDMA-assisted psychotherapy in MAPS-sponsored clinical trials conducted over the past seventeen years. The overarching lessons I take from Stan, on which all the others depend, are lessons of courage, scientific honesty, and the ethical responsibility of physicians and therapists to the well-being of their patients, and of all of us to the well-being of each other. It takes these qualities to explore one's own psyche deeply; to remain intensely curious in the face of discoveries that don't fit with prevailing dogma about the nature of reality; to follow what patients are telling you they need, even if it means straying from rigid professional boundaries regarding touch and length of sessions as well as concepts about pathology and the nature of healing; and to recognize the fallacy in thinking there is some essential difference between therapist and patient. During our MAPS Therapist trainings we show videos of MDMA-assisted research sessions using this kind of non-pathologizing approach. Sometimes during the videos, experienced psychiatrists start crying and tell us, "This is why I went into psychiatry, this is the kind of work I've always wanted to do!" Stan Grof's contributions to psychiatry and psychology, to humanity, will continue to be felt, and are a gift to all of us as we support each other in healing and growth.

Light at the End of the Tunnel

Stanislav Grof's Vision for the End of Life

Charles S. Grob

I first became aware of Stanislav Grof's groundbreaking research in the early 1970s, when a friend invited me to attend a talk Stan was giving in New York City at an early meeting of the Association of Humanistic Psychology. Up to that time, I had some interest in psychedelics, from experiences in college in the late 1960s and from having recently discovered the voluminous literature on the topic of psychedelic research. On that cold and dark winter evening at the AHP meeting in New York, Stan presented a fascinating overview of his remarkable range of investigations with psychedelics, with particular emphasis on the application of the psychedelic treatment model with patients who had severe existential depression and anxiety reactive to their diagnosis of terminal cancer. Hearing Stan present the rationale, cross-cultural correlates, history of prior research with psychedelics and his findings of his most recent treatment studies with dying cancer patients profoundly changed my career path, and no doubt, my life.

In those early years, as I adapted to a rigorous pre-medical program and later to medical school and post-graduate training, I would encounter Stan again during pivotal moments in my life when I struggled to find my way. When I began medical school in 1975, psychedelic research had descended into the Dark Ages. Virtually all approved studies in the United States had been terminated, not because of any lack of promise of these novel treatments, but rather because of the political and cultural turmoil of the era.

Indeed, I found out in a rather disconcerting way that psychedelics were no longer deemed an acceptable topic for professional discourse, outside of known adverse effects ("bad trips") and fanciful outcomes ("chromosomal damage") that were later disproven but at the time achieved sensationalized media coverage.

In my second year of medical school, we were each assigned the task of identifying an important research article in the medical literature, distilling its essence, and presenting the study to the class. Without hesitation, I chose Stan's 1973 article in the *International Journal of Pharmacopsychiatry,* describing his investigations into the clinical application of two psychedelic treatment models using LSD and DPT respectively, with patients suffering from end-stage cancer who were approaching death. This study, demonstrating an impressive degree of psychological and spiritual healing in sixty terminal cancer patients, was in my mind one of the most moving, inspirational, and important research reports I had ever come across. Stan demonstrated that after one psychedelic treatment, over 70% of his patients sustained significant improvement in mood, anxiety, quality of life, and fear of death, 30% of whom were described as having "dramatically improved." I enthusiastically read and organized my thoughts and prepared my presentation, keenly looking forward to whatever questions and comments my classmates and instructors might have. On the day of my presentation, I delivered my talk as clearly and as cogently as I could and waited for the response. And waited. In the end, there were no comments, and there were no questions. Not a single hand was raised. No one had anything to say. I realized then that, even in medical circles, the topic of psychedelics and their innate capacity under optimal conditions to facilitate healing and transformation was now off the table. Psychedelics had become a taboo topic, not to be discussed openly. And I got the message, keeping my thoughts and interests in psychedelics to myself, and waiting for a more propitious time.

Finding myself disillusioned with mainstream psychiatry as it was constituted in the late 1970s, I initially trained in internal medicine after graduating from medical school, following which I began my training in neurology. I quickly found myself, however, sinking into a depressed state, not finding the inspiration or direction in my work that I was looking for. One Friday during this period of personal trial, I found myself browsing in the Stanford Medical School bookstore where I unexpectedly came across

Stan's first two recently published books, *Realms of the Human Unconscious* and *The Human Encounter with Death.* I purchased the books, raced back home and over the weekend carefully read both of them from cover to cover. By Monday morning, I knew what I had to do. Reading Stan's books had jogged my memory as to why I had made the decision to go to medical school and become a physician in the first place. From that moment, I was back on the path that I had first envisioned after hearing Stan speak in New York. And now, almost ten years later, prompted by Stan's fascinating first two books, I was re-inspired to take up the call once again and prepare myself for the day when psychedelics would return to their deserved and respected position among the panoply of effective and permitted treatments in the world of medicine.

Following my transfer into a psychiatry residency program in Los Angeles, I was excited to begin the process of establishing a foundation that would one day allow me to work with psychedelics as an approved investigator. On my first day, I was invited into the office of the residency training director, who enthusiastically told me that I was entering the field (this was 1980) at an exciting moment in time, with new developments and exciting state-of-the-art treatment models in psychiatry abounding. He pointed to a large book on his desk and told me that the book represented the vanguard of the coming revolution in psychiatry. My excitement at that moment could barely be contained, as I eagerly pondered what the book could be, even considering briefly that Stan might have published still another book on psychedelics that I would now be introduced to. I snapped out of my trance as my training director handed me the book, whereupon my spirits sagged, as I read on the cover that it was the *DSM III, the Diagnostic and Statistical Manual for Psychiatric Disorders*, which would be heralding in this new era in psychiatry. Not exactly what I was expecting.

My first several months as a psychiatric resident were grim, and uninspiring to the point that I began to formulate plans to leave psychiatry and return to internal medicine. I had yet to capture the spirit of psychiatry, and feeling dispirited, I began to make plans to change fields once again. But, before finalizing my plans, I decided to take one more shot in the dark and wrote a long letter to Stan, in care of the publisher of his just-released book, *LSD Psychotherapy.* In my letter to Stan, I described my predicament, of having been inspired to enter the field of medicine and specialize in psychiatry in large part to conduct work on the healing potentials of psychedelics, having been inspired by his own research in this area, and now finding myself in what I felt was a wasteland of stale and ineffectual treatment models, while the paradigm then starting to dominate the field, lavishly sponsored by the pharmaceutical industry, was the psychobiological perspective. This new psychiatry emphasizing mechanisms of brain and behavior, for all its notable advances, still in my view, did not approach the heart of the matter, which was psychiatry's alienation from being true physicians of the soul. I told Stan that I was disillusioned with the field of psychiatry, as it then stood, and asked him if he thought there was any hope that psychedelics could ever be a viable psychiatric treatment.

A month later, I received Stan's reply, written from on board his intercontinental flight to Australia where he was traveling to participate in one of the meetings of the International Transpersonal Association. Stan said that while it remained "difficult to assess the future of psychedelics at this time," he continued to be, as he put it, "mildly optimistic" that the future would yet find a useful and respected place for psychedelics. This reassurance from Stan at such a vulnerable time helped to keep me in place and focused on completing my training in psychiatry, along with learning the basics of how to become a doctor and a healer. I read on the topic as much as I could, and during this time, I made frequent trips to the Bodhi Tree Bookstore in West Los Angeles, where on a regular basis, I perused the small section reserved for psychedelic books, hoping to come across new contributions from Stan and his colleagues in the field, and progressively enlarging my library. Later, when I trained and joined the full-time faculty at The Johns Hopkins Hospital in Baltimore, I systematically explored the basement stacks of the venerable Welsh Medical Library and "discovered" the remarkable writings of early psychonaut explorers from the turn of the last century, including William James, Havelock Ellis, Weir Mitchell, and others describing their astonishing experiences and reflections on their personal experimentation with mescaline, only having been recently isolated from the peyote cactus by the great German chemists Arthur Heffter and Louis Lewin in the late 1890s. This was all in the 1980s at Johns Hopkins, where the study of psychedelics had been deemed off-limits by the intransigent chairman of psychiatry. It was a time once again of keeping my views to myself, but I nevertheless used the opportunity to more fully acquaint myself with some of the fascinating early medical literature on psychedelics. This was time well spent for me, not only in developing an affinity for and mastering various elements of contemporary psychiatric practice, but also in building my foundation of historical knowledge and preparing for the day when it would become permissible once again to conduct the kind of rigorous studies of psychedelics that Stan had been engaged in and had set the template for as an active psychedelic investigator some years previous.

Stan's work with psychedelics began in the mid-1950s while a medical student in Prague, Czechoslovakia, and continued through the late 1960s to mid-1970s as director of the Maryland Psychiatric Research Center in Spring Grove. In an interview I conducted in the late 1990s for my book *Higher Wisdom: Eminent Elders Explore the Continuing Impact of Psychedelics*, Stan estimated that over these two decades, he had personally administered psychedelics in legal settings in both Prague and the United States, constituting upwards of 4,000 sessions with select patients and normal volunteers, spending a minimum of five hours during each of these sessions in the actual treatment room. This, I have no doubt, represents a greater degree of first-hand experience administering psychedelics than perhaps any other investigator, past or present.

What has always impressed me about Stan, from my very first encounter attending his talk at AHP in late 1972, has been his intellectual rigor, the remarkable length and breadth of his knowledge, and his capacity to tie together in a seamless thread his medical identity with his compassionate, careful, and scrupulous work with patients and his outstanding communication skills. Stan's sensitivity to the historical record with its many contributions from cultures far and wide, from the earliest time of shamanic presence on this planet to the contemporary world of his peers, has instructed his worldview. Through his vast experience studying psychedelics, Stan has developed a highly original and compelling theoretical framework for understanding non-ordinary states of consciousness. While mainstream psychiatry has by and large eschewed meaningful discussion of the clinical implications of Stan's theories based on powerful intrauterine and birth experience, it remains a compelling theory ripe for future explorations. With the tool of psychedelic drugs now legally available again, investigators may well have a fascinating opportunity to rigorously examine the validity and application of these theoretical positions developed more than half a century ago, during the Golden Age of psychedelic research.

Psychedelics have also returned to the medical world. Since the early 1990s, it has become feasible once again to obtain regulatory approval to conduct psychedelic research with human subjects. And, since the early 2000s, clinical treatment research has resumed. Recently, interest within mainstream psychiatry has in fact soared. While limited research funds from a few philanthropic donors sustained the field for many years, with federal granting agencies continuing to turn their backs as they have for the past half-century, over the past couple of years very large amounts of money have been infused into the field by the pharmaceutical industry and by entrepreneurs intent on profiting from what they envisage as a new enterprise centered around psychedelic treatment. Companies of relatively recent origin are suddenly and shockingly being valued in the billions of dollars. It remains to be seen whether the commercialization of psychedelics can be implemented in an optimally safe and effective way, but there is no doubt that the field of psychedelics is back, and perhaps with a vengeance, 50 years after a previous generation of research was shut down in the late 1960s and early 1970s. All the more reason, then, to look back at Stan's tremendous accomplishments and teachings from his many years as a distinguished and greatly respected researcher of this now resurgent field of psychedelic treatment. It will also benefit the next generations to study not only the great body of his work but also Stan's personal example, one of steadfastly according himself with the highest standards of probity.

From 2004 to 2008, at the Harbor-UCLA Medical Center, I had the distinct privilege to conduct the first approved research investigation of a psychedelic (psilocybin) in the treatment of severe existential anxiety and depression in individuals with advanced-stage cancer, since Stan was forced to shut down his treatment program many years before. Stan's work at Spring Grove, in addition to that of several of his colleagues elsewhere, including Eric Kast, Gary Fisher, Sidney Cohen, and Walter Pahnke, provided valuable lessons and instruction in how to create a safe and effective treatment model, and most importantly, how to optimize the opportunity to achieve a highly therapeutic and even transformative outcome, with a population of patients approaching the end of life. I have always felt and taught that it is critical for us to learn the lessons of those who came before.

Stan and his contemporaries were the pioneers who created the foundation of an entirely new and novel form of treatment that, after many years of repression, has now reemerged. To these early visionaries, we owe a great debt of gratitude for their early discoveries and contributions, without which we would be denied the stable foundation that this evolving field will need to rely on in order to effectively engage the challenges that no doubt lie in our collective futures.

One particularly intriguing finding from Stan's early work treating terminal cancer patients with potent psychedelic drugs was the direct correlation of a mystical experience with a positive therapeutic outcome. Weeks and even months following a single treatment session conducted under optimal conditions, this powerful psychospiritual epiphany was strongly associated with sustained improvement of mood, lowered anxiety, decreased pain, and greater spiritual well-being. This mystical experience, catalyzed by the profound effects of the psychedelic treatment, appears to serve the function of restoring a sense of meaning and purpose to the lives of these highly vulnerable individuals, even to the point of recovering their identity, their sense of self that had so eroded under the onslaught of the relentless progression of their disease. What Stan and his colleagues explored, and what we are now allowed to employ once again today, is an existential medicine that potentially reveals and heals the great existential crisis that is an inevitable part of human existence, a crisis that is often most evident during our final days.

As I read in one of Stan's lesser-known yet stellar books, *Beyond Death: The Gates of Consciousness*, "*Mors certa, hora incerta:* the most certain thing in life is that we will die; the least certain is the time when it will happen." The great human dilemma is that we do all die. Inevitably, it is our fate—for all of us. And sadly, the modern world we inhabit has poorly prepared us for this eventuality. Modern men and women are often ill-prepared for this terminal chapter. Serious medical illness can be treated far more effectively than ever before, yet when hope and viable medical options run out, what then? Stan and his colleagues have handed down to us important models to facilitate healing and transformation, even during the final days of our stay on earth.

From Stan's knowledge of the discoveries of his peers and those far and wide who preceded him, along with his advanced knowledge of world cultures, he has evolved a model treatment of conscious dying that may give us an anticipation of light at the end of the tunnel. Thanks to the pioneers of the field of psychedelics, most notably Stan Grof, we now have the opportunity and the tools to once again open up the psychedelic treatment model and determine if the early studies using psychedelics to treat the existential crisis of advanced medical illness can continue to be replicated, and if so, be made available in safe, sanctioned, and skilled treatment settings to those profoundly in need of psychological and spiritual healing at the end of life. This is the hope and the promise of psychedelics. To Stan Grof goes much of the credit for our now having this remarkable opportunity before us. It is now up to us to implement an optimal use of these mysterious and remarkable compounds in as respectful and careful a manner as those who came before would expect.

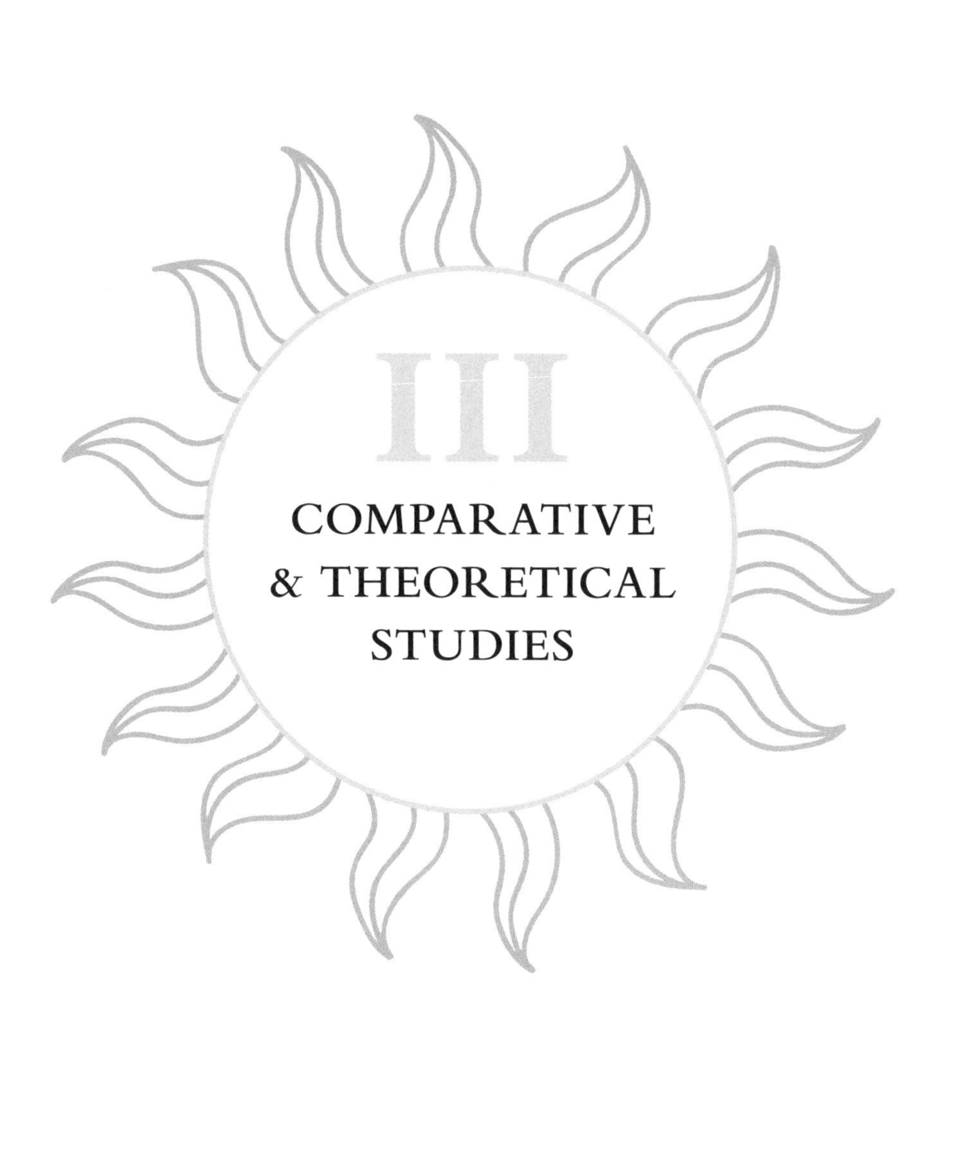

III
COMPARATIVE & THEORETICAL STUDIES

Seekers of a Second Birth

James, Grof, and the Varieties of Perinatal Experience

Sean Kelly

There are towering figures we have not met personally with whom we nevertheless feel a profound karmic connection and who have enormous impacts on our lives, both internally and externally. In my case, Hegel and Jung are such figures. Any gratitude we might feel can only be expressed inwardly, as a kind of silent prayer. Occasionally, there are equally consequential figures that one has the great good fortune to know personally. Stanislav Grof is such a figure in my life, and I am delighted that I have been given this chance to express my deep gratitude for the precious gift of his life's work.

I have written elsewhere on the centrality of Grof's contributions to transpersonal psychology and to what many refer to as the New Paradigm.[1] Most recently, I have suggested how Grof's understanding of the perinatal process and the death/rebirth archetype can illuminate the current planetary initiation and its associated collective Near-Death Experience into which the Earth community is being drawn.[2] For this offering, I want to bring Grof's revolutionary findings into dialogue with the work of William James, another great pioneer in the study of extraordinary experiences. I have two goals in what follows: first, to highlight the extent to which the central conceptual analogies of James's *Varieties of Religious Experience* both foreshadow and are amplified by Grof's transpersonal-theoretical constructs; and secondly, to explore the implications of James's observations, in tandem with those of Grof, for a key issue in transpersonal theory—namely, the relationship between the perinatal and the transpersonal. I am assisted in this enterprise by a sympathetic reading of Hegel as well as by some of Ken Wilber's helpful distinctions, though I take issue with Wilber on the nature of the perinatal process (and more generally on the relation of the *pre* to the *trans*).

Transmarginal and Holotropic

I have yet to track down the first use of the field analogy for the nature of consciousness, though I suspect it might have originated with Gustav Fechner, whom James greatly admired. In any case, James remarks in the *Varieties* that "the expression 'field of consciousness' has but recently come into vogue in the psychology books."[3] "The important fact which this field formula commemorates," James adds, "is the indetermination of the margin."[4] "I cannot but think," he continues, "that the most important step forward that has occurred in psychology since I have been a student of that science is the discovery, first made in 1886, that, in certain subjects at least, there is not only consciousness of the ordinary field, with its usual centre and margin, but an addition thereto in the shape of a set of memories, thoughts, and feelings which are extra-marginal and outside the primary consciousness altogether, but must be classed as conscious facts of some sort, able to reveal their presence by unmistakable signs."[5] The author of this discovery was James's good friend and fellow psychical researcher, Frederic Myers, who coined the term "subliminal Self." While James also refers to the "subliminal," he often uses "transmarginal" instead, probably because the latter is more consonant with the field analogy with its associated distinction between focal point and margin. The advantage of "subliminal" is that it appeals to the depth dimension ("sub-") and to the notion—this time definitely associated with Fechner—of a variable "threshold" of consciousness.

Grof is clearly in the depth-psychological tradition which, though most familiarly associated with Freud and Jung, actually stretches back through James and Myers all the way to Fechner and ultimately, it would seem, to the Romantics (especially Schelling). Though he does not make explicit use of the threshold analogy, Grof's depth-oriented map of the psyche, in fact, postulates a series of margins or boundaries, starting with the "sensory barrier," followed by the boundaries which separate, in order of increasing depth, the biographical from the perinatal, and the latter from the transpersonal "realms" of the psyche. At the same time, however, Grof, like James, invokes the field analogy—appealing, as James obviously could not, to the new physics—to account for some of the more challenging findings to emerge from contemporary consciousness research (that is, findings which, from a variety of sources—including shamanic studies, parapsychology, and psychedelic research—challenge the dominant assumptions of the so-called Cartesian-Newtonian paradigm).[6]

Focusing in particular on the "non-ordinary" states of consciousness that typically accompany deep experiential work, Grof proposes the distinction between two fundamental "modes" of consciousness—the hylotropic and the holotropic. Whereas the hylotropic "involves the experience of oneself as a solid physical entity with definite boundaries and a limited sensory range, living in three-dimensional space and linear time....," the holotropic "involves identification with a field of consciousness with no definite boundaries [recall James on the "indetermination" of the field's margin] which has unlimited experiential access to different aspects of reality without the mediation of the senses."[7] From the perspective of depth, one could say that the individual, in the holotropic mode, is subject to an indeterminate lowering of the threshold of consciousness. One of the paradoxes of this topography, however, is that once the perinatal and transpersonal boundaries are crossed, the depth metaphor begins to fail—or at least becomes less serviceable—since there is now the possibility of the experience of, or identification with, increasingly ego-alien realities which may strike the individual as "higher," more encompassing, or radically "other" in ways less obviously related to an increase in personal "depth." Such is the case, for instance, in encounters with celestial deities, or in states of identification with Gaia or the cosmos at large, or in experiences of what Grof calls the Supracosmic and Metacosmic Void.[8]

Appealing to the perennial-philosophical metaphor of the Great Chain of Being—though he came to prefer the more complex concept of a "nested hierarchy," or "holarchy"—Ken Wilber has made much of the distinction between the "lower" (personal and prepersonal) and "higher" (transpersonal) domains. While he has retained[9] the notion that (transpersonal) Spirit has greater depth than (merely personal or prepersonal) Mind—and that both Mind and Spirit have greater depth than Matter, he has more or less abandoned the depth-psychological orientation along with the "bi-polar" distinction (to use Michael Washburn's expression) between two fundamental kinds or "modes" of consciousness. This orientation and this distinction, according to Wilber, militate against a proper appreciation of the unconditional difference between the prepersonal and the transpersonal, or at the very least do not afford the opportunity to distinguish between the two. In Wilber's estimation, the problem is particularly acute with respect to the status of the perinatal,[10] to which we shall turn shortly.

James, for his part, is quite aware of the difference between the pre and the trans, though he does not employ these terms. He recognizes, for example, that in "delusional insanity, paranoia, as they sometimes call it, we may have a *diabolical* mysticism, a sort of religious mysticism turned upside down."[11] He goes on to say, however, "that from the point of view of their psychological mechanism, the classic mysticism and these lower mysticisms spring from the same mental level, from the great subliminal or transmarginal region of which science is beginning to admit the existence, but of which so little is really known. That region contains every kind of matter: 'seraph and snake' abide there side by side."[12] James is not making the (anti) metaphysical claim that there are no ontological realities or "higher powers," as he elsewhere puts it, corresponding to the "classic mysticism." He is, rather, making the purely psychological claim that, "[i] f there be higher powers able to impress us, they may get access to us only through the subliminal door."[13]

Grof, similarly, has no doubt as to the existence of "higher" metaphysical realms. What his research suggests, however, is that in fact—that is, from the point of view of the actual phenomenology of holotropic states of consciousness—it is often difficult, if not impossible, to make an absolute distinction between the pre and the trans.[14] In so stating, Grof argues with James that "'seraph and snake' abide ...side by side." For example:

> Through observing clients in non-ordinary states we discover that their neurotic or psychosomatic symptoms often involve more than the biographical level of the psyche. Initially, we may find that the symptoms are connected to traumatic events that the person suffered in infancy or childhood, just as described in traditional psychology. However, when the process continues and the experiences deepen, the same symptoms are found to

be also related to particular aspects of the birth trauma. Additional roots of the same issue can then be traced even further to transpersonal sources, for example, an experience in a past life, an unresolved archetypal theme, or the person's identification with an animal.[15]

The Natal, the Transnatal, and the Metanatal

I turn now from topographical considerations to those of process and transformation. In the *Varieties*, James is especially concerned with those transformations of consciousness that involve the emergence of a new, religiously informed "centre of personal energy."[16] Such transformations, or "conversions" (*metanoia*), involve "the process, gradual or sudden, by which a self hitherto divided, and consciously wrong, inferior, and unhappy, becomes unified and consciously right, superior, and happy, in consequence of its superior hold upon religious realities."[17] In more contemporary terms, one could say that James is interested in the phenomenon of the integration (or individuation, in the Jungian sense of the word) of personality in association with, or even as a consequence of, encounters with the transpersonal domains of consciousness. In specifically Grofian terms, James's view of "conversions" would correspond to the successful negotiation of spiritual emergences ("gradual") and emergencies ("sudden").[18]

Following the Christian tradition of discourse concerning the experience of conversion, James characterizes these transformations as second "births," and refers to those who have been so transformed as the "twice born." It is an open question to what extent the second birth of consciousness into the transpersonal involves the (re)processing of material from the first. What James's treatment suggests, however—though this really only becomes apparent in the light of Grof's subsequent findings—is that the process of biological birth is a primary analog—if not the prototypical occasion—for the emergence or creation of consciousness in general, whether personal or transpersonal.[19] The concept of the perinatal, in other words, is complexly "overdetermined" and, like the notions of the subliminal or transmarginal and the holotropic, straddles the biological and the psychospiritual, as it does the pre and the trans. One could say, therefore, that the concept of the perinatal involves not only a natal but a para- or *trans*natal dimension as well. Though the transnatal dimension of the perinatal process involves, in its most developed form, a *transpersonal* (James's "religious") death/rebirth experience, the experience nevertheless remains a (second) *birth*. One would expect it, therefore, to manifest the same kinds of themes already associated with the four phases, or "matrices," of the strictly natal dimension of the perinatal process. Such indeed is the case, as I will illustrate with a few representative passages from the *Varieties*.

Starting with the first matrix, perhaps the most evocative, if unsophisticated, passage comes from an account by Mrs. Jonathan Edwards, who describes an evening of "heavenly sweetness of Christ's excellent love... with an inexpressibly sweet calmness of soul in an entire rest in him.... At the same time my heart and soul all flowed out in love to Christ, so that there seemed to be a constant flowing and reflowing of heavenly love, and I appeared to myself to float or swim, in these bright, sweet beams, like the motes swimming in the beams of the sun, or the streams of his light which come in at the window."[20] The often repeated "sweetness," the "rest" and "calm," and especially the image of floating or swimming in warm light, are all typical descriptors of the undisturbed intra-uterine state of the first perinatal matrix (BPM I). In more abstract conceptual terms, we have Vivekananda's description of *samadhi* as a state where there is "no feeling of *I*, and yet the mind works, desireless, free from restlessness, objectless, bodiless."[21]

In the actual birth process (natal dimension), the second matrix signals the onset of labor, and therefore the end of the "heavenly sweetness" of BPM I. James himself provides, in the guise of a letter from a fictitious French correspondent, an autobiographical example of a transpersonally modulated experience of BPM II. He describes how, during a period of prolonged "philosophic pessimism and depression of spirits," he was suddenly struck with a panic fear of his own existence in association with the image of an epileptic patient he had seen in an asylum. "That shape am *I*, I felt, potentially."[22] "There was such a horror of him," writes James, "and such a perception of my own merely momentary discrepancy from him, that it was as if something hitherto solid within my breast gave way entirely, and I became a mass of quivering fear.... I awoke morning after morning with a horrible dread at the pit of my stomach, and with a sense of the insecurity of life that I never knew before, and that I have never felt since."[23] Another striking example of a transpersonally modulated experience of BPM II is taken from the memoirs of one "Father Gratry." At one point, he recounts how, following a period of mental isolation and excessive study, he was beset by images of what Grof describes as "cosmic engulfment" followed by a foretaste of his inevitable lot in Hell, which he characterizes as the complete loss of the ability to experience any positive feeling whatsoever. "I had such a universal terror," he writes, "that I awoke at night with a start, thinking that the Pantheon was tumbling on the Polytechnic school, or that the school was in flames, or that the Seine was pouring into the Catacombs, and that Paris was being swallowed up. And when these impressions were past, all day long without respite I suffered an incurable and intolerable desolation, verging on despair.... But what was perhaps still more dreadful is that the very idea of heaven was taken away from me.... It was like a vacuum.... I could conceive no joy, no pleasure in inhabiting it."[24]

The third matrix corresponds to the second stage of birth, where the fetus typically experiences crushing pressures—often with near suffocation—a generalized hyper-arousal, and an active struggle for life itself. The following example is from an unidentified patient in a debilitating state of agitated depression who feels mercilessly persecuted by God or the devil, he knows not which. He complains that "fear, atrocious fear, presses me down, holds me without respite, never lets me go.... Eat, drink, lie awake all night, suffer without interruption—such is the fine legacy I have received from my mother! What I fail to understand is this abuse of power. There are limits to everything, there is a middle way. But God knows neither middle way nor limits. I say God, but why? All I have known so far has been the devil.... I am defenseless against the invisible enemy who is tightening his coils around me."[25]

The next example captures the transition from the third to the fourth matrix—that is, to the phase corresponding to the final stage of delivery where, if all goes well, there is a sudden release from the agony of the passage through the birth canal. This release, moreover, is typically associated with intense light and a flood of positive feeling. Once again, the person reporting the experience is unidentified, though we know that it took place at a Christian revival meeting.[26]

> I know not how I got back into the encampment, but I found myself staggering back to Rev. ______'s Holiness tent—and as it was full of seekers and a terrible noise inside, some groaning, some laughing, and some shouting, ... I fell on my face by a bench and tried to pray, and every time I would call on God, something like a man's hand would strangle me by choking.... I thought I should surely die if I did not get help, but just as often as I would pray, that unseen hand was felt on my throat and my breath squeezed off.... I don't know how long I lay there or what was going on.... When I came to myself, there was a crowd around me praising God. The very heavens seemed to open and pour down rays of light and glory. Not for a moment only, but all day and night, floods of light and glory seemed to pour through my soul, and oh, how I was changed, and everything became new. My horses and hogs and even everybody seemed changed.[27]

Richard Tarnas was the first to offer a Hegelian reading of the perinatal process,[28] assimilating the four matrices to the triphasic rhythm of the dialectic (or speculative reason, to be more precise). This same rhythm, which I have shown[29] is equally evident in Jung's concept of individuation, is invoked by James to describe the fundamental pattern of conversion. The three phases or "moments" of universality, particularity, and individuality—or identity, difference, and identity-in-difference—apply in equal measure to the natal and transnatal dimensions of the perinatal process (in this case, the second and third matrices refer together to the second moment of difference or particularity). We could say that it is a question here of the *meta*natal, in the sense of a generative logic (in Hegel's sense of the term) for the birth (and death/rebirth) of consciousness as such, whether personal or transpersonal.

I would point out that the natal and transnatal dimensions are prefigured in the second and third moments, respectively, of the metanatal. The first moment of identity/universality corresponds to the metanatal itself as pure potential or archetypal possibility. The moment of particularity or difference corresponds to the first, biospsychic birth—what I am calling the natal dimension of the perinatal process—which, though it involves the death of the participatory consciousness of the fetus, signals, by virtue of this death, the emergence of the initial form of the separate-self sense. Once constellated, this separate-self sense will, in an average lifetime lived to full psychophysical maturity, undergo a series of developmental metamorphoses through the various "fulcra."[30] The passage through each fulcrum, as described by Wilber, transpires according to the threefold (metanatal) rhythm of identity, differentiation, new identity, so that the normal course of personal development is marked by a series of mini deaths and rebirths. Perhaps the most significant of these involves the emergence of a second, super- (and ambiguously[31] supra-) ordinate self-sense (the mental ego, with the prior body ego tending to fall more or less into the "shadow"). While many individuals will experience at least some transient non-ordinary states of consciousness of the transpersonal varieties, relatively few will undergo the radical ego-death associated with the second birth into the transnatal.[32] (See figure 1)

In terms of the distinctions I have introduced above, what I am arguing against is the collapsing of the natal and transnatal dimensions of the perinatal process into the sole dimension of the metanatal. While the latter indeed prefigures the former, it does so only as logical or archetypal possibility. The full actualization of the second and third moments of the metanatal, in other words, can only come about in the context of the first and second births, in whose experiential phenomenologies the metanatal moments are embedded, and out of which, after the fact, they are theoretically abstracted.

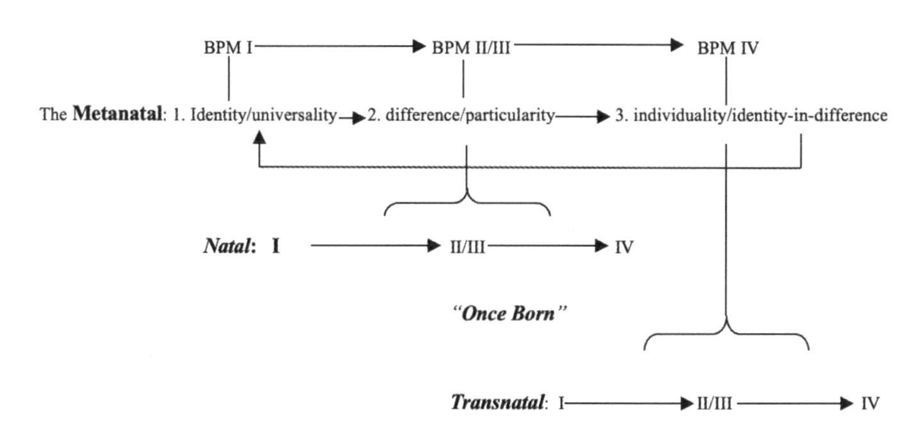

FIGURE 1 The Perinatal Process

Two Transpersonal Worldviews

Alongside his distinction between the once-born and the twice-born, James proposes two basic personality types—the "healthy-minded" and the "sick-souled"—with their two associated religious, or transpersonal, worldviews. Both worldviews, as we shall see, can be thought of as flowing from two possible biases or positions of the transnatal relative to the natal. In its most pronounced manifestations, the healthy-minded personality involves the "systematic expulsion...of all contractile elements."[33] In some cases, "[t]he capacity for even a transient sadness or a momentary humility seems cut off from them as by a kind of congenital anaesthesia."[34] James quotes a woman friend of his, whose words, in describing "the feeling of continuity with the Infinite Power, by which all mind-cure disciples are inspired," may serve as a summary of the healthy-minded worldview.[35] "The first underlying cause of all sickness," she writes, "weakness, or depression is the human sense of separateness from that Divine Energy which we call God. The soul which can affirm in serene but jubilant confidence, as did the Nazarene: "I and the Father are one," has no further need of healer, or of healing. This is the whole truth in a nutshell, and other foundation for wholeness can no man lay than this fact of impregnable divine union."[36]

By contrast, the sick-souled individual, in James's evocative encapsulation, tends to the belief that, "[b]ack of everything is the great spectre of universal death, the all-encompassing blackness...."[37] "Unsuspectedly from the bottom of the fountain of pleasure...something bitter rises up: a touch of nausea, a falling dead of the delight, a whiff of melancholy...."[38] "In short, life and its negation are beaten up inextricably together.... Yet the two are equally essential facts of existence; and all-natural happiness thus seems infected with a contradiction. The breath of the sepulchre surrounds it."[39] While James often gives the impression that the healthy-minded corresponds to the once-born and the sick-souled to the twice-born, one must be careful not to collapse all four terms into a single distinction. It is true that James considers that nothing less than a full-fledged second birth can satisfy the spiritual cravings of the sick-souled individual. However, he admits that, in his consideration of healthy-mindedness, "we found abundant examples of regenerative process. The severity of the crisis in this process is a matter of degree.... so that in many instances it is quite arbitrary whether we class the individual as a once-born or a twice-born subject."[40] James, in fact, proposes a congenital variability in the pain threshold,[41] along with "a certain discordancy or heterogeneity in the native temperament of the subject"[42] as significant factors in predisposing individuals to one worldview or the other. Both worldviews, in other words, have their natal (congenital) and transnatal expressions, depending upon whether or not the individual has experienced the radical restructuring of personality associated with the second birth.[43] James gives Bunyan and Tolstoy as examples of two sick-souled individuals who eventually experienced such a second birth. "They had drunk too deeply of the cup of bitterness," he writes, "ever to forget its taste, and their redemption is into a universe two stories deep. Each of them realized a good which broke the effective edge of his sadness; yet the sadness was preserved as a minor ingredient in the heart of the faith by which it was overcome."[44]

Despite his appreciation for the way in which the Mind Cure movement drew attention to the healing potential of intentionally focused consciousness, James leaves no question as to which of the two transpersonal worldviews he considers most satisfying. There is no doubt, he says, "that healthy-mindedness is inadequate as a philosophical doctrine, because the evil facts which it refuses positively to account for are a genuine portion of reality; and they may after all be the best key to life's significance, and possibly the only openers of our eyes to the deepest levels of truth."[45] From the perinatal, and specifically metanatal, perspective, it is clear that the healthy-minded worldview, in its "systematic expulsion...of all contractile elements," involves a problematic relationship to the second moment of difference and particularity—to "negativity," in the Hegelian sense of the term—along with its associated cluster of such existential facts as suffering, evil, and death. Such a worldview, however pragmatically useful in certain contexts (and granting the possibility that it might even be metaphysically true), nevertheless bears the mark of an incomplete perinatal gestalt. Correlative to the muting or bracketing of the second moment of the metanatal is a repression, from the point of view of what could be described as a premature identification with the transnatal, of the natal dimension as a whole (which, as we have seen, involves the developmental—and, one might add, incarnational—drama of the separate-self sense, with its series of mini deaths and rebirths). The liberation from pain, suffering, and death which such a worldview promises is therefore contingent upon a radical denial, and the spiritual monism which it proffers as a metaphysical solace is correspondingly flat and monochromatic.[46] (See figure 2.)

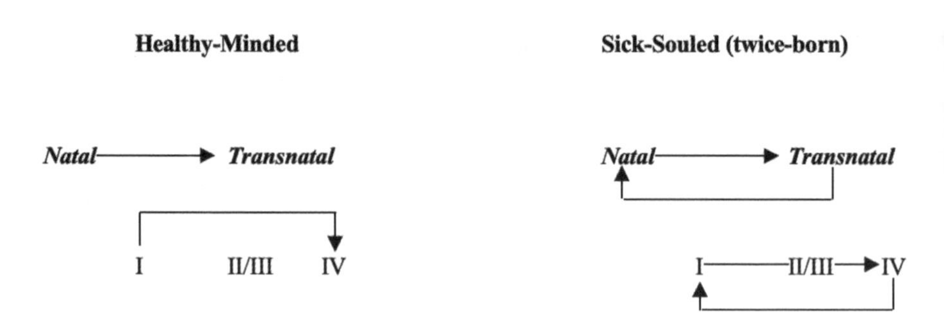

FIGURE 2 Two Transpersonal Worldviews

Saintliness and a Higher Sanity

James uses the word "saintliness" to describe the character of those twice-born individuals for whom "spiritual emotions are the habitual centre of the personal energy."[47] He lists several typical traits of the saintly character, including the "feeling of being in a wider life than that of this world's selfish little interests;"[48] an "immense elation and freedom, as the outlines of the confining selfhood melt down;"[49] and more generally, a "shifting of the emotional centre toward loving and harmonious affections, towards 'yes, yes,' and away from 'no,' where the claims of the non-ego are concerned."[50] In the most exemplary of cases, these traits tend to manifest themselves in asceticism (the subordination of personal, lower order, needs to spiritual ones), "strength of soul" (where "new reaches of patience and fortitude open out"), and an increase of "charity" (in the form of "tenderness for fellow creatures").[51]

At the same time, however, James recognizes that the saintly character is typically prone to various forms of "excess," including fanaticism, indiscriminate "tenderness," and masochistic self-mortification. Once again, we see the impossibility, in the actual state of affairs, of making a clean break between the trans and the pre, between the spiritual ("seraph") and the pathological ("snake"). From a Grofian perspective, it seems clear that the saintly character, as James describes him or her, is particularly open to the holotropic mode of consciousness (the "feeling of being in a wider life," an openness "where the claims of the non-ego are concerned"). From the examples he gives, it is clear as well that the "spiritual emotions" which typically dominate the saint's field of consciousness are informed by, or continue to draw upon, the perinatal dimensions of the psyche. While James's encapsulation of the "faith state" as "a paradise of inward tranquility"[52] is suggestive of BPM I, and the "immense elation and freedom" of the saint points to BPM IV, the dynamics of BPM III are equally prominent. And it is here that the line between the pre and the trans, or the natal and the transnatal, is especially hard to draw. The extremes to which Suso, the fourteenth-century mystic, went to maintain a state of self-inflicted torture,[53] or the declaration of the founder of the Sacred Heart order that "Nothing but pain makes my life supportable,"[54] would strike most contemporaries as unambiguously pathological. Nevertheless, James stresses the point that such extremes—or at least the potential for such extremes—not only often accompany, but in some way seem to feed, the capacity for the same individuals to practice the unambiguously saintly virtues of patience, fortitude, and charity.[55]

Grof, as we have seen, has stressed the impossibility of effecting an absolute separation of the transpersonal from the perinatal, or of what I have called the transnatal from the natal. He even goes so far as to recognize the existence of transpersonal counterparts (Hell realms, for example, or various wrathful deities) to those negative perinatal dynamics which, in the realm of personal and social psychology, manifest themselves as debilitating pathology. In the case of saintly asceticism, however, it seems to be not so much a matter of such transpersonal counterparts (though the example of Saint Anthony's desert visitations do point in this direction), as of a situation where negative natal dynamics are reframed within a transnatal context. As with the healthy-minded worldview, there is, with the prototypical saint, a problematic relation to the natal dimension as a whole, with BPM III being particularly charged. In contrast to the healthy-minded worldview, however, saintly asceticism does not attempt a simple negation of the natal dimension, but actively, and even strenuously, engages it as something only too real which must, nevertheless, be sublimated or otherwise overcome. In its spiritual meaning, as James puts it, "asceticism stands for nothing less than for the essence of the twice-born philosophy. It symbolizes, lamely enough no doubt, but sincerely, the belief that there is an element of real wrongness in this world, which is neither to be ignored nor evaded, but which must be squarely met and overcome by an appeal to the soul's heroic resources, and neutralized and cleansed away by suffering. As against this view, the ultra-optimistic form of the once-born philosophy [I would say, more precisely, the healthy-minded worldview] thinks we may treat evil by the method of ignoring."[56]

The virtues of asceticism notwithstanding, James by no means condones its excesses, nor does he have much use for its more moderate, traditional expressions. Is it not possible, he asks, "for us to discard most of these older forms of mortification, and yet find saner channels for the heroism which inspired them?"[57] I would suggest that the kind of deep experiential work pioneered by Grof must be considered preeminent among such possible "saner channels." Such work, where the individual is allowed to confront, and eventually release, some of the most painful of psychic materials (whether personal or collective)—and all in a safe, supportive, and indeed sacred, setting—leaves ample opportunity for the exercise of the kind of heroism which James so rightfully admires. In this context, instead of saintliness, Grof speaks of a "higher sanity" displayed by "individuals who have achieved a balanced interplay of both complementary modes [i.e., the hylotropic and the holotropic] of consciousness."[58] Echoing James on the "shifting of the emotional centre," in the personality of the saint, "toward loving and harmonious affections, towards 'yes, yes,' and away from 'no,' where the claims of the non-ego are concerned," Grof says the following on his notion of a higher sanity.

> Balanced integration of the two complementary aspects of human experience tends to be associated with an affirmative attitude toward existence—not the status quo or any particular aspects of life, but the cosmic process in its totality, the general flow of life. An integral part of healthy functioning is the ability to enjoy simple and ordinary aspects of everyday life, such as elements of nature, ...as well as eating, sleeping, sex, and other physiological processes of one's body.
>
> ... The generally affirmative attitude toward existence creates a metaframework that makes it possible to integrate in a positive way even the difficult aspects of life.[59]

Mysticism and the Cosmic Game

James considers mystical states of consciousness to be "the root and center" of religious, or transpersonal, experiences.[60] Here I am not concerned with the experiential qualities of such states,[61] but rather with the metaphysical "overbeliefs," as James would say, with which mystical states tend to be associated. In this respect, James discerns the existence of a perennial consensus, or "eternal unanimity,"[62] which points "in definite philosophical directions," one of which is optimism, the other, monism.[63] The "overcoming of all the usual barriers between the individual and the Absolute," he writes, "is the great mystic achievement. In mystic states we both become one with the Absolute and we become aware of our oneness. This is the everlasting and triumphant mystical tradition, hardly altered by differences of clime or creed. In Hinduism, in Neoplatonism, in Sufism, in Christian mysticism, in Whitmanism, we find the same recurring note, …which brings it about that the mystical classics have, as has been said, neither birthday nor native land."[64]

Just a few pages later, however, James admits that he oversimplified the matter and that, on closer inspection, "the supposed unanimity largely disappears."[65] In contrast to the monism of Shankara's Vedanta, for instance, or that of Neoplatonism, James notes that Sankhya, by contrast, is dualistic and that with the great Spanish mystics, union with the Absolute "is for them much more like an occasional miracle than like an original identity."[66] There is more here at issue than the simple question of a mystical consensus. It is a question, rather, of the coherence of monism—or negatively stated, of unqualified Non-dualism—as a metaphysical overbelief. While James, in the end, comes down in favor of a "pluralistic universe" consisting of a collection of "larger and more godlike" selves, "of different degrees of inclusiveness, with no absolute unity realized in it at all,"[67] he is nevertheless attracted to, and arguably fascinated by, the monism he is ultimately unable to affirm. This situation parallels his estimation of the healthy-minded worldview which, though appreciated for its therapeutic efficacy in the case of those individuals temperamentally suited to it, he nonetheless considers, as we have seen, "inadequate as a philosophical doctrine."

Although Grof has, in the bulk of his published research, generally eschewed metaphysical pronouncements, in *The Cosmic Game* he focuses on "those parts of the records that described experiences and observations related to basic ontological and cosmological questions."[68] In his assessment of these records, I find an analogous tension to that of James's estimation of monism or unqualified Non-dualism. With Grof, however, the tension is inversely polarized, as it were. On the one hand, he underscores the fact that, in deep experiential work, "we realize that we have to experience in our life a certain amount of physical and emotional pain and discomfort that is intrinsic to incarnate existence in general,"[69] and that, in such work, the "experience of agony gives a new dimension to the experience of ecstasy," just as "the knowledge of darkness enhances the appreciation of light...."[70] Like Jung before him, Grof does not minimize the reality of evil. Refusing to consider it as a mere "privation" of the good, he sees evil—along with suffering and death—as "an intrinsic part of creation."[71]

On the other hand, Grof at several points characterizes the metaphysical implications of his research in terms that clearly suggest a form of unqualified Non-dualism. We must recognize, he states, that "all boundaries that we normally perceive in the universe are arbitrary and ultimately illusory. The entire cosmos is in its deepest nature a single entity of unimaginable dimensions, Absolute Consciousness."[72] Grof amplifies such statements by likening the cosmos to an "infinitely sophisticated 'virtual reality'"[73] and, updating Plato's analogy of the cave, to the projection of a particularly compelling film.

Once again, though the metaphysical doctrine of unqualified Non-dualism may, in fact, be true, it becomes problematic to maintain without minimizing, if not implicitly denying, the existence of evil, suffering, and death. These three marks of finitude cannot exist without boundaries, yet we are told that all boundaries are ultimately illusory. Grof is well aware of this quandary, and his book is rich in allusions to various religious, mythological, and philosophical options. However, I would like to suggest that, if there is to be Non-dualism, it be qualified in a way that makes it consistent with the deeper logic of the perinatal process.

Metaphysically considered, Grof understands evil—and the illusion of separateness—as a necessary consequence of the intentional "splitting" or "partitioning" of the Absolute in the interests of the birthing, or creation, of "Absolute [Self-] Consciousness." From the perspective of the perinatal process, as we have seen, such splitting or partitioning corresponds to the second moment of the metanatal and, correlatively, to the natal dimension as a whole (i.e., to the realm of literal birth, sex, and death). To insist on the illusory character of all boundaries in the name of the Non-dual Absolute, therefore, would seem, as in the case of the healthy-minded worldview, to suggest an abortive perinatal gestalt. This is not to deny the reality, or the profound therapeutic potential, of those relatively rare experiences where all boundaries dissolve in the "mystical bath," as James describes it, of the first perinatal matrix (whether natally or transnatally conceived). It is, however, to guard against an illegitimate conflation of such experiences with the third metanatal moment (corresponding to BPM IV) of the completed perinatal gestalt. Such a conflation could be seen as an instance of what Wilber calls the "pre/trans fallacy," though not, as Wilber defines it, involving an elevation of the prepersonal to the status of the transpersonal, since in this case it is a question of two actually transpersonal epiphanies and their associated metaphysical conceptions of the nature of the Absolute. The fallacy, rather, here consists in mistaking the first metanatal moment—corresponding, as Tarnas puts it, to the "initial state of undifferentiated unity"—for the third moment which "both overcomes and fulfills the intervening alienated state—restoring the initial unity but on a new level that preserves the achievement of the whole trajectory."[74]

Concluding Remarks

James is one of the few figures Grof does not consider in his extensive review of the history of depth and transpersonal psychology.[75] As I hope the preceding will have demonstrated, however, the two have much to say to one another. I began with the field analogy to which both James and Grof appeal in their respective topographies of the psyche, in each case underscoring the complex character of the relation between the pre and the trans, and more particularly between the pathological and the spiritual. I then turned to the concept of the perinatal, where the complexity in question is most in evidence. Here I proposed, with the help of James's notion of the "twice born," the distinction between the *natal* and *transnatal* (or personal and transpersonal) dimensions of the perinatal process, both of which share the same triphasic, *metanatal* structure or pattern. It is from this meta-perspective that the concept of the perinatal can be seen, not so much topographically as a "realm" of the psyche, but instead as an expression of the generative logic for the birth, or creation, and ongoing transformation of consciousness in general, whether personal or transpersonal.

With these theoretical distinctions in place, I offered some reflections on the relative merits of James's two typical and mutually incompatible transpersonal worldviews. The focus here, as in the ensuing discussion of saintly asceticism, was the problematic character of the second metanatal moment and the demands of finite, incarnate existence with which this moment is associated. This focus was carried over to the final section where not only the idealism of the healthy-minded worldview but the metaphysical position of unqualified Non-Dualism as well were seen, from the metanatal perspective, to imply an incomplete perinatal (or rather, metanatal) gestalt.

Much more, obviously, remains to be said. I will be satisfied if, in initiating a dialogue between these two pioneers of transpersonal psychology, I have contributed something to a theoretical articulation of the "metaframework" alluded to by Grof in his discussion of the ideal of a "higher sanity." For it is to the actualization of such an ideal that both James and Grof, and we who continue to be inspired by their research and speculations, have devoted the better part of our labor, a labor which, though not without its own travails, is sweet to those who seek a second birth.

Grof and Whitehead

Visions of a Postmodern Cosmology

JOHN BUCHANAN

Introduction

Over the last half of the twentieth century, the research findings and theories of transpersonal psychology in general, and the work of Stanislav Grof, in particular, helped raise to consciousness elements of experience that have been increasingly obscured by the relentless trajectory of modern science and Western civilization. Scientific experimentation, clinical research, and experiential explorations in the areas of parapsychology, psychedelic studies, thanatology, shamanistic and ritual practices, and other types of exceptional experiences provide strong evidential justification for including the spiritual dimension of human existence among the regions of reality that must be taken into account when constructing our cosmological models. This is important not only because these experiences reflect important elements of reality, but also for the practical and compelling reason that this spiritual dimension holds our best hope for the kind of psychospiritual renaissance necessary to guide and sustain us through the impending global crises. In *Beyond the Brain,* Grof envisions this fundamental change in our understanding of the universe in terms of a major *paradigm shift:* "The central message of this book is that Western science is approaching a paradigm shift of unprecedented proportions, one that will change our concepts of reality and of human nature, bridge the gap between ancient wisdom and modern science, and reconcile the difference between Eastern spirituality and Western pragmatism."[1]

In several of his writings, Grof has explored the parameters of an overarching framework that might serve to integrate the discoveries of contemporary science, as well as other modern domains of knowledge, with the findings of transpersonal psychology, the wisdom of premodern religious traditions, and other spiritual sources. In *Beyond the Brain,* Grof concludes his analysis of this problem by suggesting that the *holonomic* approach—where every part in some important way embodies the whole system—holds great promise in this regard, even though it does not in itself constitute a paradigm as such.[2] In *The Cosmic Game,* Grof takes a somewhat different tack by describing and extrapolating upon "the extraordinary philosophical, metaphysical, and spiritual insights that have emerged in the course of this work."[3] In these two books—which represent Grof's most complete statements of his general philosophical views—many important ideas and proposals are put forth concerning the nature of reality, the relationship between science and religion, the meaning of existence, the problem of evil, and ultimate cosmic principles. These ideas are the general subject matter of this essay. My purpose, however, is to suggest that Alfred North Whitehead's *process philosophy* offers a compelling metaphysical and cosmological framework for the new paradigm that Grof is seeking to articulate—a framework which can faithfully incorporate the theories and findings of transpersonal psychology and integrate them into the emerging worldview of a postmodern science and civilization. While Whitehead's ideas have exerted considerable influence on many important twentieth-century thinkers, his philosophy is still relatively unknown and underappreciated outside of specialized philosophical and theological circles. This essay is an attempt to convey to a wider audience some of Whitehead's insights that are particularly relevant to transpersonal theory, and also to suggest that transpersonal psychology has a significant contribution to make to a Whiteheadian cosmology.

Whitehead's Speculative Philosophy

Whitehead characterizes his mature philosophical system as *speculative philosophy*—that is, "the endeavor to frame a coherent, logical, necessary system of general ideas in terms of which every element of our experience can be interpreted."[4] (Whitehead himself called his final system the *philosophy of organism;* today his position is more generally referred to as *process* philosophy, which is the term I use in this essay.) The purpose of such an all-encompassing theory is *not*—at least for Whitehead—to reduce or limit the bounds of reality, but rather to provide a glimpse into the fundamental structure of the universe. This in turn creates a general scheme of ideas that can be applied across disciplines and fields of experience, facilitating cross-fertilization of thought and supplying a shared language of interpretation.[5] The possibility of coherent communication between disciplines is critical to a field such as transpersonal psychology, whose theories and research are so easily perceived as inherently foreign to more traditional areas of study. If a framework of ideas is going to emerge that can serve effectively as the foundation for the New Paradigm envisioned by Grof and others, I would argue that *some* type of speculative philosophy, as defined by Whitehead, is exactly what is required.[6]

While some may find such an overarching theory distastefully monolithic and contrary to an open or pluralistic understanding of the broad range of spiritual experiences described in transpersonal psychology's historical, cross-cultural, experimental, and experiential investigations, it seems to me that an integrative approach of this kind is precisely what is called for to fulfill Grof's anticipation of a future paradigmatic synthesis. Grof writes: "when this new vision of the cosmos is completed, it will not be a simple return to prescientific understanding of reality, but an overarching creative synthesis of the best of the past and the present. A worldview preserving all the achievements of modern science and, at the same time, reintroducing into the Western civilization the spiritual values that it has lost, could have profound influence on our individual, as well as collective life."[7] To accomplish a "creative synthesis" of this scope—a cross-cultural and historical integration of scientific and spiritual values, methods, and theories—demands as powerful and far-reaching a system of ideas as can possibly be found. Whitehead's philosophy is just such a system, and is intended for this precise purpose. "Philosophy," writes Whitehead, "frees itself from the taint of ineffectiveness by its close relations with religion and with science, natural and sociological. It achieves its chief importance by fusing the two, namely, religion and science, into one rational scheme of thought."[8] Although the feasibility of this type of synthetic, philosophic project has been held in ever-increasing disrepute since the Kantian critique of metaphysics, Whitehead was able not only to show a way around Hume's and Kant's enormously influential attacks on the very possibility of metaphysics, but also to master enough fields of thought so as to be able to create a twentieth-century synthetic vision of the nature of reality and the cosmos.

With a grounding in logic, mathematics, the sciences, aesthetics, and William James' psychology and philosophy of experience, along with its primary mission to coherently integrate science and religion, Whitehead's philosophy certainly seems well-positioned to make a substantial contribution to Grof's New Paradigm. Of course, all the above would be moot if Whitehead's process philosophy was not sympathetic to the worldview of transpersonal psychology, so let me immediately turn to that issue.

Grof and Whitehead: Limits of Modern Thought

Grof's critique of the foundations of modern science helps to highlight a number of important perspectives held in common by Grof and Whitehead, as well as some crucial nuances of difference.[9] In the first place, both are critical of the scientific establishment for retarding the advancement of ideas because of prior *philosophical* commitments to a particular worldview, as is illustrated in this indictment by Grof: "Many scientists use the conceptual framework of contemporary science in a way that resembles a fundamentalist religion more than it does science. They mistake it for a definitive description of reality and authoritatively implement it to censor and suppress all observations that challenge its basic assumptions."[10] Whitehead is equally adamant in taking modern science to task for its obstruction of progress towards truth: "Nothing is more curious than the self-satisfied dogmatism with which mankind at each period of its history cherishes the delusion of the finality of its existing modes of knowledge. Skeptics and believers are all like. At this moment scientists and skeptics are the leading dogmatists. Advance in detail is admitted: fundamental novelty is barred. This dogmatic common sense is the death of philosophic adventure."[11]

More particularly, Grof and Whitehead both point to John Locke's empiricism as a key juncture in the history of ideas. Locke's sensationalist credo, in Grof's rendering: "There is nothing in the intellect that had not first been processed through the senses,"[12] especially as developed and extended by Hume, leads to a series of unfortunate consequences for both understanding the spiritual dimension of reality and for speculative and metaphysical endeavors in general.[13]

Furthermore, when conscious sense perception is determined to be the only avenue of evidence for science and philosophy, whole domains of human experience become scientifically unimportant, or epiphenomenal at best, and thus carry no weight in serious formulations of the nature of reality. This generates the "thin" view of the universe that James rails against in *A Pluralistic Universe* and, according to Grof's analysis, leads to an explaining away and a pathologizing of mystical and other spiritual experiences.[14] Thus a major point of agreement between Grof and Whitehead is that *all* human experience must be considered evidence about the nature of reality and weighed carefully in our philosophic and scientific attempts to construct models of the universe.

It is important to note that this evidence from human experience is not taken merely to reveal facts about human nature itself; the position being presented here is revolutionary only if it argues that human experience has direct access to data that convey real information about the universe beyond our local selves (our individual bodies and "minds")—and this is exactly what Grof and Whitehead are claiming. We are speaking here of information *beyond* that conveyed by ordinary sense perception, which implies a type of perception denied in principle by most of modern science. Oddly enough, this Humean-based science itself has no way of philosophically accounting for the data of *normal* sensory perception regarding either its arrival in conscious awareness or its ability to provide information about the physical universe. All we have, according to Hume, are images and ideas in the mind with no way of knowing how or even whether they are connected to something beyond our experience. We have the *data* of sense perception, but no objective relations. In the following, Whitehead draws out the disastrous implications of the "empirical" method of modern science as based solely on observations through normal sensory perception: "Science conceived as resting on mere sense perception, with no other source of observation, is bankrupt, so far as concerns its claim to self-sufficiency. Science can find no individual enjoyment in nature: Science can find no aim in nature: Science can find no creativity in nature; it finds mere rules of succession. ...The reason for this blindness of physical science lies in the fact that such science only deals with half the evidence provided by human experience."[15] Both transpersonal psychology and process philosophy are keenly interested in accounting for this *other half* of the evidence.

Grof's extensive research into exceptional states of awareness has led him to conclude that *holotropic* states—that is, certain types of highly, psychically-open or spiritually-attuned non-ordinary experiences—offer access to a heightened range and greater depth of consciousness, which theoretically can provide experience and information about any other aspect of the universe.[16] Grof's theory of the psychical and spiritual potential of non-ordinary states is highly complementary with process philosophy's cónceptualization of a *second mode of perception,* one more primary than conscious sense perception, which Whitehead refers to as *perception in the mode of causal efficacy*. This alternative mode of perception represents a *direct inner connection* to the feelings and energies of all past actualities, thus providing not only an explanatory basis for the exceptional information received in holotropic states of awareness, but more generally accounting for how we have a real and direct experiential interaction with the rest of the universe. As a direct intuitive connection between the psyche and the brain's neural events (and the rest of the events in the universe as well), "perception in the mode of causal efficacy" is foundational for both normal sensory perception and the kind of psychical perceptions found in parapsychological and transpersonal experiences, thereby explaining both sensory *and* non-sensory perception with a single set of categories.

By providing an *experiential* mode of connection between all actual events in the universe, Whitehead's novel theory of perception bears immediately on such vexing philosophical problems as mind-body interaction and escapes what Santayana referred to as the "solipsism of the present moment," both of which arise unavoidably out of Hume's doubt and Kant's answer (that is, that we have no knowledge of things-in-themselves and thus of the world outside us).[17] Whitehead's notion of a more fundamental, psychical mode of perception receives greater attention later in this essay, but first, it is important to understand how Whitehead and Grof compare in their approaches to the nature of consciousness or subjectivity.

Grof's critique of modern science's views on consciousness focuses on two related points: its claims that consciousness evolved out of material or non-mental substances and that human consciousness is completely derivative from neurological processes.[18] Whitehead agrees with Grof that it is metaphysically incoherent to suggest that experiential events arise out of totally non-experiential substances, be it within the evolutionary process, the embryonic process, or within the processes of the human brain. To claim that this shift from "matter to consciousness"—from physical in the sense of insentient to mental in the sense of experiential—occurs via "emergence" of new properties, is to confuse the appearance of new qualities *within* a single type of metaphysical substance with the sudden appearance of an *entirely new kind* of metaphysical substance. For example, for water to emerge out of the combination of hydrogen and oxygen atoms is categorically distinct from an experiencing event arising out of the combination of many neurological elements that are in themselves devoid of any subjective content. Subjective experience is not a quality that can emerge out of non-experiencing substances: either "physical matter" is in some sense experiential in its fundamental make-up (Whitehead's approach) or "consciousness" has always been constitutive of reality (Grof's approach). While these two positions may sound similar, Whitehead's method places emphasis on the primacy of relatively *unconscious* experience, while much transpersonal theory tends to describe reality as a derivative of, or emanating from, a pervasive or depth source of Consciousness (which, however, is normally inaccessible to ordinary consciousness in the "hylotropic" mode).

Thus, despite their shared belief that experience is central to the metaphysical nature of reality, there are some important theoretical differences concerning the manner in which Whitehead and Grof understand the *subjectivity* that lies at the heart of each of their systems. Grof conceives of consciousness as existing in both an everyday mode (hylotropic) and an expanded mode (holotropic)—within which "human beings can also function as vast fields of consciousness, transcending the limitations of the physical body, of Newtonian time and space, and of linear causality."[19] At "its farthest reaches," the human psyche "is essentially commensurate with all of existence and ultimately identical with the cosmic principle itself."[20] Furthermore, Grof argues that modern consciousness research supports the so-called perennial philosophy in its view that "all of existence is permeated by superior intelligence." A Whiteheadian could, with certain qualifications regarding terminology, readily support Grof's conclusions up to this point. However, in the following statement, Grof seems to argue that only the Universal Mind (Absolute Consciousness) is fully actual and thus our experience of individuality is, in the final analysis, illusory insofar as we are ignorant of our true Identity as the One Consciousness: "All the roles in the cosmic drama;" as Grof puts it, "have ultimately only one protagonist, Absolute Consciousness. This is the single most important truth about existence revealed in the ancient Indian Upanishads."[21]

While process philosophy's understanding of reality is compatible with most, if not all, of the non-ordinary experiences described by Grof to support his endorsement of the Upanishadic view of consciousness, it would question the philosophic validity and wisdom of such monistic conclusions. Having saved human experience, and the experience of the universe at large, from the epiphenomenal clutches of a materialistic, mechanistic scientism, do we really want to immediately relegate human experience to the realm of illusion, to the status of an epiphenomenal manifestation of Universal Mind/Absolute Consciousness?

If this *is* the Truth, then we must face it boldly. But what if there is another way of looking at the evidence of transpersonal psychology that not only gives it full fare in the cosmological order, but is also able to preserve more fully the value, individuality, and freedom of all entities in the universe? Whitehead's process philosophy provides us with an understanding of reality that recognizes the central metaphysical role of subjectivity, and of God, but that also assigns ontological reality and value to all individuals, or experiences, as they are in and for themselves—as well as for their importance to others and to God. Let me next try to briefly outline how this is so with a rudimentary explication of some relevant aspects of Whitehead's philosophy.

Whitehead's Panexperientialism

David Ray Griffin has characterized Whitehead's metaphysics as a version of *panexperientialism,* one holding that all actualities are made up of moments (pulses) of experience or "feeling." Both human experience and human feeling have been abstracted from and generalized by Whitehead in order to describe the basic kind of experience or subjectivity that he believes is pervasive throughout the universe. The relation between these moments or "occasions" of experience is described in a famous dictum: "The many become one, and are increased by one."[22] The *many* past events, or moments of experience, are creatively conjoined to form a new *one* (that is, a new "pulse" of experiential unification), which then joins the array of previously created experiential events as *one more* objective experience waiting to be synthesized into future occasions of experience. This process of creating new actualities out of the accomplished past, referred to by Whitehead as the *creative advance,* is the core principle of his mature philosophy: "The ultimate metaphysical principle," Whitehead asserts, "is the advance from disjunction to conjunction, creating a novel entity other than the entities given in disjunction."[23] This deceptively simple formulation entails many fascinating implications that, however, require some exegesis, as Whitehead's philosophy is both highly original and complex. I will be able to touch on only a few central ideas that are most critical for understanding Whitehead's relevance to transpersonal psychology.

First, for the concerns of this essay, it is important to realize that these moments of experience are composed of mostly *unconscious* experiential activity. Electronic, atomic, molecular, and cellular "actual occasions" (moments or pulses of experience) operate at a level of experience that, by human standards of feeling, would be considered *completely* unconscious. Even in the case of highly complex entities, such as human beings, many of their moments of experience involve no normally perceptible dimension of consciousness; when consciousness *is* a factor, it accounts for only a small portion of the overall experience, most of which occurs at an unconscious level. For in Whitehead's scheme, a moment of experience begins with a flood of unconscious feelings of the past that must be selected, highlighted, and integrated into a unique "drop" of reality. Consciousness results from comparing or contrasting feelings from different stages of this integrative process. Thus conscious awareness can arise only near the end of this synthetic process of experiential unification. Consciousness is the "crown of experience."[24]

I think the vast majority of psychologists today would agree with Whitehead's analysis of human experience as functioning primarily at an unconscious level. This has become standard psychological theory. I believe that even many transpersonal psychologists would concur with this account, while at the same time agreeing with Grof and Wilber that, *ultimately,* the universe is made up entirely of pure or absolute consciousness. They would share Grof's and Wilber's view that while unconscious experience is real at one level of analysis, at a higher level of reality "everything is consciousness" in some sense. Whitehead takes seriously the notion of unconscious feeling or experience and does not want to diminish reality by suggesting that it is in some way illusory or ultimately "something else." While placing experience and subjectivity at the center of his theory of the nature of reality, his metaphysical cosmology also holds that most of the universe is composed of experience lacking any significant conscious awareness.

Whitehead's panexperientialist position presents a challenge to some of Grof's characterizations of transpersonal experiences, such as where his language seems at times to attribute some form of "consciousness" to the plant world or the realm of inorganic matter, or when he states that consciousness is "a primary principle of existence" that "plays a critical role in the creation of the phenomenal world."[25] And process philosophy must take serious exception to Ken Wilber's tendencies towards characterizing *consciousness* as the general exemplification of the "interiority" or "the within of things,"[26] or of equating "Consciousness" with the "pure Emptiness" or Openness of enlightenment,[27] as all of these interpretations misconstrue the metaphysical significance of conscious feeling. Overall, however, I believe Whitehead's understanding of the nature of experience is highly compatible with the evidence of transpersonal psychology—allowing for certain modifications in how this evidence is interpreted—while at the same time being more congruent with our everyday experience of the world as composed largely of objects that do not seem to possess consciousness in the human sense of the word.

To help account for the admixture of experiencing subjects and seemingly subjectless objects that populate our world, Whitehead distinguishes between momentary events—which arise through synthesizing past moments of experience—and the social groupings ("societies") that such events form. For example, an electron is considered a 'serially-ordered society'—a single, temporal strand composed of momentary electronic events. The situation is more complex in the case of a cell, where each cellular event is the novel synthesis of especially the atomic and molecular events within the special environment of the cell. Here we have the cell as an "organism of organisms," which also is a serially-ordered society when considered as a temporal strand of cellular events, and which may also be a part of a larger organismic society, such as an animal body.

Social groupings of actualities may also be less hierarchical or intrinsically interconnected. While a chair does constitute a social grouping of atomic and molecular events, process philosophy argues that there is no special event that emerges out of this society to become a synthetic event giving overall subjectivity to the chair. A chair is a mere *aggregate* of events and thus has no "soul," so to speak. This may seem an obvious point, but it illustrates that Whitehead's metaphysics is able to describe how a chair may be composed of experiential events—the atoms and molecules are active centers of unified feelings—yet not be conscious or have a soul, an important point if process philosophy, or the New Paradigm, is to avoid being dismissed as primitive animism.[28] The same argument holds for rocks, rivers, and plants: while there *is* a social feeling among and between the momentary events constituting these societies, there is *not,* in Whitehead's view, a "central occasion" that arises as a unique, higher-order synthesis of the many lower-order events, as is the case in a cell, an animal, a human being, or in other spiritual entities. Thus, while transpersonal experiences may be able to tap into the "social feelings" of a mountain or a tree, and also to access the meaning of the archetype or Idea exemplified by them, from a process perspective it would be misleading to attribute "consciousness" (or a soul) to such entities in-themselves, even though *their* generalized social feelings may appear in *our* conscious awareness. Thus the universe might be thought of as a vast field of *feeling* or experiential activity, in contrast to those theories that consider consciousness the primary feature of reality.[29]

That said, there is another way within the process perspective that Consciousness *does* permeate all of the universe. God, as the Ultimate Actuality, may be seen as *pervading* all of reality with its consciousness, both in the sense of entering into the formation of every occasion, and in receiving and integrating every accomplished event. Thus, in this sense, the universe can be understood as a *field of consciousness,* but it is also an *ocean of feeling* events possessing limited or no consciousness experience. Feeling and experience are pervasive, as is God's conscious synthesis of all actuality—which floods the entire universe—but consciousness is not the defining quality of every subjective process.

Feeling the World

To really enter into the heart of Whitehead's metaphysical system, an understanding of his notion of "prehension" is critical. Prehension is Whitehead's technical term for the kind of fundamental feeling that pervades the cosmos. Whitehead chose the term *feeling* to suggest both the human emotions and bodily sensations that we commonly refer to with this term, and which serve as experiential guideposts to his more generalized metaphysical use of the word. Used in this full metaphysical sense, *prehension* refers to the flow of experience from past entities into the newly arising event. In an evocative phrase, Whitehead writes that prehension can be "conceived as the transference of throbs of emotional energy."[30] Prehension may also be thought of as the "grasping" of past experiences or, more simply, as feeling the feelings of others—the new occasion of experience creates itself by incorporating (feeling) the experiences (feelings) of past events. Thus there is a fundamental *sympathetic* relation that lies at the heart of reality: every entity co-creates itself out of its feelings of the other beings in its world. Prehension also refers to the feelings involved in the synthesizing and integrating activity carried out as the new moment of experience selects, highlights, and harmonizes the influences of the past into a new unity of feeling.

A good analogy for this process of creative synthesis is found in the experience of *insight*. Try to recall the feelings one has during a moment of insight into some problem or issue: the sense of something rising up from the depths of one's experience, the tingling feeling of excitement over something new about to happen, and finally the sense of satisfaction as seemingly unrelated data are pulled together into a novel whole that sheds new light on all its parts. Whitehead uses the term "concrescence" to indicate this "pulling together" of many feelings into a new whole, which is the basic structure of every new moment of experience. In process philosophy, concrescence, "actual occasion," or, in less technical language, the pulsating moment of experience, all refer to this fundamental creative process whereby the newly-arising occasion feels the influences of past actualities and then integrates these feelings into a novel moment of experiential unity. "The many become one, and are increased by one."

This act of feeling, or prehending, past experiential events—perhaps it is better to imagine these past events *flowing into* the new occasion[31]—forms the basis for the new mode of perception articulated by Whitehead's philosophy: *perception in the mode of causal efficacy*.[32] In this mode of perception, elements from past events are felt as flowing directly into the internal constitution of the new event. A transpersonal example of this would be telepathy, understood as one person's thoughts or feelings flowing directly into another person's experience. Whitehead, by the way, acknowledges that telepathy—"feeling at a distance"—is a real possibility within the metaphysical parameters of his philosophy.[33]

Whitehead's *evidence* for perception in the mode of causal efficacy has an interesting correlation within the field of psychology. He argues that very short-term memory and bodily sensations provide a direct sense of the causal influence of other actualities—precisely the kind of evidence whose existence is denied by Hume. What Whitehead is suggesting is that we are not limited to mere passive awareness of sensory data. Not only is our immediate perceptual experience endowed with a sense of meaning and reality that exceeds a sheer presentation of sensory data, our here-and-now awareness includes *within it* a sense of derivation from both our immediate past experience and from our body. Our thoughts, and words, flow along together into coherent phrases that derive their meaning and impetus from real past intentions inherent within them; we feel auditory sounds as *coming through* our ears, and visual images as *received through* the eyes; we feel the pain in our stubbed toe as *arising within* the toe and flowing into our consciousness; we *share* our body's sexual excitement. It is not mere coincidence that psychology, especially depth psychology, has been focused largely on memories, repressed feelings, and the body's energies. As the primary experiential sources of perception in the mode of causal efficacy, these regions of experience provide access to the depths of our being—and to the universe beyond us as well.

Whitehead's philosophical innovation in proposing the category of *prehension* is revolutionary: it brings together, under one general concept, the notions of perception, energy, causality, memory, and emotion[34]—all as manifestations of past events flowing into a new occasion of experience. Particularly significant for transpersonal psychology, this "prehensive perception" allows for the possibility of *direct access* to realities that lie beyond our everyday sensory ways of knowing the world.

Prehending past events, that is, perception in the mode of causal efficacy, represents an ideal *mode of access* for the kind of information and experiences that occur in non-ordinary or holotropic states. In a Whiteheadian cosmology, each new moment of experience is fundamentally *open* to *all* past experience.[35] As Griffin makes clear, however, this mode of perception is not a type of special psychical intuition that is "higher" than conscious sensory awareness, but rather is a more *fundamental* mode of perception.[36]

The reason this type of perception is viewed as extraordinary—aside from its complete theoretical divergence from mainstream modern philosophy and science—relates to the difficulties of bringing information obtained through the mode of causal efficacy clearly into conscious awareness. To see exactly why this is so, we must understand a little bit about Whitehead's theory of the *phases of concrescence.*

The Structure and Depth of Human Experience

When examined genetically, that is, from inside the event itself, the newly arising moment of experience (the 'concrescence') can be broken down into several phases of integrative activity. In the first phase of concrescence, feelings from past events flow in, or are "grasped," by the newly-forming occasion. Here, the past events exert a *causal* influence: thus, Whitehead's term "perception in the mode of causal efficacy." This first phase of experience is necessarily *unconscious,* as it precedes the later phases of synthesis that include the kind of integrative activity that alone can provoke consciousness. Consciousness requires the selective and synthetic activities of the later phases of concrescence and, in particular, a special *contrast* of feelings derived from comparing primitive causal feelings with complex feelings relating to the same primitive elements.[37] Thus conscious sensory perception and intentional thought and initiative occur primarily in the highest or latest phases of experience, while the feelings that produce a sense of connection to the aliveness of nature, to the full presence of others, to the vitality of the human body, and to God and the spiritual realm, occur in the unconscious depths of the earliest phases of experiential integration. Therefore, these primitive causal feelings of aliveness and connection tend to be obscured and trivialized by the higher mental activity and thought processes that dominate the later phases. This process, whereby conscious thought and sense perception obscure or override our deeper and more fundamental intuitions of reality, is facilitated and reinforced by the habitual feeling patterns of the ego structure that try to canalize the flow of primitive feeling into safe and familiar channels of awareness.

In short, consciousness tends to neglect those elements of experience that are embedded within our primitive but direct contact with reality, and to highlight late-arising and derivative aspects such as normal sense perception and "thinking."[38] From a Whiteheadian perspective, holotropic states are precisely those experiential patterns that tend to reverse this situation, moving awareness away from its preoccupation with the sensory world and ego concerns and shifting the illumination of consciousness "back" into the depths of experience, thus allowing feelings from the mode of causal efficacy to flow more fully into awareness. In essence, I believe that holotropic or non-ordinary experiences are produced by this enhanced flow of primitive feeling into conscious awareness.

A key difference between non-ordinary states in general, and the holotropic states that Grof defines as revelatory of deeper levels of reality, may relate to how successfully this heightened flow of feeling is synthesized in the higher levels of experience. A sheer flood of feeling may be overwhelming, delusory, or simply confusing, unless it is *recanalized* into integrative patterns and symbols that are appropriate forms for the novel flow of feelings and the data they carry. This enhanced flow itself is related to a relaxing of habitual ego structures, which normally act to filter, distort, and obstruct the natural flow of unconscious feeling and the information contained therein—in other words, there is a lifting of *repression*. However, it should be pointed out that from a process perspective, some kind of selection process—and thus some type of "repression" or limitation of feeling—*must* occur in all actualities (except perhaps for God), since no finite occasion could synthesize the entire past universe into full conscious awareness.

In a process cosmology, the *depth unconscious* is conceived as the whole past universe of events: *the psyche's depth is the past.* Whitehead provides a philosophical grounding for Grof's similar conclusions about the nature of the human unconscious, which Grof sees as extending from the biographical-recollective, to the perinatal, and finally into the collective and transpersonal levels of the unconscious.

A Whiteheadian theory of the human unconscious would start with the fact that while the entire past universe forms the ultimate range of the depth unconscious, human moments of experience tend to be dominated by two factors: feelings from the neurons of the central nervous system as they mediate bodily feelings and sensory input; and past moments of experience of the human psyche. The latter appears both as memories of past events and, perhaps more significantly, as the inherited structural patterns of feeling that define ego identity and all the associated habitual modes of thought, emotion, and behavior—not to mention patterns of repression, which also involve chronic holding patterns within the bodily matrix. Just as the body retains its habitual repression patterns through chronic muscle tension, the unconscious psychic components of the ego structure are recreated moment-to-moment as the new human-level occasion of experience *prehends* its past moments and recasts itself in similar patterns of feeling (or non-feeling). These unconscious habitual psychic patterns can be envisioned as relatively stable "templates" or "matrices" operating within the concrescent structure that *mould* the raw experience of the past into predictably bounded forms of conscious awareness. Memories of past personal moments, inherited psychical structural patterns, and the bodily and neural feelings, together constitute a personal level of the unconscious that correlates with Grof's biographical-recollective and perinatal realms of the unconscious.

Of course, the perinatal realm as described by Grof is not limited to the biographical components of the birth process, but also tends to include, or open up into, the archetypal and transpersonal realms. As I have already tried to indicate, when the personal level of the unconscious is transcended—that is, when the habitual unconscious patterns of ego structure, and the related preoccupation with sensory data, personal memories, and normal bodily sensations are relaxed and released as primary organizing principles for moment to moment awareness—then the Whiteheadian metaphysical insight that every occasion of experience is in its essence *fully open* to the entire past comes into play. For when the field of attention is partially freed up in this manner, the vague but direct intuitions of feelings from beyond the personal realm can begin to flow up from the unconscious depths into conscious awareness. And in a process cosmology, the thoughts and feelings of other minds, the vibrant and living energy of the world around us, events from any time or place in the past, and the presence and feelings of spiritual beings, and of God, are all flowing in through the depths of our unconscious synthesis of the universe in every moment. Here, then, is the crux of the Whiteheadian metaphysical basis for Grof's transpersonal realm of the unconscious: the inherent openness of each moment of human experience to the entire past universe of feeling through perception in the mode of causal efficacy.

The relationship between transpersonal "openness" and Grof's notion of ego depotentiation and ego death is apparent within this Whiteheadian analysis. The ego's ongoing sense of identity is created largely by its habitual style or pattern of regulating and shaping the flow of unconscious feeling so as to produce familiar forms of conscious awareness. Transpersonal experiences tend to arise through either a breakdown, or an intentional relaxing, of these habitual patterns of unconscious feeling that are the underlying source of one's sense of ego identity. Thus the same type of shift in concrescent structure that makes possible the kind of openness necessary for transpersonal experiences also produces a correlative perception of a "loss of self," ranging from a pleasant sense of ego-lessness to a terrifying feeling of ego death or destruction. Waiting on just the other side of this breakdown of ego defenses or self-structures, however, is a much deeper and richer connection to the universe.

This view of the human unconscious as being capable of receiving feelings from all past events and exerting an influence upon future occasions means that the human psyche itself should be considered a *psychical organ of perception and projection,* intrinsically endowed with the potential for a multitude of parapsychological and other exceptional abilities. Psychical and mystical experiences, therefore, can be understood to be *natural* phenomena rather than supernatural; they represent the unfamiliar extremes of a single unified field or experiential range. The task, then, is to learn how to understand and properly access these potentials. Non-ordinary states of consciousness, especially as researched and mapped by Grof, represent a major tool in this ongoing investigation.

Conclusion

In concluding *The Cosmic Game,* Grof speaks of the ideas he has presented as providing "an overarching metaframework for the phenomena constituting consensus reality."[39] I have argued here that Whitehead's process philosophy can contribute significantly to this "metaframework" or New Paradigm by offering a sophisticated and complementary metaphysical system capable not only of grounding transpersonal psychology within a coherent theory of experience, but also of providing a systematic connection between transpersonal thought and the social and physical sciences, religion, and our everyday life. Whitehead's belief that "there are wider truths and finer perspectives within which a reconciliation of a deeper religion and a more subtle science will be found"[40] comes close to a definition of transpersonal psychology itself; and Grof's description of the New Paradigm fits process thought to a tee: "The updated model shows the universe as a unified and indivisible web of events and relationships; its parts represent different aspects and patterns of one integral process of unimaginable complexity."[41] Together, Grof's transpersonal psychology and Whitehead's process philosophy form the basis for a new cosmological vision for our world.

Let me summarize some of the ways that I have found Whitehead's ideas to be extremely congenial to a transpersonal orientation. This summary also serves as a recapitulation of this essay's introduction to Whiteheadian thought:

1. The Whiteheadian universe is composed of "actual occasions" or moments of synthetic experience: thus experience lies at the *heart* of reality; the universe is "alive" and subjective; all actualities are composed metaphysically of the same "stuff," yet have their own individual reality.
2. Whitehead's highly original notion of "prehension" means that the universe has an essential *interconnectedness,* that *feeling* is primary, and that there is a fundamental *openness* to Being.
3. Whitehead's hypothesis of "perception in the mode of causal efficacy" implies that there exists a more primary mode of intuition than physical sense perception; that we are in touch with a *real world;* that all actualities possess a basic, psychic connection to all other existing entities; that certain parapsychological powers are inherent potentials arising out of the metaphysical structure of reality.
4. The "creative advance" introduces *process,* and an evolutionary trend, into the very nature of the universe, thereby reversing Newton's bias toward the static and Eternal. This process view—whereby the momentary subjective events "perish" on their attainment of actuality, only to become the past "objects" of experience for the next generation of newly-arising subjective moments—overcomes many previously intractable dualisms, including the subject-object dichotomy and the bifurcation of nature into mind and matter.
5. *Creativity,* as the metaphysical ultimate of ultimates, means that novelty and freedom are central to the nature of reality. *Every* experience is inherently valuable just as it is—as well as for its contributions to future actualities and to God.

Whitehead's process philosophy offers a strong and sympathetic foundation for Grof's transpersonal psychology, providing independent philosophical and metaphysical support and grounding for many of Grof's theories, such as a form of memory existing independently of neural storage mechanisms, a mode of transpersonal intuition transcending normal temporal and spatial boundaries, and a universe composed entirely of modes of experience. Also, and perhaps most importantly, Whitehead's epistemology—which shows how at a very deep level of experience there exists a direct, *unmediated* perception of reality—demonstrates how it is philosophically conceivable for transpersonal experiences to provide real evidence about the nature of the universe, evidence that is not simply reducible to individual and cultural factors or "filters."

On the other hand, Grof's research has produced empirical evidence that appears to support a number of Whitehead's more controversial notions, including "perception in the mode of causal efficacy," God's dual nature, the reality of a pervasive "aliveness" or "subjectivity" throughout nature, and the existence of "memory without a material substrate"[42] that parallels the Whiteheadian interpretation of memory as the direct prehension of past experiences of the human psyche or soul. More generally, Grof's vast evidence from observations of holotropic states has demonstrated scientifically that mystical and transpersonal experiences are not limited to a few extraordinary individuals, but rather are a widespread and generally accessible phenomenon of human nature. This evidence goes a long way towards addressing Whitehead's reticence concerning the inclusion of mystical experience in constructing a cosmology, on the grounds that speculative philosophy must be founded on stable and widespread features of human experience.[43] Finally, Grof's revisioning of the universe in terms of a vaster, spiritual cosmology shares a great sympathy with Whitehead's ideas, despite some differences in emphasis. Grof's view of the cosmos is much richer in its variety of spiritual beings and realms, however, and this is an area where a process cosmology can productively deepen itself by drawing on Grof's cartography of non-ordinary states. For example, Grof's cartography of the human psyche reveals a universe characterized by parapsychological phenomena, an enspirited nature, multiple realms of spiritual beings, something like a collective unconscious, and the possibility of a wealth of mystical encounters with God and other ultimate levels of reality.

On the other hand, a Whiteheadian would question both the interpretation of certain types of transpersonal experiences and some of the broader implications drawn by Grof concerning the ultimate nature of things. For example, even though reliable intuitions of future events are conceivable within a Whiteheadian metaphysics, the future here is understood as only *partially* determined—thus it would be impossible to have absolute knowledge of a future event. Perfect precognition, as in apodictically knowing the future ahead of time, would conflict with Whitehead's basic notion of temporal flow. The future is *defined* as that which is not yet actual. Past influences may conspire to generate a probable future, but the actual future remains mere possibility (or probability) until it actually happens. While future events may be imperfectly intuited by various means, they cannot be known absolutely.

More generally speaking, process philosophy is diametrically opposed to interpretations or theories that entail complete predestination, total determinism, or any other metaphysical position that contradicts our overwhelming awareness of the reality of freedom and responsibility in our everyday lives. This, of course, does not close the book on the subject. There are many factors and possibilities that make the topic of freedom and responsibility a complex one, especially when viewed from a transpersonal perspective. But I believe that our starting point, at least, should be a philosophy that recognizes the fundamental importance of individual creativity, responsibility, and freedom—both existentially and metaphysically.

Another point, which we have already encountered, is that Whitehead's panexperientialism describes most actualities as consisting totally, or primarily, of *unconscious* feeling, versus the transpersonal community's tendency to speak of consciousness as a ubiquitous phenomenon. Thus a process interpretation would refrain from using categories such as "Consciousness of Inorganic Matter," preferring Grof's formulations such as "Plant Identification" or "Animal Identification,"[44] which avoid the appearance of attributing a human-like consciousness to lower-level entities that might better be described in terms of more primitive feeling[45] (which is not to say that many animals may not be capable of significant levels of conscious awareness).[46]

My reason for pursuing this point with such force is that I believe this metaphysical distinction between *feeling* and *consciousness* is critical to our understanding of reality and ourselves. Let me summarize why, in non-technical language. First, it draws our attention to the underlying *depths* of experience as yet unrealized in conscious awareness, and away from the narrow boundaries of everyday consciousness. Second, "feeling" provides a sense of the process that is central to *all* actualities in the universe: *feeling* is our fundamental mode of interconnection with others and is the core of our true nature. And, third, the notion of feeling is highly suggestive of the aliveness and vibrancy that we encounter in our emotions, in our bodies, and in our vital experience of other entities and the universe.

Finally, Grof's framing of ultimate matters in terms of a *cosmic drama*, where Absolute Consciousness seeks solace, entertainment, creative possibility, and variety through the act of dividing undifferentiated Being into the many isolated bits of consciousness that constitute the universe,[47] holds more affinities to Whitehead's perspective than I have so far indicated. Whitehead describes God's involvement with the universal process in terms of quite similar kinds of motivations, and every moment of experience does in a sense separate itself from its initial connection with God and the universe through the process of its self-creative activity. However, a process metaphysics invests full reality and value in each moment of becoming and thus takes a more cosmologically concrete approach to the problem of evil and the importance of the adventure of the universe in its evolution. In Griffin's excellent phrase: the world is a "locus of value-realization."[48] All actualities are of value in themselves and for others, as well as for God. This is not illusion.

Part and parcel with a mystical connection to God comes a sense of one's place in the unfolding "creative advance" and the importance of one's contributions to the future of *all* entities in the universe. Religious experiences of God instill *world loyalty*. Process philosophy understands the *illusory* elements of reality not so much in terms of losing track of one's true Identity with Absolute Consciousness, but rather in terms of our imperfect perception of the universe as it truly is, and our valuing the enduring elements of ego more highly than our deepest metaphysical identity: the creative moment of experience itself—harmoniously aligned with the divine aims.

Henri Corbin writes: "A philosophy that does not culminate in a metaphysic of ecstasy is vain speculation; a mystical experience that is not grounded on a sound philosophical education is in danger of degenerating and going astray."[49] Process philosophy furnishes an intellectual basis for consciousness to expand beyond its limited perspective and into its own aboriginal depths of experience; transpersonal psychology provides various experiential modes of access capable of illuminating these normally unconscious elements. A fusion of Whiteheadian and transpersonal theories could provide a foundation for a new cosmology for our world and offer social direction and orientation through a unified perspective on philosophy, science, and religion. By joining the depth insights and psychospiritual methodology of transpersonal psychology with the metaphysical system and cosmological vision of process philosophy, we may obtain at least part of what will be needed to carry us through the impending cultural, environmental, and spiritual crises and into an entirely new mode of relating and being in the universe—and with each other.

Subtle Connections

Grof, Jung, and the Quantum Vacuum

ERVIN LASZLO

Are human beings entirely discrete individuals, their organism enclosed by the skin and their minds enclosed by the cranium housing the brain? Or are there effective, if subtle, interconnections between humans—and between humans and the world at large? This study argues that the latter assumption is likely to be true. Though the evidence for "subtle connections" is not in the form of incontrovertible "hard data," it is nevertheless cogent and significant. The directly pertinent findings are generated by research on psi-phenomena and the practice of psychotherapists. Possible explanations for the findings can be traced to the ideas of C.G. Jung and are now pursued at the leading edge of the physical sciences.

The Findings: Psi Experiments

Controlled experiments concerning subtle connections between subjects removed in space, and occasionally also in time, date back to the 1930s, to J.B. Rhine's pioneering card- and dice-guessing work at Duke University. Since then, experimental designs have become sophisticated and experimental controls rigorous; physicists have often joined psychologists in carrying out the tests. Explanations in terms of hidden sensory cues, machine bias, cheating by subjects, and experimenter error or incompetence have all been considered, but they were found unable to account for a number of statistically significant results.[1]

Relevant work began in the 1970s, when Russell Targ and Harold Puthoff carried out some of the best-known experiments on subtle connections among distant subjects in regard to the transference of thoughts and images.[2] They examined the possibility of telepathic transmission between individuals, one of whom would act as "sender" and the other as "receiver." The receiver was placed in a sealed, opaque, and electrically shielded chamber, while the sender was in another room where he or she was subjected to bright flashes of light at regular intervals. Electroencephalograph (EEG) machines registered the brain-wave patterns of both. As expected, the sender exhibited the rhythmic brain waves that normally accompany exposure to bright flashes of light. But, after a brief interval the receiver also began to produce the same patterns, although he or she was not exposed to the flashes and was not receiving sense-perceivable signals from the sender.

Targ and Puthoff also conducted experiments on remote viewing. In these tests, sender and receiver were separated by distances that precluded any form of sensory communication between them. At a site chosen at random, the sender acted as a "beacon;" the receiver then tried to pick up what the beacon saw. To document his or her impressions, the receiver gave verbal descriptions, at times accompanied by sketches. Independent judges found that the descriptions of the sketches matched on the average 66% of the time the characteristics of the site that was actually seen by the beacon.[3]

Remote viewing experiments reported from other laboratories involved distances from half a mile to several thousand miles. Regardless of where they were carried out, and by whom, the success rate was generally around fifty percent—considerably above random probability. The most successful viewers appeared to be those who were relaxed, attentive, and meditative. They reported that they received a preliminary impression as a gentle and fleeting form, which gradually evolved into an integrated image. They experienced the image as a surprise, both because it was clear and because it was clearly elsewhere.

Images could also be transmitted while the receiver was asleep. Over several decades, Stanley Krippner and his associates carried out "dream ESP experiments" at the Dream Laboratory of Maimonides Hospital in New York City.[4] The experiments followed a simple yet effective protocol. The volunteer, who would spend the night at the laboratory, would meet the sender and the experimenters on arrival and had the procedure explained to him or her. Electrodes were then attached to the volunteer's head to monitor brain waves and eye movements; there was no further sensory contact with the sender until the next morning. One of the experimenters threw dice that, in combination with a random number table, gave a number that corresponded to a sealed envelope containing an art print. The envelope was opened when the sender reached his or her private room in a distant part of the hospital. The sender then spent the night concentrating on the print.

The experimenters woke the volunteers by intercom when the monitor showed the end of a period of rapid-eye-movement (REM) sleep. The subject was then asked to describe any dream he or she might have had before awakening. The comments were recorded, together with the contents of an interview the next morning when the subject was asked to associate with the remembered dreams. The interview was conducted double-blind—neither the subject nor the experimenters knew which art print had been selected the night before.

Using data taken from the first night that each volunteer spent at the dream laboratory, the series of experiments between 1964 and 1969 produced 62 nights of data for analysis. They exhibited a significant correlation between the art print selected for a given night and the recipient's dreams on that night. The score was considerably higher on nights when there were few or no electrical storms in the area, and sunspot activity was at a low ebb—that is, when the Earth's geomagnetic field was relatively undisturbed.

A particularly striking example of transpersonal contact and communication can be found in the work of Jacobo Grinberg-Zylverbaum at the National University of Mexico.[5] In more than fifty experiments performed over five years, Grinberg-Zylberbaum paired his subjects inside sound and electromagnetic radiation-proof "Faraday cages." He asked them to meditate together for twenty minutes. Then he placed the subjects in separate Faraday cages where one of them was stimulated and the other not. The stimulated subject received stimuli at random intervals in such a way that neither he nor she nor the experimenter knew when they were applied. The non-stimulated subject remained relaxed, with eyes closed, and was instructed to feel the presence of the partner without knowing anything about his or her stimulation.

In general, a series of one hundred stimuli were applied—flashes of light, sounds, or short, intense but not painful electric shocks to the index and ring fingers of the right hand. The EEG of both subjects was then synchronized and examined for "normal" potentials evoked in the stimulated subject and "transferred" potentials in the non-stimulated subject. Transferred potentials were not found in control situations where there was no stimulated subject, where a screen prevented the stimulated subject from perceiving the stimuli (such as light flashes), or when the paired subjects did not previously interact. However, in experimental situations with stimulated subjects and with interaction, the transferred potentials appeared consistently in some 25% of the cases. A particularly poignant example was furnished by a young couple, deeply in love. Their EEG patterns remained closely synchronized throughout the experiment, in consonance with their report of deep feelings of oneness.

In a limited way, Grinberg-Zylberbaum could also replicate his results. When a subject exhibited the transferred potentials in one experiment, he or she usually exhibited them in subsequent experiments as well.

A related experiment investigated the degree of harmonization of the left and right hemispheres of the subject's neocortex. In ordinary waking consciousness the two hemispheres—the language-oriented, linearly thinking, rational "left brain" and the gestalt-perceiving intuitive "right brain"—exhibit uncoordinated, randomly diverging wave patterns in the electroencelograph. When the subject enters a meditative state of consciousness, these patterns become synchronized, and in deep meditation, the two hemispheres fall into a nearly identical pattern. In deep meditation, not only the left and right brains of one and the same subject, but also the left and right brains of different subjects manifest identical patterns. Experiments with up to twelve subjects simultaneously showed an astonishing synchronization of the brain waves of the entire group.[6]

In the past few years, experiments such as these have been matched by hundreds of others. They provide significant evidence that identifiable and consistent electrical signals occur in the brain of one person when a second person—especially if he or she is closely related or emotionally linked—is either meditating, provided with sensory stimulation, or attempts to communicate with the subject intentionally.[7]

Interpersonal connection beyond the sensory range can also occur outside the laboratory; it is particularly frequent among identical twins. In many cases, one twin feels the pain suffered by the other and is aware of traumas and crises, even if he or she is halfway around the world. Besides "twin pain," the sensitivity of mothers and lovers is equally noteworthy: countless stories are recounted of mothers having known when their son or daughter was in grave danger or was actually involved in an accident.

Interpersonal connection is not limited to twins, mothers, and lovers—the kind of closeness that a therapeutic relationship creates between therapist and patient seems also to suffice. A number of psychotherapists have noted that, during a session, they experience memories, feelings, attitudes, and associations that are outside the normal scope of their experience and personality. At the time these strange manifestations are experienced, they are indistinguishable from the memories, feelings, and related sentiments of the patients themselves. It is only later, on reflection, that they come to realize that the anomalous items stem not from their own life and experience but from their patient.

It appears that in the course of the therapeutic relationship, some aspect of the patient's psyche is projected into the mind of the therapist. In that location, at least for a limited time, it integrates with the therapist's own psyche and produces an awareness of some of the patient's memories, feelings, and associations. Known as "transference," this phenomenon can be useful in the context of therapy: it can permit the patient to view what was previously a painful element in his or her personal consciousness more objectively, as if it belonged to somebody else.

Actual physical manifestations seem also capable of being transmitted from one individual to another. Transmissions of this kind came to be known as "telesomatic": they consist of physiological changes that are triggered in the targeted person by the mental processes of another. The distance between the individuals involved seems to make little or no difference. William Braud and Marilyn Schlitz carried out hundreds of trials regarding the impact of the mental imagery of senders on the physiology of receivers—the latter were distant and unaware that such imagery was being directed to them. They claim that the mental images of the sender can "reach out" over space and cause changes in the physiology of the distant receiver—effects comparable to those one's own mental processes produce in one's own body. People who attempt to influence their own bodily functions are only slightly more effective than those who attempt to influence the physiology of others from a distance. In several cases involving a large number of individuals, the difference between remote influence and self-influence was almost insignificant: "telesomatic" influence by a distant person proved to be nearly as effective as "psychosomatic" influence by the same person.[8]

The Findings: Grof's Experience with Altered States of Consciousness

Complementing psi-experiments in regard to the ability of the human mind to penetrate beyond the limits of personal sensory experience are the findings of modern psychotherapists. The pertinent evidence comes clearly to the fore in the work of Stanislav Grof. In reviewing findings gathered in the course of many decades, Grof suggests that standard cartographies of the human mind need to be complemented with additional elements. To the standard "biographical-recollective" domain of the psyche, we should add a "perinatal" and a "transpersonal" domain. The transpersonal domain, it appears, can mediate connection between our mind and practically any part or aspect of the phenomenal world.[9]

Grof's experience derives from work with non-ordinary states of consciousness (NOSCs), induced in his patients either by psychedelic drugs or holotropic breathing. NOSCs embrace a large part of the human psyche; the states of normal waking consciousness are but the tip of the iceberg. As over a hundred years ago William James had noted, "Our normal waking consciousness...is but one special type of consciousness, whilst all about it, parted from it by the filmiest of screens, there lie potential forms of consciousness entirely different. We may go through life without suspecting their existence; but apply the requisite stimulus, and at a touch they are all there in all their completeness."[10] People in "primitive" and classical cultures knew how to apply the requisite stimulus—some tribes, such as the !Kung Bushmen of the Kalahari desert, could enter altered states all at the same time.

In many parts of the world, ancient peoples combined chanting, breathing, drumming, rhythmic dancing, fasting, social and sensory isolation, even specific forms of physical pain to induce altered states. The native cultures of Africa and pre-Colombian America used them in shamanic procedures, healing ceremonies and rites of passage; the high-cultures of Asia used them in various systems of yoga, Vipassana or Zen Buddhism, Tibetan Vajrayana, Taoism, and Sufism. The Semitic cultures used them in Cabala, the ancient Egyptians in the temple initiations of Isis and Osiris; the classical Greeks in Bacchanalia, and the rites of Attis and Adonis as well as in the Eleusinian mysteries. Until the advent of Western industrial civilization, almost all cultures held such states in high esteem for the remarkable experiences they convey and the powers of personal healing and interpersonal contact and communication they render accessible.[11]

Today, at the leading edge of the contemporary sciences, research on non-ordinary states of consciousness is becoming accepted as a legitimate part of the new discipline known as "consciousness research." The insight that surfaces is, as Charles Tart noted, that these states tend to make our connections to each other and to our environment more evident. Grof's records of the verbal reports of his patients make this very clear.[12]

In the experience of "dual unity," a patient in a NOSC experiences a loosening and melting of the boundaries of the body ego and a sense of merging with another person in a state of unity and oneness. In this experience, despite the feeling of being fused with another, the patient retains an awareness of his or her own identity. Then, in the experience of "identification with other persons," the patient, while merging experientially with another person, has a sense of complete identification to the point of losing the awareness of his or her own identity. Identification is total and complex, involving body image, physical sensations, emotional reactions and attitudes, thought processes, memories, facial expression, typical gestures and mannerisms, postures, movement, and even the inflection of the voice. The "other" (or others) can be someone in the presence of the patient or someone absent; he or she can be part of an experience from the subject's childhood, his or her ancestry, or even of a previous lifetime.

In "group identification and group consciousness" there is a further extension of consciousness and melting of ego boundaries. Rather than identifying with individual persons, the patient has a sense of becoming an entire group of people who share some racial, cultural, national, ideological, political, or professional characteristics. The depth, scope, and intensity of this experience can reach extraordinary proportions: people may experience the totality of suffering of all the soldiers who have ever died on the battlefield since the beginning of history, the desire of revolutionaries of all ages to overthrow a tyrant, or the love, tenderness, and dedication of all mothers in regard to their babies. Identification can focus on a social or political group, the people of an entire country or continent, all members of a race, or all believers of a religion.

"Identification with animals" goes beyond the human transpersonal dimension: it involves a complete and realistic identification with members of various animal species. The experience can be authentic and convincing, including body image, specific physiological sensations, instinctual drives, unique perceptions of the environment, and the corresponding emotional reactions. The nature and scope of these experiences distinguish them from ordinary human experiences; they often transcend the scope of fantasy and imagination.

While less frequent, "identification with plants and botanical processes" occurs as well. On occasion, patients have a complex experience of becoming a tree, a wild or garden flower, a carnivorous plant, kelp, *Volvox globator*, plankton in the ocean, a bacterial culture, or an individual bacterium. In the still more embracing experience of "oneness with life and all creation," an individual expands his or her consciousness to such an extent that it encompasses the totality of life on this planet, including all of humanity and all flora and fauna of the biosphere. Instead of identification with one living organism, the patient identifies with life itself as a cosmic phenomenon.

Experience in NOSCs can also penetrate beyond the sphere of life: it can include the macroscopic and microscopic phenomena of the inorganic world. In the "experience of inanimate matter and inorganic processes," patients report experiential identification with the waters of rivers and oceans, with various forms of fire, with the earth and with mountains, and with the forces unleashed in natural catastrophes such as electric storms, earthquakes, tornadoes, and volcanic eruptions. They can identify with specific materials, such as diamonds and other precious stones, quartz crystals, amber, granite, iron, steel, quicksilver, silver, and gold. The experiences extend into the microworld and may involve the dynamic structure of molecules and atoms, Brownian motions, interatomic bonds, electromagnetic forces, and subatomic particles. Grof concludes that every process in the universe that in an ordinary state of consciousness can be objectively observed, can also be subjectively experienced in a non-ordinary state.

The cosmic dimensions of the experiences in NOSC can encompass all of the planet Earth. In "planetary consciousness," the subject's consciousness expands to the Earth's geological substance with its mineral kingdom and its biosphere with all its life forms. The Earth as a whole appears to be one complex organism, oriented toward its own evolution, integration, and self-actualization. In "extraterrestrial experiences"—a further expanded form of consciousness—other celestial bodies and astronomical processes are included. The subject can experience traveling to the moon, sun, other planets, stars, and galaxies; he or she can experience explosions of supernovas, contraction of stars, quasars and pulsars, even passage through black holes. The experience can occur in the form of simply witnessing such events, or of actually becoming them, experiencing them intimately, as if being a part of the experienced thing or event. At the widest (and comparatively rare) form of this experience—"identification with the entire physical universe"—the subject has the feeling that his or her consciousness encompasses the entire cosmos. All its processes are experienced as part of the organism and psyche of the all-encompassing universe-system.

In addition to the spatially expanded forms of consciousness, there are experiences that recall OBEs (out-of-body experiences), clairvoyance, clairaudience, and telepathy. More relevant for our purposes are experiences involving a displacement in time. Time-displacement experiences range from "embryonal and fetal experiences," where the subject recalls his or her intrauterine life as a fetus, through "ancestral experiences" involving identification with one's biological ancestors, "racial and collective experiences" where those involved are not one's direct ancestors but members of the same race, or sometimes the entire human species (suggestive of Jung's "collective unconscious" of which more will be said later), all the way to "past incarnation experiences." The essential characteristic of the latter is a convincing sense of remembering something that had already happened in the history of the same person. Subjects maintain their sense of individuality and personal identity, but experience themselves in another form, at another place and time, and in another context. In these reincarnation-type experiences, the birth of the individual appears as a point of transformation, where the enduring record of multiple lifetimes enters the bio-psychological life of the individual.

According to Grof, the memories that surface in past incarnation experiences share with other transpersonal experiences the capacity to provide instant and direct extrasensory access to information about some aspect of the world. If such experiences are veridical, all divisions and boundaries in the universe are in some sense illusory and arbitrary; in the last analysis, it is only a cosmic consciousness that actually exists.[13]

Toward an Explanation: Jung's Unus Mundus

What explanation can we give for the varied yet remarkably consistent phenomena unearthed in controlled psi-experiments and in the work of Grof and other psychotherapists with patients in altered states of consciousness? Just what is the nature of the "cosmic consciousness"—or similar factor—that would connect our psyche with the world at large?

Carl Jung, fascinated with this seemingly esoteric aspect of the human psyche, attempted an explanation in terms of a higher or deeper reality that would connect human minds with each other as well as with physical reality. He was led to his explanatory concept by a comparison of unconscious processes in individuals with the myths, legends, and folktales of a variety of cultures at various periods of history. Jung found that the individual records and the collective material contain common themes. This prompted him to postulate the existence of a collective aspect of the psyche: the "collective unconscious." The dynamic principles that organize this material are the "archetypes." Archetypes are irrepresentable in themselves, but have empirical manifestations: these are the archetypal images and ideas. "The archetype as such," Jung wrote, "is a psychoid factor that belongs, as it were, to the invisible, ultraviolet end of the psychic spectrum. It does not appear, in itself, to be capable of reaching consciousness."[14]

While in the realm of the spirit, at the upper, "ultraviolet" end of the psychic spectrum, archetypes are dynamic organizers of ideas and images, at the lower, "infrared" end of the spectrum, the biological instinctual psyche shades into the physiology of the organism, merging with its chemical and physical conditions. As Jung noted, "the position of the archetype would be located beyond the psychic sphere, analogous to the position of physiological instinct, which is immediately rooted in the stuff of the organism and, with its psychoid nature, forms the bridge to matter in general."[15]

Jung formulated his concept of the archetype in collaboration with quantum physicist Wolfgang Pauli. He was struck by the fact that while his own research into the human psyche led to an encounter with such "irrepresentables" as the archetypes, research in quantum physics had likewise led to "irrepresentables": the micro-particles of the physical universe, entities for which no complete description appeared possible.

Jung concluded: "When the existence of two or more irrepresentables is assumed, there is always the possibility—which we tend to overlook—that it may not be a question of two or more factors but of one only."[16] The single factor that underlies the irrepresentables of physics and of psychology may be the same as that which underlies the synchronicities Jung had investigated: meaningful coincidences that tie together in an acausal connectedness the physical and the psychological worlds. Jung named the common factor that would underlie and connect these worlds the *unus mundus* ("unitary world"). The foundation for the *unus mundus* is "that the multiplicity of the empirical world rests on an underlying unity, and that not two or more fundamentally different worlds exist side-by-side or are mingled with one another."[17] As Charles Card summarized, "The realms of mind and of matter—*psyche* and *physis*—are complementary aspects of the same transcendental reality, the *unus mundus*. Archetypes act as the fundamental dynamical patterns whose various representations characterize all processes, whether mental or physical. In the realm of the *psyche*, archetypes organize images and ideas. In the realm of *physis*, they organize the structure and transformations of matter and energy, and they account for acausal ordering as well. Archetypes acting simultaneously in both the realms of *psyche* and *physis* account for instances of synchronistic phenomena."[18]

Jung relates the subtle connections that appear in synchronistic events involving the psyche of different individuals, as well as the psyche of one person and the physical world around that person, to the underlying archetypal reality of the *unus mundus*, which is itself neither psychic nor physical—it stands above or lies beyond, both *psyche* and *physis*.

Toward an Explanation: The Quantum Vacuum

Jung's concept points the way toward a fruitful avenue of research: a deeper reality that connects mind and mind, and mind and matter. This approach should enter the current stream of consciousness research. For the present, most researchers seek an explanation of mental events mainly in terms of physical processes in the brain. But henceforth, the mental events needing explanation should include not only the workings of the individual brain but, in light of the findings of psi-experimenters and psychotherapists, the subtle connections that link human brains with each other and with the world at large.

It seems likely that world and brain—and more importantly, cosmos and consciousness—are interconnected by a continuous information-conserving and transmitting *field*. Such a field cannot be postulated in an *ad hoc* manner—science must respect the law laid down by William of Occam in the 14th century: entities are not to be multiplied beyond necessity. New entities—which can also be forces or fields—can only be postulated when doing so is the *simplest*, the most *economical,* and the most *rational* way of explaining a given set of findings and observations.

A field that constitutes the simplest, the most economical, and rational explanation of the current findings may exist. In line with my own thinking, David Bohm suggested that it is the as yet imperfectly understood "zero-point field" (ZPF) that seems present throughout the quantum vacuum. In the following, we shall explore what is known about this field of the vacuum, what is currently hypothesized about it, and how it could account for the subtle interconnections noted above.

Received knowledge about the vacuum—In quantum physics, the quantum vacuum is defined as the lowest energy state of a system of which the equations obey wave mechanics and special relativity. It is considerably more than just the state of a system, however. It is the locus of a vast energy field that is neither classically electromagnetic nor gravitational, nor yet nuclear in nature. Instead, it is the originating source of the known electromagnetic, gravitational, and nuclear forces and fields. It is the originating source of matter itself.

The technical definitions of the quantum vacuum point to a continuous energy sea in which particles of matter are specific substructures. According to Paul Dirac's calculation, all particles in positive energy states have negative-energy counterparts (by now, such "antiparticles" have been found experimentally for all presently known particles). The zero-point field of the quantum vacuum is a "Dirac-sea": a sea of particles in the negative energy state. These particles are not observable—physicists call them "virtual." But they are not fictional for all that. By stimulating the negative energy states of the ZPF with sufficient energy (of the order of 10^{-27} erg), a particular region of it can be "kicked" into the real (that is, observable) state of positive energy. This is the process known as pair-creation: out of the vacuum emerges a positive energy (real) particle, with a negative energy (virtual) particle remaining in it. Thus, the Dirac-sea is everywhere; the observable universe floats, as it were, on its surface.

The quantum vacuum contains a staggering density of energy. John Wheeler estimated its matter-equivalent at 10^{94} gram per cm^3—and that is more than all the matter in the universe put together. Compared with this energy density, the energy of the nucleus of the atom—the most energetic chunk of matter in the known universe—seems almost minuscule: it is "merely" 10^{14} gram/cm^3.

The vacuum itself is not material: its zero-point energies—which, according to David Bohm, exceed all the energies bound in matter 10^{40} times—are in the negative state. This is fortunate, for if they were not, the universe would instantly collapse to a size smaller than the radius of an atom. (This follows from $E = mc^2$, Einstein's celebrated mass-energy equivalence relation: energy corresponds to mass, and mass, in turn, entails gravitation.)

Because the "real" world of matter—that is, of energy bound in mass—is so much less energetic than the vacuum, the observable universe is not a solid condensate floating on top of the vacuum, but like a set of bubbles suspended in it. In terms of energy, the material world is not a *solidification* of the quantum vacuum, but a *thinning* of it.

Speculations on the vacuum—A thin line divides what is already known and accepted about the quantum vacuum and what is still speculative and controversial. Here we review the relevant explorations: those that concern interactions between the observable world of matter-energy and the vacuum's zero-point energies.

The world of matter and the quantum vacuum are known to interact. For example, under certain conditions the vacuum's zero-point energies act on electrons orbiting atomic nuclei. The effects occur when electrons "jump" from one energy state to another: the photons they emit exhibit the so-called Lamb-shift (a frequency slightly shifted from its normal value). Vacuum energies also create a radiation pressure on two closely spaced metal plates. Between the plates some wavelengths of the vacuum field are excluded, thereby reducing its energy density with respect to the field outside. This creates a pressure—known as the Casimir effect—that pushes the plates inward and together.

Other interactions may exist as well. Some years ago, Hungarian physicist Lajos Jánossy assigned "relativistic effects" (such as the slowing down of clocks when accelerated close to the speed of light, or the increasing of the mass of objects at those velocities) to the interaction of real-world objects with the vacuum's energy field. Close to the speed of light the matter-particles of objects rub against the force-particles (bosons) of the vacuum, and this friction slows down their processes and increases their mass. In this concept, the ZPF of the vacuum is a physical field that interacts with the objects that move in space and time.

Currently, another Hungarian, maverick theoretician László Gazdag, developed Jánossy's concept into a full-fledged "post-relativity theory."[19] In his theory, the vacuum's energy field has the properties of a superfluid. It is known that in supercooled helium, all resistance and friction ceases; it moves through narrow cracks and capillaries without loss of momentum. Conversely, objects move through the fluid without encountering resistance. (Since electrons also move through it without resistance, superfluids are also superconductors.) Thus, in a sense, a superconducting superfluid is not "there" for the objects or electrons that move through it—they get no information about its presence. This could explain why we, and even our most sensitive instruments, fail to register its presence.

In Gazdag's reinterpretation of Einstein's relativity theory, the celebrated formulas describe the flow of bosons in the superfluid ZPF. This flow is what determines the geometrical structure of spacetime, and hence the trajectory of real-world photons and electrons. When particles of light and matter move uniformly, spacetime is Euclidean; when they are accelerated, the ZPF interacts with their motion. Then spacetime appears curved. (As Russian physicist Piotr Kapitza noted, in a superfluid only those objects move without friction that are in constant quasi-uniform motion. If an object is strongly accelerated, vortices are created in the medium, and these vortices produce resistance: the classical interaction effects surface.)

Front-line research in physics confirms the basic notion that underlies these assumptions. Current work follows up a suggestion made by physicists Paul Davies and William Unruh in the mid-1970s. Davies and Unruh, like Jánossy and Gazdag, based their argument on the difference between constant-speed and accelerated motion in the vacuum's zero-point field. Constant-speed motion would exhibit the vacuum's spectrum as isotropic (the same in all directions), whereas accelerated motion would produce a thermal radiation that breaks open the directional symmetry. The "Davies-Unruh effect," too small to be measured with physical instruments, prompted scientists to investigate whether accelerated motion through the vacuum field would produce incremental effects. This expectation has borne fruit. It turned out that the inertial force itself could be due to interactions in that field.

In 1994 Bernhard Haisch, Alfonso Rueda, and Harold Puthoff gave a mathematical demonstration that inertia can be considered a vacuum-based Lorentz-force.[20] The force originates at the subparticle level and produces opposition to the acceleration of material objects. The accelerated motion of objects through the vacuum produces a magnetic field, and the particles that constitute the objects are deflected by this field. The larger the object, the more particles it contains, hence the stronger the deflection—and greater the inertia. Inertia is thus a form of electromagnetic resistance arising in accelerated frames from the distortion of the zero-point (and otherwise superfluid) field of the vacuum.

More than inertia, mass also appears to be a product of vacuum inter-action. If Haisch and collaborators are right, the concept of mass is neither fundamental nor even necessary in physics. When the massless electric charges of the vacuum (the bosons that make up the superfluid zero-point field) interact with the electromagnetic field, beyond the already noted threshold of energy, mass is effectively "created." Thus mass may be a structure condensed from vacuum energy, rather than a fundamental given in the universe.

If mass is a product of vacuum energy, so is gravitation. Gravity, as we know, is always associated with mass, obeying the inverse square law (it drops off proportionately to the square of the distance between the gravitating masses). Hence, if mass is produced in interaction with the ZPF, then also the force that is associated with mass must be so produced. This, however, means that *all* the fundamental characteristics we normally associate with matter are vacuum field-interaction products: inertia, mass, as well as gravity.

Regarding the full scale of interactions between vacuum energies and the micro- as well as macro-world of matter-energy, the work of a group of Russian physicists is of particular significance. Anatoly Akimov, G.I. Shipov, V.N. Binghi, and co-workers developed a sophisticated theory of what they call the "physical vacuum."[21] In their theory, the vacuum is a real physical field extending throughout the universe; it registers and transmits the traces of both micro-particles and macro-objects. The theory, which at the time of writing has not been published outside Russia, is important and fascinating enough to merit some further details.

In standard theories, the energetic properties of the quantum vacuum are generally considered in the framework of quantum electrodynamics. This framework gives rise to elegant and relatively simple mathematics. But such formulas, though highly sophisticated, can be misleading. They may not provide the best possible account of physical reality. Stochastic electro-dynamics, for example, produces a "messier" math, but its tenets about the real world may be closer to realistic assumptions about the nature of reality. In any case, quantum electrodynamics, as other scientific theories, can always be reconsidered or extended.

The Russian physicists do not hesitate to undertake this step. They take their cue from earlier work by Einstein. In a seminal treatment, G.I. Shiphov showed that in accordance with the Clifford-Einstein program of the geometrization of spacetime, the vacuum can be described not only in terms of Riemannian (four-dimensional) curvature but also in terms of Cartan torsion. In the 1920s, studies carried out by Albert Einstein, and E. Cartan laid the foundation of the theory that became subsequently known as the ECT (Einstein-Cartan Theory). The idea stemmed originally from Cartan, who at the beginning of the century speculated about fields generated by angular momentum density. This idea was later elaborated independently by a number of Russian physicists, including N. Myshkin and V. Belyaev. They claim to have discovered the natural manifestations of enduring torsion fields.

Presently Akimov and his team consider the quantum vacuum as a universal torsion wave-carrying medium. The torsion field is said to fill all of space isotropically, including its matter component. It has a quantal structure that is unobservable in non-disturbed states. However, violations of vacuum symmetry and invariance create different and, in principle, observable states.

The torsion field theory takes a modified form of the original electron-positron model of the "Dirac-sea": the vacuum's energy field is viewed as a system of rotating wave packets of electrons and positrons (rather than a sea of electron-positron pairs). Where the wave-packets are mutually embedded, the field is electrically neutral. If the spins of the embedded packets have the opposite sign, the system is compensated not only in charge, but also in classical spin and magnetic moment. Such a system is said to be a "phyton." Dense ensembles of phytons are said to approximate a simplified model of the physical vacuum field.

When the phytons are spin-compensated, their orientation within the ensemble is arbitrary. But when a charge q is the source of disturbance, the action produces a charge polarization of the vacuum, as prescribed by quantum electrodynamics. When a mass m is the source of disturbance, the phytons produce symmetrical oscillations along the axis given by the direction of the disturbance. The vacuum then enters a state characterized by the oscillation of the phytons along their longitudinal spin-polarization; this is interpreted as a gravitational field (G-field). The gravitational field is thus the result of vacuum decompensation arising at its point of polarization—which is an idea that was originally introduced by Sakharov. Given that the gravitational field is characterized by longitudinal waves, it cannot be screened, which is in accordance with observation and experiment. Hence m-disturbance produces the G-field, much as q-disturbance produces the electromagnetic field.

Akimov *et al.* go further. Following a thesis advanced by Roger Penrose, they represent the vacuum equations in the spinor form and thereby obtain a system of nonlinear spinor equations where two-component spinors represent the potentials of torsion fields. These equations can describe charged, as well as neutral, quantum, and classical particles. They thus allow that the vacuum field is disturbed not only by charge and mass, but also by classical spin. In that event the phytons oriented in the same direction as the spin of the disturbance keep their orientation. Those opposite to the spin of the source undergo inversion; then the local region of the vacuum transits into a state of transverse spin polarization. This gives the "spin field" (S-field), viewed as a condensate of fermion pairs.

As a result, Akimov *et al.* view the vacuum as a physical medium that can assume various polarization states. Given charge polarization, the vacuum is manifested as the electromagnetic field. Given matter-polarization, it is manifested as the gravitational field. And given spin-polarization, the vacuum manifests as a spin-field. All fundamental fields known to physics correspond to specific vacuum polarization-states.

The above "torsion-field theory of the physical vacuum" can claim that all objects, from quanta to galaxies, create vortices in the vacuum. The vortices created by particles and other material objects are information carriers, linking physical events quasi-instantaneously. The group-speed of these "torsion-waves" is of the order of 10^9 C—one billion times the speed of light. Since not just physical objects, but also the neurons in our brain create and receive torsion-waves, not only particles are "informed" of each other's presence (as in the famous EPR experiments), but also humans can be so informed: our brain, too, is a vacuum-based "torsion-field transceiver." This suggests a physical explanation not only of quantum non-locality but also of telepathy, remote viewing, and the other telesomatic effects discussed above.[22]

Torsion waves are both superluminal and enduring. Meta-stable "torsion-phantoms" generated by spin-torsion interaction can persist even in the absence of the objects that generated them. The existence of these phantoms has been confirmed in the experiments of Vladimir Poponin and his team at the Institute of Biochemical Physics of the Russian Academy of Sciences.[23] Poponin, who has since repeated the experiment at the Heartmath Institute in the US, placed a sample of a DNA molecule into a temperature-controlled chamber and subjected it to a laser beam. He found that the electromagnetic field around the chamber exhibits a specific structure, more or less as expected. But he also found that this structure persists long after the DNA itself has been removed from the laser-irradiated chamber: the DNA's imprint in the field continues to be present when the DNA is no longer there. Poponin and his collaborators conclude that the experiment shows that a new field structure has been triggered from the physical vacuum. This field is extremely sensitive; it can be excited by a range of energies close to zero. The phantom effect is a manifestation, they claim, of a hitherto overlooked vacuum substructure.

Theories such as those we have cited here foreshadow a major leap in the scientific world picture: the physical foundations of the universe acquire an active role in all its functions and processes. Life, and even mind, is a manifestation of the constant if subtle interaction of the wave-packets classically known as "matter" with the underlying physically real zero-point vacuum field.

If the emerging world picture is to be completed, we must evolve an explicit hypothesis to describe the basic dynamics of the overall range of matter-vacuum interaction. In my "quantum-vacuum interaction (QVI) hypothesis," the non-classical energy field of the vacuum (consisting of scalar as well as electromagnetic wave propagations) registers the spacetime behavior and evolution of matter-energy systems in the form of interfering wavefronts. The conserved interference patterns form a holographic information field accessible to systems with a stereodynamic pattern isomorphic to the systems that produced the patterns.

The applicable process can be described as forward and reverse Fourier (more exactly, Gabor) transforms. Hence matter-energy systems ranging from quanta to complex atomic, molecular, cellular, and multicellular structures, including human brains, decode ("read out") the information they and analogous systems have encoded ("read into") the field. Given that wavefronts superpose in multiple dimensions, the ZPF of the vacuum acts as an information-conserving and transmitting universal holofield, interconnecting systems with each other, as well as with their subsidiary systems (internal parts) and suprasystems (external environments).

Conclusions

The astonishing psi-phenomena that come to light in controlled experiments and the equally astonishing findings of expert psychotherapists cannot be dismissed as mere chimera, figments of a fertile but undisciplined imagination. The findings are part and parcel of the manifestation of human consciousness, the subconscious domains of which extend far beyond the confines of both brain and organism.

The findings may be real, yet their acceptance hinges critically on discovering ways to connect them with the received frameworks of knowledge. As long as there is no conceivable tie between an anomaly and the basic paradigm that frames knowledge in the pertinent field, the anomaly will remain just that: a paradoxical, uncomprehended item, relegated to the back shelf of the science establishment. Recognition of a conceivable tie could, however, make for a significant difference—it could open up feasible avenues of conceptual analysis, theory formulation, and experimental testing. For that reason, likely hypotheses of brain-brain and brain-universe (or, in an alternative terminology, consciousness-consciousness, and consciousness-world) interaction need to be seriously scrutinized, for intrinsic meaningfulness, consistency with observations, as well as mesh with the currently known frameworks of explanation.

In the case discussed above, the scientific validation of the findings would have an additional bonus. Not only would it introduce greater coherence into our world picture—binding together the hitherto anomalous findings of consciousness research with our knowledge of the physical world—it would also introduce greater coherence into human affairs. As thoughtful observers have frequently remarked, many of our current ills are due to the sense of separateness and lack of empathy we experience vis-à-vis our fellow humans and the nonhuman realms of nature (in modern societies, as Woody Allen quipped, "nature and I are two"). The scientist's recognition that we do have deeper ties to each other and to the natural environment could make a significant cultural impact and therewith on the dominant attitudes of the public.

T.S. Eliot asked, "What are the roots that clutch, what branches grow out of this stony rubbish? Son of man you cannot say, or guess, for you know only a heap of broken images."[24] Perhaps the exploration of our subtle ties with each other and with nature could enable us to know more than a heap of broken images. It could help us to recognize Bateson's "pattern that connects": the subtle connecting pattern present in the cosmos and in the biosphere—and likewise in our brain and consciousness. In this regard, Grof's contribution to the understanding of the deeper reality we have good reason to believe underlies our perceptions and apprehensions is both fundamental and monumental.

Grof's Perinatal Matrices and the Sāṃkhya *Guṇas*

Thomas Purton

In this chapter, I would like to explore a connection between the characteristics associated with the *guṇas* of Hinduism's Sāṃkhya school and those of the perinatal matrices discovered by Stanislav Grof. While much of Grof's cosmology and metaphysics is evocative of Hindu thought, his discovery of the perinatal matrices was made independently of any association with Hinduism: it emerged from the identification of basic patterns of experience in deep experiential work, which Grof and his patients then began to relate to the stages of biological birth. I will try to show that Grof's model may provide a corroboration, arrived at through entirely independent means, of a core tenet of the metaphysics of Sāṃkhya. I will not assume any prior knowledge either of Grof's work or of Hinduism.

The *Guṇas* in Sāṃkhya

No concept in Hindu thought has perhaps been as difficult to accommodate into a Western mindset as that of the three *guṇas* (*triguṇa*) of *tamas*, *rajas*, and *sattva*, although such a difficulty is not unique to the West. In its originating school the substratum of an entire metaphysics, *triguṇa* is not to be found in any non-Hindu school of thought.[1] From the perspective of comparative religion, *triguṇa* has always remained isolated as an exclusively Hindu concept, failing to undergo any significant osmosis into other schools. This in itself is worthy of note: it is consistent with the suspicion that a transmission of its true nature and importance was never fully successful.

The metaphysics under which the three *guṇas* find their first full exposition appears to a Westerner no less strange. The Sāṃkhya school in which the *guṇas* emerge divides the universe most fundamentally into *puruṣa*, pure consciousness,[2] and *prakṛti*, what consciousness can be conscious of. *Puruṣa* is what sees and *prakṛti* is what is seen. In this fascinating form of dualism, what is seen is not only what the eyes see but what the mind experiences; not only, even, what the mind experiences but what the mind is. *Puruṣa* is consciousness in its purest possible form, possessing such soteriological significance that it constitutes an entire pole of a dualism.

This is something very different from the split between mind and body Western thought is familiar with, and while an identification with *puruṣa* as the ultimate goal of the spiritual journey might perhaps evoke the Jungian Self, *puruṣa* in itself contains no psyche and no strivings. According to the *Sāṃkhya Kārikā*,[3] in *puruṣa* the inactive appears (through *prakṛti*) as though active; in *prakṛti* the non-conscious appears (through *puruṣa*) as though conscious. *Puruṣa* might be compared to a transparent jewel, *prakṛti* to a colored surface upon which the jewel is placed: no matter what surface the jewel is placed on, it retains its fundamental identity. In the illustration below the bottom half of the jewel appears to be colored, but the color does not belong to the jewel itself.[4]

FIGURE 1 First visual analogy of *puruṣa*, represented by the jewel, and *prakṛti*, represented by the colored surface.

Yet in a certain sense *puruṣa* and *prakṛti* are intertwined in a way that Western mind and matter are not. "For the purpose of *puruṣa*'s aloneness," says the *Sāṃkhya Kārikā*, "the two [come together] like the blind and the lame; that conjunction is creation, emergence."[5] *Prakṛti* is blind, unable to see; *puruṣa* is lame, unable to move; the experience of creation is the result of their cooperation. This mutual dependence strongly implies that we cannot consider *prakṛti* a part of a realist metaphysic. Mikel Burley has argued for Sāṃkhya as instead being a "metaphysics of experience,"[6] in which *prakṛti* is not the matter of materialism but the matter of experience itself.[7] The colored surface here represents the permutations of matter as it is subjectively experienced by consciousness, not an independent material reality.

Every part of this surface can then be understood as specifically some combination of the three *guṇas* mentioned earlier: *tamas*, *rajas*, and *sattva*. Just as every possible color in the spectrum must be some blending of the three primary colors, so too, in this reading of Sāṃkhya, every possible subjective experience, all of the different ways in which the jewel can be illusorily colored, must be of some blending of the three. Partitioning our subjective experience in this way seems highly unintuitive, but part of this chapter's intention is to evoke how, through Grof's work, it may become a little easier to understand how these fundamental building blocks can combine to produce every possible facet of our experience.

The *Guṇas* Elsewhere in Hindu Scripture

The earliest known reference to the three *guṇas* is in the *Vedas,* the texts considered *śruti* (revealed scripture) in Hinduism and orally transmitted since at least the second millennium BCE. The *Atharva Veda Samhita* states:

> Behold now the lotus with nine gates [the orifices of the physical body], encircled by the three *guṇas*.[8]

In the late Vedic *Chāndogya Upaniṣad*, various aspects of the natural world are each decomposed into red (associated with heat), white (associated with water), and black (associated with food), with the following maxim repeated: "The three colors (forms) alone are true."[9] These three forms are stated to comprise all possible permutations of what is thinkable or understandable; having known them, the learned state, "No one can now mention to us anything which we have not heard, thought of, or known."[10]

The *Upaniṣad* notes that "Whatever appeared to be unknown they knew to be the combination of these three deities."[11] The analogy of the colored surface above, with all possible combinations of the color spectrum derived from the three primary colors, is striking here.

The second explicit use of the word *guṇa* that we know of is in the *Śvetāśvatara Upaniṣad*, which (in a separate chapter) also contains the same division of red, white, and black:

> The sages, absorbed in meditation through one-pointedness of mind, discovered the (creative) power, belonging to the Lord Himself and hidden in its own *gunas*.[12]

> There is one unborn [*prakṛti*, or possibly "female"][13]—red, white, and black—which gives birth to many creatures like itself.[14]

The later *Maitrāyaṇīya Upaniṣad* contains the following passage:

> Verily, in the beginning this world was *tamas* alone....When impelled [*īrita*] by the Supreme, that goes on to differentiation [*viṣamatva*]. That form, verily, is *rajas*. That *rajas*, in turn, when impelled, goes on to differentiation. That, verily, is the form of *sattva*.[15]

And in the *Bhagavad Gītā*, to be discussed later in this chapter, Krishna states:

> Whatever forms are produced, O Kaunteya [Arjuna], in all the wombs whatsoever, the great Brahman (*mūlaprakṛti*) is their womb, and I am the seed-giving father. *Sattva*, *rajas*, and *tamas*—these *guṇas*, O! mighty-armed, born of *prakṛti*, bind the indestructible embodied one, fast in the body.[16]

The *Gītā* here contains an explicit allusion to birth. But we might note that evocations of birth seem to be implicit in the above texts also. In the above passage from the *Maitrāyaṇīya* we can consider that in the form described as *rajas,* the baby firstly differentiates from the womb in response to the impelling of the waters breaking; in the form described as *sattva*, it then further differentiates from the maternal body in the moment of actual birth. The *Śvetāśvatara* talks of a female (or *prakṛti* itself) giving birth to many creatures. And if asked to associate the specific colors we might intuitively associate with the entrapment in the dark prison of the womb, the bloody fight through the birth canal, and the emergence into light, we might well arrive at the black, red, and white of the *Chāndogya* and *Śvetāśvatara Upaniṣads*.[17]

It is the Sāṃkhya school, however, that marks the first point in which the *guṇas* find their first systematic placement in a coherent metaphysics, and no explicit or even implicit connections with birth seem to be made here.[18] The *guṇas* in Sāṃkhya are "of the nature of gladness [*sattva*], perturbation [*rajas*] and stupefaction [*tamas*]; serving to illuminate [*sattva*], activate [*rajas*] and restrain [*tamas*]; [they] subjugate, support, generate and combine with one another. *Sattva* is light and illuminating; *rajas* is impelling and moving; *tamas* is heavy and delimiting."[19] These characteristics are intriguing but not very intuitive either as (following the trail of the above *Upaniṣads*) possible allusions to birth, or even as the basic building blocks of the metaphysics of experience discussed above. Under such a metaphysics, *sattva* must surely be *the experience of* light and illumination, *rajas* the experience of being impelled and of movement, and *tamas* the experience of heaviness and delimitation.[20] But the descriptions the *Sāṃkhya Kārikā* provides of the three do not sound particularly intuitive or plausible in this context.[21]

If it could somehow be shown that these characteristics described above are indeed credible as experience's most fundamental constituents, this would both justify and elucidate a metaphysics of experience. Such a demonstration would also do much to explain the pivotal and comprehensive functioning of *triguṇa* in an entire cosmology.[22] As philosophical concepts, the *guṇas* can always be appreciated; but as the three partitions of the entire universe, they can be difficult to apply to our everyday experience of it. This is an appropriate point to introduce Grof's discovery.

Grof's Perinatal Matrices

Grof began his career as a psychoanalytically-trained psychiatrist in Europe in the 1950s, working initially with lysergic acid diethylamide (LSD-25) and, when the substance was banned, a form of accelerated breathing evocative of the breath of fire (*agni prāṇāyāma*) technique in Kuṇḍalinī Yoga. Both techniques appeared to take the experiencer on a journey into scenes from their own personal history, scenes from past lives, and other non-ordinary phenomena. As he continued his work, Grof began to notice that these experiences seemed ultimately to decompose into four basic experiential patterns:

> Pattern One:
>
> Subjects frequently talk about timelessness of the present moment and say that they are in touch with infinity. They refer to this experience as ineffable and emphasize the failure of linguistic symbols and the structure of our language to convey the nature of this event and its significance....An individual can, for example, talk about this experience as being content-less and yet all-containing; everything that he can possibly conceive of seems to be included. He refers to a complete loss of his ego and yet states that this consciousness has expanded to encompass the whole universe.[23]
>
> Pattern Two:
>
> This experience is characterized by a striking *darkness* of the visual field and by ominous colors. Typically, this situation is absolutely unbearable and, at the same time, appears to be endless and hopeless; no escape can be seen either in time or in space....The deepest levels are related to various concepts of hell, to situations of unbearable physical, psychological, and metaphysical suffering that will never end, as they have been depicted by various religions. In a more superficial version of the same experiential pattern, the subject...is selectively aware only of the ugly, evil, and hopeless aspects of existence.[24]
>
> Pattern Three:
>
> The individual experiences sequences of immense condensation of energy and its explosive release and describes feelings of powerful currents of energy streaming through his whole body. The visions typically accompanying these experiences involve scenes of natural disasters and the unleashing of elemental forces....[25]
>
> One important experience...is the encounter with consuming *fire*, which is perceived as having a purifying quality. The

individual who, in the preceding experiences, has discovered all the ugly, disgusting, degrading, and horrifying aspects of his personality finds himself thrown into this fire or deliberately plunges into it and passes through it. The fire appears to destroy everything that is rotten and corrupt in the individual and prepares him for the renewing and rejuvenating experience of rebirth.[26]

Pattern Four:

> After the subject has experienced the very depth of total annihilation and "hit the cosmic bottom," he is struck by visions of blinding *white* or golden light and has the feelings of enormous decompression and expansion of space. The general atmosphere is that of liberation, redemption, love, and forgiveness....He experiences overwhelming love for his follow men, appreciation of warm human relationships, solidarity, and friendship.... Feelings of joy and relief are accompanied by deep emotional and physical relaxation, serenity, and tranquility.[27]

The resonances of the second, third, and fourth patterns with the characteristics of *tamas*, *rajas* and *sattva*, and with the colors of black, red, and white discussed above should be clear. We can see the importance, in making this connection, of understanding Sāṃkhya as a metaphysics of experience: both the *guṇas* and these patterns that Grof identifies are patterns not of an external reality independent of the individual, but of consciousness' very experience of the world. The implication of what Grof discovered here, one consistent with this metaphysics, is that it is *in* such experience that reality is constituted at its most profound and fundamental level.

Regarding Grof's first pattern, it is accepted in all schools of thought in which the *guṇas* are present that there exists a state beyond them. In the dualist Sāṃkhya school the ultimate aim is to liberate *puruṣa* from the *guṇas* completely, an achievement known as *kaivalya*, meaning aloneness or solitariness. In non-dual schools such as Advaita Vedānta the place beyond the *guṇas* is variously referred to as *mokṣa*, *mukti*, or *nirvāṇa*: transcendental states free from the incessant cycles of death and rebirth and characterized by freedom, egolessness, bliss, freedom from sorrow, and the absence of a distinction between self and other.

Importantly for the point of connection being made here, then, the identification by Grof of only three basic experiential patterns would immediately be challenged by those familiar with how the *guṇas* are understood in Hindu scripture. In all known scriptures referencing the *guṇas* there exists a place that is beyond them and transcends them. Including the experience of non-duality or *mokṣa* in addition to the experience of the three *guṇas* is therefore not a contrived addendum performed to accommodate Grof's first pattern, but necessary for an authentic metaphysics of experience based upon the three.[28]

With these four basic patterns identified, Grof continued his journey:

> It was actually people themselves who started relating them to the specific stages of the biological birth process, and so I extracted somehow the experiential patterns from people's accounts, and started referring to them as basic perinatal matrices.[29]

This realization was consistent with what Grof already knew to be the case, which was that in the sessions, patients regressed into earlier and earlier scenes from their lives:

> When our process of deep experiential self-exploration moves beyond the level of memories from childhood and infancy and reaches back to birth, we start encountering emotions and physical sensations of extreme intensity, often surpassing anything we previously considered humanly possible.[30]
>
> Various forms of experiential psychotherapy have amassed convincing evidence that biological birth is the most profound trauma of our life and an event of paramount psychospiritual importance. It is recorded in our memory in miniscule details down to the cellular level and it has a profound effect on our psychological development....
>
> Conscious reliving and integration of the trauma of birth plays an important role in the process of experiential psychotherapy and self-exploration. The experiences originating on the perinatal level of the unconscious appear in four distinct experiential patterns, each of which is characterized by specific emotions, physical feelings, and symbolic imagery. These

patterns are closely related to the experiences that the fetus has before the onset of birth and during the three consecutive stages of biological delivery. At each of these stages, the child experiences a specific and typical set of intense emotions and physical sensations. These experiences leave deep unconscious imprints in the psyche that later in life have an important influence on the life of the individual.[31]

Grof describes these basic perinatal matrices in terms of their relationship to the actual experience of birth as follows:

First Perinatal Matrix (BPM I):

The biological basis of this matrix is the experience of the original symbiotic unity of the fetus with the maternal organism at the time of intrauterine existence. During episodes of undisturbed life in the womb, the conditions of the child can be close to ideal. However, a variety of factors of physical, chemical, biological, and psychological nature can seriously interfere with this state.[32] ... Pleasant and unpleasant intrauterine memories can be experienced in their concrete biological form.[33]

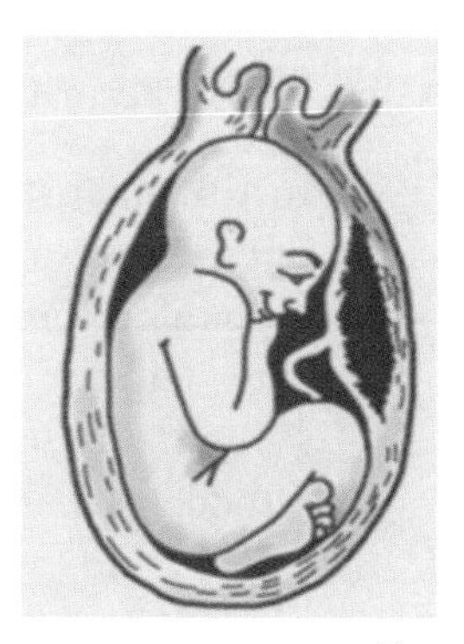

FIGURE 2 BPM I[34]

Second Perinatal Matrix (BPM II):

This experiential pattern is related to the very onset of biological delivery and its first clinical stage. Here the original equilibrium of the intrauterine existence is disturbed, first by alarming chemical signals and then by muscular contractions. When this stage fully develops, the fetus is periodically constricted by uterine spasms; the cervix is closed and the way out is not yet available.[35]

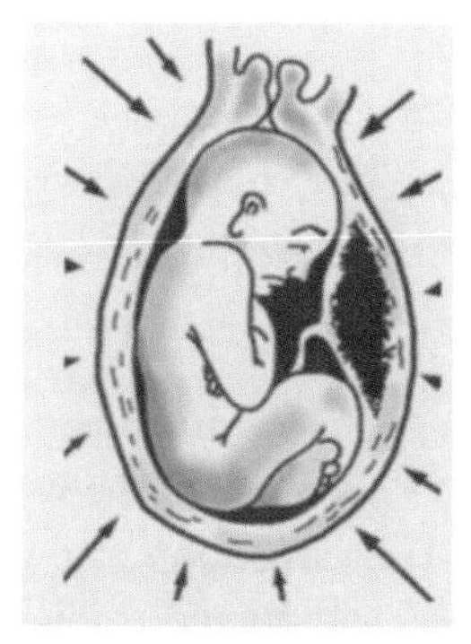

FIGURE 3 BPM II[36]

> Third Perinatal Matrix (BPM III):
>
> In this stage, the uterine contractions continue, but unlike in the previous stage, the cervix is now dilated and allows a gradual propulsion of the fetus through the birth canal. This involves an enormous struggle for survival, crushing mechanical pressures, and often a high degree of anoxia and suffocation.[37]

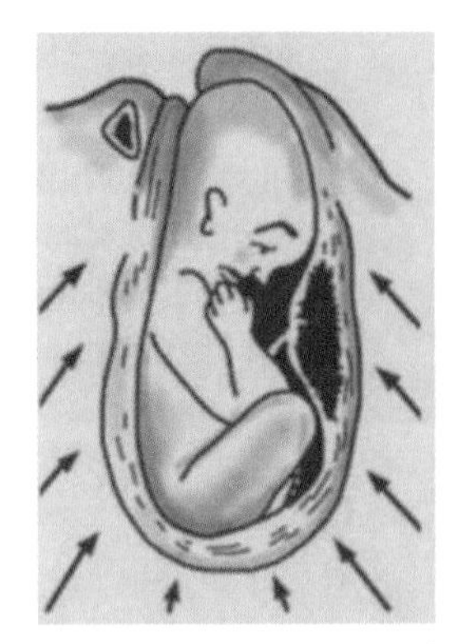

FIGURE 4 BPM III[38]

> Fourth Perinatal Matrix (BPM IV):
>
> In this final stage, the agonizing process of the birth struggle comes to an end; the propulsion through the birth canal culminates and the extreme build-up of pain, tension, and sexual arousal is followed by a sudden relief and relaxation. The child is born and, after a long period of darkness, faces for the first time the intense light of the day.... After the umbilical cord is cut, the physical separation from the mother has been completed and the child begins its new existence as an anatomically independent individual.[39]

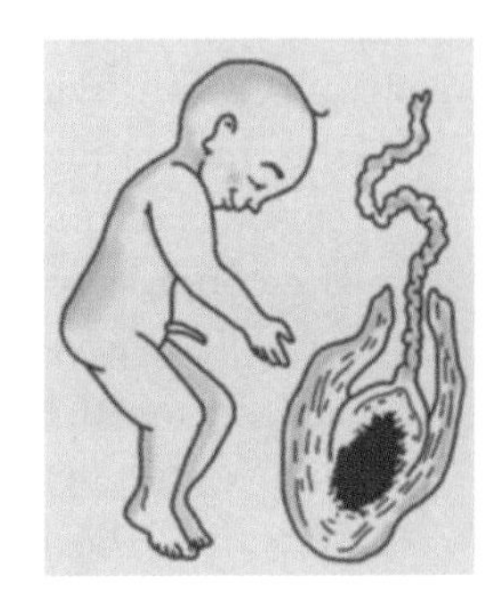

FIGURE 5 BPM IV[40]

The characteristics of *tamas*, *rajas*, and *sattva* described by Sri Aurobindo below can be considered in the light of what is taking place as the baby is born:

> Dominated by *tamas*, man does not so much meet the rush and shock of the world-energies whirling about him and converging upon him as he succumbs to them, is overborne by them, afflicted, subjected ... the tamasic man seeks only somehow to survive, to subsist so long as he may, to shelter himself in the fortress of an established routine of thought and action in which he feels himself to a certain extent protected from the battle....[41]

> The *rajasic* doer of action…is one eagerly attached to the work, bent on its rapid completion, passionately desirous of fruit and reward and consequence, greedy of heart, impure of mind, often violent and cruel and brutal in the means he uses; he cares little whom he injures or how much he injures others so long as he gets what he wants, satisfies his passions and will, vindicates the claims of his ego.[42]

> When into all the doors in the body there comes a flooding of light, as if the doors and windows of a closed house were opened to sunshine, a light of understanding, perception and knowledge—when the intelligence is alert and illumined, the senses quickened, the whole mentality satisfied and full of brightness and the nervous being calmed and filled with an illumined ease and clarity…one should understand that there has been a great increase and uprising of the *sattvic guṇa* in the nature.[43]

For the purposes of clarity, I have isolated the descriptions Grof provides specifically of the experience of birth in his description of each perinatal matrix, but it is important to appreciate that it is only rarely that the experience is entirely that of the individual undergoing birth—or rather that this very experience commingles with the experience of the respective pattern in many other forms. For example, the identification with the fetus floating blissfully in amniotic fluid can also be the identification with the whole universe in BPM I; the experience of the trapped and terrified fetus in BPM II is so intense that it can become that of the concentration camp victim; the fight through the birth canal in BPM III can be of the same bloodlust and frenzy as the soldier engaging in battle; and an experience out in nature with friends can carry the same joy and liberation as the moment of birth in BPM IV.

The following material from Grof's own work, describing a progression from a rajasic state (BPM III) to a sattvic (BPM IV) one contains multiple evocations of the scriptures quoted from in this chapter:

> The session started with an incredible upsurge of instinctual forces. Waves of orgasmic sexual feelings alternated or combined with aggressive outbursts of immense power. I felt trapped by steel-like machinery threatening to choke me to

death, yet mesmerized and carried along by this irresistible outpouring of life energies. My visual field was glowing with a spectrum of red colors that had an awesome and numinous quality. I somehow sensed that it symbolized the mystical power of blood, uniting humanity in strange ways throughout the ages. I felt connected with the metaphysical dimensions of cruelties of all kinds—torture, rape and murder—but also to the mystery of the menstrual cycle, birth, delivery, death... The underlying theme behind all this seemed to be a profound identification with an infant's struggle to free himself from the clutches of the birth canal....

All of a sudden, I seemed to be losing all my connections to reality, as if some imaginary rug was pulled from under my feet....Filled with indescribable horror, I saw a gigantic figure of a deity towering over me in a threatening pose. I somehow instinctively recognized that this was the Hindu god Śiva in his destructive aspect. I felt the thunderous impact of his enormous foot that crushed me, shattered me to smithereens, and smeared me like an insignificant piece of excrement all over what I felt was the bottom of the cosmos.[44]

In the next moment, I was facing a terrifying giant figure of a dark goddess whom I identified as the Indian Kālī. My face was being pushed by an irresistible force toward her gaping vagina that was full of what seemed to be menstrual blood or repulsive afterbirth. I sensed that what was demanded of me was absolute surrender to the forces of existence and to the feminine principle represented by the goddess. I had no choice but to kiss and lick her vulva in utmost submission and humility. At this moment, which was the ultimate and final end of any feeling of male supremacy I had ever harbored, I connected with the memory of the moment of my biological birth. My head was emerging from the birth canal with my mouth in close contact with the bleeding maternal vagina.

I was flooded with the divine light of supernatural radiance and beauty whose rays were exploded into thousands of exquisite peacock designs. From this brilliant golden light emerged a figure of a Great Mother Goddess who seemed to embody love and protection of all ages. She spread her arms and reached toward me, enveloping me into her essence. I merged with this incredible energy field, feeling purged, healed, and nourished.[45]

What takes place here evokes a movement from the emanation of Kālī as she is more generally known, as ferocious and frightening, to the emanation known in the Śaivite Trika tradition as Parā, loving and nourishing. In this tradition, Kālī exists in three forms. Parā is white, beautiful, and benevolent: "She is [white] like the best crystal and she pours out nectar everywhere."[46] Parā is worshipped both as one of the three, and as their sum and source as the "essence of the mothers."[47] Aparā is black; Parāparā, between Parā and Aparā, is red;[48] both have the same appearance: "Her tongue flickers in and out like lightning....Her mouth yawns wide and at its corners are terrible fangs. Ferocious, with her brows knitted in rage, wearing a sacred thread in the form of a huge snake, adorned with a string of human corpses round her neck...she seems to swallow space itself."[49] Whether responded to passively (in BPM II or *tamas*) or actively (in BPM III or *rajas*), this description is a typical experience of the torturing interior of the maternal body reported in Grofian sessions.

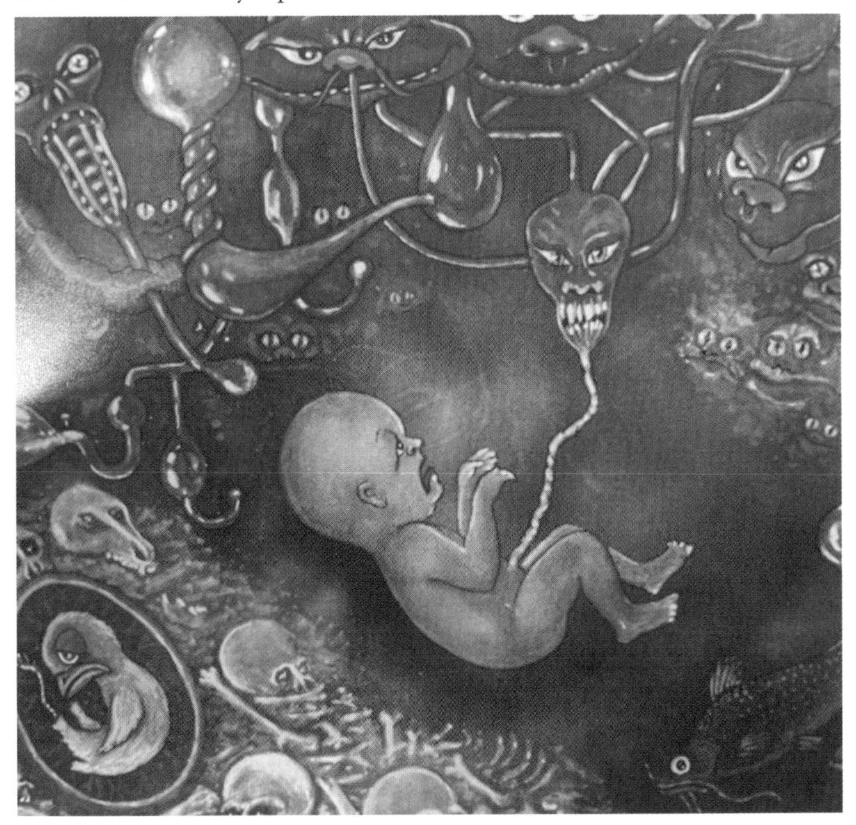

FIGURE 6 The horrors of the birth trauma.[50]

Parā is the power of the subject-element, Aparā the power of the object-element, and Parāparā the field of the subject-object relation:[51] we can consider that in Grof's BPM II (*tamas*) the baby is completely objectified, in BPM III (*rajas*) there are a series of violent reversals between subject and object,[52] and that BPM IV (*sattva*) sees the emergence of the subject. In the *Niruttara Tantra* of the related Kaula tradition, Śiva states: "The vulva (*bhaga*), the blessed goddess (*bhagavatī*), is to be understood as Dakṣiṇā [Kālī], lady of the three *guṇas*."[53]

We can see that Grof's work may contribute to an understanding of the *guṇas*[54] in at least two respects:

1. They are phenomena that are literally experienced. What the *Sāṃkhya Kārikā* describes as the "delimiting" quality of *tamas*, for example, can be experienced in multiple interlocking ways: as the baby trapped in the womb before the cervix has opened, as a prisoner, or simply as a state of despair from which there appears to be no exit.
2. As phenomena literally experienced, they undergo progressive degrees of distillation in deep experiential work, to the point where the experience is that purely of one of the three.[55] The temporal regression to the time of one's birth in the Grofian session can be understood as an effect of this distillation.

The *Bhagavad Gītā*

The hierarchy in which *rajas* overpowers *tamas* and *sattva* overpowers *rajas* is another remarkable point of connection between how the *guṇas* are understood in Hinduism and the progression through the matrices described by Grof. This hierarchy is described in the *Bhagavad Gītā*,[56] and I will end by exploring the *Gītā* in the light of this point of connection.[57]

Two of the questions that have haunted the *Gītā* as its popularity has inexorably grown over the centuries are the extent and nature of its allegorical quality, and whether it should be understood as an *Upaniṣad* in its own right. In relation to the latter, there is no tangible evidence that the *Gītā* is a separate scripture inserted by a different seer into the *Mahābhārata*:[58] it is but a part of the sixth chapter of this vast epic. Why, then, did it come to be one of the three canonical texts (*prasthānatrayī*) of the Vedānta schools, and why does this specific fragment of the *Mahābhārata* speak so particularly profoundly and deeply to us? An understanding of the *guṇas* as finding their purest expression in the experience of birth may throw some light on these questions.

The *Gītā* begins with Arjuna's sense of gravitas and sorrow around the portending Kurukṣetra War fought against his own cousins: the need to take up arms against members of his family, in the name of righteousness but not, he senses, necessarily in the service of it. Krishna's response is to impel Arjuna, without equivocation, to fight. When Arjuna presents Krishna with creditable reasons to cast down his arms—"Of what avail is dominion to us, O Govinda?[59] ...They, *for whose sake* we desire kingdom, enjoyment and pleasures, stand here in battle...."[60]—Krishna rejects them. While Krishna does not explicitly use the words *tamas* or *rajas* in his words to Arjuna at this point, there are strong evocations of the need for a movement from a powerless tamasic state into a forceful rajasic one:

> Whence is this perilous condition come upon thee, this dejection, unlike of Āryan, heaven-excluding, disgraceful, O Arjuna? Yield not to impotence, O Pārtha! It does not befit thee. Cast off this mean weakness of heart! Stand up, O Parantapa (scorcher of foes)![61]

Yet *rajas*, Krishna will say later,[62] *binds us to* action, rather than being action in itself. The implication is that one can engage in action, as Arjuna now must, but in renouncing its fruits,[63] not be bound to it. Here then is a sophisticated and nuanced reading of how one can engage in physical battle without succumbing to the *rajas guṇa*, the *guṇa* that, in a later *śloka*, Krishna will declare to be the very source of evil:

> Arjuna said:
>
> ...Impelled by what does man commit sin, though against his wishes, O Vārṣṇeya, constrained, as it were, by force?
>
> The Blessed Lord said:
>
> It is desire, it is anger born of the *rajas guṇa*, all-devouring, all-sinful; know this as the foe here (in this world).[64]

Arjuna is to fight, but not selfishly, and not in such a manner that *rajas* will bind him and destroy him. We are now taught that this is true no less of an internal struggle than of an external one. There are in fact two separate contexts in the *Gītā* in which Arjuna is explicitly impelled to kill. The first is in the external battle of the war itself:

> ...Stand up, O son of Kuntī, resolved to fight. Having made pleasure and pain, gain and loss, victory and defeat the same, engage in battle for the sake of battle...."[65]

The second is with the internal enemy of desire referenced above:

> Enveloped, O son of Kuntī, is 'wisdom' by this constant enemy of the wise in the form of 'desire,' which is difficult to appease, like fire....O best of the Bhāratas, controlling first the senses, kill this sinful thing, the destroyer of knowledge and wisdom.[66]

Most readings of the *Gītā* can be classified according to which of the two forms of fight are stressed: whether we are presented here with external reality as we must confront it or with an allegory for our own internal battles. The following is from Sri Aurobindo's revolutionary career:

> Under certain circumstances a civil struggle becomes in reality a battle and the morality of war is different from the morality of peace. To shrink from bloodshed and violence under such circumstances is a weakness deserving as severe a rebuke as Sri Krishna addressed to Arjuna when he shrank from the colossal civil slaughter on the field of Kurukṣetra.[67]

The *Gītā's* explicit injunction to take up arms[68] has been taken absolutely literally,[69] and it cannot be said that here is an injunction explicitly contradicted by anything that comes later in the text. But the layering and profundity of what is taught is visible in the way it can be read as an implicit injunction not to take up arms no less.[70] Kurukṣetra, while a real place in Northern India, translates literally as *kuru-kṣetra*, field of the Kurus[71], and in Krishna's pronouncement "This body, O Kaunteya, is called *kṣetra*,"[72] he employs the same word. If here then is an allegory for the body, the very teaching may be that battles seemingly external to this body must be understood as internal to it. Yet, unusually for an allegory, Arjuna seems at times close to questioning its very narrative frame: "If it be thought by you that 'knowledge' is superior to 'action,' O Janārdana, why then, do you, O Keśava, engage me in this terrible action?"[73] Krishna appears to have set Arjuna the particularly difficult task of simultaneously fighting an allegorical battle and a real one.

One senses that what is asked of us here is more than just to fall on one side of the debate. If it could somehow be shown that our task, as Arjuna's, is to alchemize *rajas* into *sattva* both internally and externally without these two battles contradicting one another, this would surely be the most consistent with what is taught here. Grofian work, in unveiling a form of encounter with *rajas* that is subjectively external but objectively internal, effectively contributes a proposal on how this might be achieved.

Grof shows that in every human life there are two, conceptually but not experientially different, encounters with *rajas*. The first is at birth itself. Here, thousands of hours of clinical work have shown how all of the characteristics associated with the *rajas guṇa*—cruelty, selfishness, war, action, struggle—are directly confronted, experienced, and worked through. One important point here is that at this point, the embracing of *rajas* is absolutely natural and necessary. The baby needed to move from a tamasic state, in which it was passively submitting to the uterine contractions, into a rajasic state, where it was quite literally engaging in a physical fight. From the perspective of the baby, birth simply is a bloody fight, and *rajas* no more and no less than the furious and desperate struggle for separation. This battle was absolutely and rightly selfish: the baby cared for nothing but its freedom, and the harder and more mercilessly it fought, the more successful it was.

The second is following birth. *Rajas* can emerge with formidable intensity in post-natal life. But Grof's case material shows that the degree of intensity with which it emerges is dependent upon the extent to which the original birth experience has been processed and worked through. There are many rajasic encounters with the external world in the individual's biography. Yet they seem to trace like a river into the ocean of an even more powerful and potent *rajas* experienced at birth,[74] which in most cases appears to have been dissociated from: the conscious knowledge and processing of it is absent. The full experiencing of this *rajas*, as though for the first time, is what facilitates the most profound and important movement out of *rajas* into *sattva*.

Hence the baby did indeed need to summon selfish desire, did need to fight, in order to be born. But if this energy of *rajas*, this energy which proved so successful and so necessary in the battle through the birth canal, is unconsciously *retained* following birth—if it remains to afflict the adult individual rather than being worked through, confronted, and dissolved—then it does indeed answer the question "impelled by what does man commit sin, though against his wishes, O Vārṣṇeya, constrained, as it were, by force?"[75] The original battle of birth was one in which *rajas* was necessarily and rightly indulged to the full. But the battle fought in the present day must become one in which *rajas* is not acted out—acted out in cruelty, selfishness, and even war—but in which its original manifestation is remembered and processed. In the former battle one embraces *rajas*; in the latter, one remembers, works through, and divests oneself of it.

This is then worth entering into dialogue with the *Gītā*'s injunctions to defeat *rajas* both internally[76] and, implicitly, externally.[77] *Rajas* is in a very real sense defeated, in a battle subjectively experienced as external, as the baby completes its birth. But this battle must then be repeatedly fought again, in what is objectively an internal struggle, by the adult in deep experiential self-exploration. Through repeated such encounters, it is alchemized into *sattva* in the present day. The original rajasic experiences are worked through and processed, and become understood and appreciated as historical events rather than anything that belongs in the here and now. The present becomes ever-more differentiated from the past, and ultimately the rajasic state no longer afflicts and distorts one's encounters with others in everyday life.

This reading retains the necessary understanding of the *Gītā* as allegory—we know that it is far more than just the story of a physical battle—while also honoring the tragedy and the suffering entailed *in* physical battle. The *Gītā* is clearly communicating a more profound truth than that war—real, tangible physical hardship and suffering—can at times be necessary. Yet what the existing allegorical readings do not account for is that it must also be saying more than that the real fight, the really difficult fight, is an internal one: because such a reading misses that the extremes of physical suffering are amongst the most difficult of all creation's permutations that we must face. In George Orwell's *1984*, as Winston begins his beatings at the hands of the prison guards, dark truths emerge about how we respond to the extremities of such suffering:

> One question at any rate was answered. Never, for any reason on earth, could you wish for an increase of pain. Of pain you could only wish one thing: that it should stop. Nothing in the world was so bad as physical pain. In the face of pain there are no heroes, no heroes, he thought over and over as he writhed on the floor....[78]

And as the extremes of *tamas* and *rajas* are confronted in deep experiential work, the tormenting birth canal melds relentlessly with the experience of torture and war. But Krishna's instructions to Arjuna are those that Winston could not bear to hear, and indeed could be said to be one of this scripture's maxims: in the face of pain, there are yet heroes. If the *Gītā's* most frightening teaching—of the terrible, unimaginable, reality of war—is diminished, so too may be some of its wisdom. It may only be by fully living through the horror of the birth trauma,[79] of enduring states in which even physical suffering melds into fathomless abysses of *rajas* and *tamas*, that *sattva* truly can show its face. Richard Tarnas summarizes the effect of Grofian work as follows:

> The price [of working under Grof's techniques] was dear—in a sense the price was absolute: the reliving of one's birth was experienced in a context of profound existential and spiritual crisis, with great physical agony, unbearable constriction and pressure, extreme narrowing of mental horizons, a sense of hopeless alienation and the ultimate meaninglessness of life, a feeling of going irrevocably insane, and finally a shattering experiential encounter with death—with losing everything, physically, psychologically, intellectually, spiritually. Yet after integrating this long experiential sequence, subjects regularly reported experiencing a dramatic expansion of horizons, a radical change of perspective as to the nature of reality, a sense of sudden awakening,[80] a feeling of being fundamentally reconnected to the universe, all accompanied by a profound sense of psychological healing and spiritual liberation.[81]

In the language of the Sāṃkhya school, only once *puruṣa* has fully faced *tamas* and *rajas* in their purest possible state can *sattva* truly blossom and grow. Hence it is no wonder that the *Gītā's* story is one that has resonated with so many millions of people—why a single part of a vast epic was chosen as representative of all humanity's struggle—because it is a story which all of us have been through. The baby really did experience its whole universe, the womb that previously nurtured and nourished it, turn against it and declare war on it, just as Arjuna experiences his own family, his own teachers, now as his enemies. It really did summon the courage to move into a physical fight; and perhaps Krishna's intervention is one found in every birth, his calm encouragement possessing its own perinatal corollary in some subtle biological realm. Yet the real victory attained, the true move from *rajas* into *sattva,* is in birth so ingenious—all of the pain becomes explicable; all of the enmity becomes a joyous biology—that it far surpasses the experience of a bloodied victor on the Kurukṣetra plains. There is something about the power of Krishna's communications to Arjuna that imply triumphing in this fight will take him to a place he cannot yet imagine. The *Gītā's* realism as a story of war is not to be discounted; yet its true allegorical nature is implicit in its beauty, and in its intimation that Kurukṣetra's battle has in fact already been fought—and won.

Conclusion

If there are clear resonances between the characteristics of the *guṇas* and the final three of Grof's matrices, the differences between the cosmologies and praxes of the *guṇas*' originating school and Grofian work are important to stress. Sāṃkhya stresses emancipation: *prakṛti's* transmigration over multiple lives in a state of bondage is to find release.[82] Grofian work, conversely, stresses confrontation: *puruṣa* takes a voyage into, and out of, a consciousness of the *guṇas*' darkest and purest extremes. A critical discovery of Western depth psychology is then that parts of the psyche determinedly resist making this journey. This implies that psychological and spiritual praxis must entail patiently steering *puruṣa*—the seer—over the most frightening and painful parts of *prakṛti*.[83] Here, the jewel does not remain static. It is moved over *prakṛti*'s terrain until it has seen permutations of its *guṇas* perhaps appropriate to the individual's *karma*; far further down the spiritual path would then lie an appreciation of its essentially liberated nature.

FIGURE 7 Second visual analogy of *puruṣa*, represented by the jewel, and *prakṛti*, represented by the colored surface.

If we consider the courage of Arjuna's journey, this may not be a reading alien to it. I will close with the following quotation by Eknath Easwaran, in which the journey through the *guṇas* and the reversal of effect into cause appear to be both integrated and placed in context:

> Evolution, according to the *Gītā*, is a painfully slow return to our native state. First *tamas* must be transformed into *rajas*—apathy and insensitiveness into energetic, enthusiastic activity. But the energy of *rajas* is self-centered and dispersed; it must be harnessed to a higher ideal by the will. Then it becomes *sattva*, when all this passionate energy is channeled into selfless action. This state is marked by happiness, a calm mind, abundant vitality, and the concentration of genius.
>
> But even this is not the end. The goal of evolution is to return to unity: that is, to still the mind. Then the soul rests in pure, unitary consciousness, which is a state of permanent joy.[84]

Sartre's Rite of Passage

THOMAS J. RIEDLINGER

The idea of "birth trauma" is highly fantastic.
–Jean-Paul Sartre

In February 1935, the existentialist philosopher Jean-Paul Sartre entered Sainte-Anne's Hospital in Le Havre, France, and was given an injection of the psychedelic drug mescaline.[1] His fame was still several years in the future; Sartre was then 29 years old and employed as a college philosophy teacher. Unpublished and unknown, he lay waiting on a hospital bed in a dimly lit room for the mescaline to take effect. He was writing a book on the imagination and hoped that the drug would induce hallucinations. It succeeded too well, reported Simone de Beauvoir, his lifelong companion, and fellow philosopher. She phoned Sainte Anne's that afternoon and heard him talking in a thick, blurred voice. He said that her phone call had rescued him from a desperate battle with octopuses. Ordinary objects seemed to change their shape grotesquely, Sarte told her: umbrellas were deforming into vultures, shoes turned into skeletons, and faces looked monstrous, while behind him, just past the corner of his eye, scrambled crabs, octopuses, and grimacing things. For the rest of the evening, he continued to see frightful apparitions: giant beetles, an orangutan's leering face pressed to the window, and huge, fat flies.[2] But he seemed to recover completely by the following day and referred to the experience with "cheerful detachment."

Beauvoir later learned that, for several days after, Sartre suffered recurring attacks of anxiety. He was in a state of deep depression, "and the moods that came upon him recalled those that had been induced by the mescaline." A doctor prescribed twice-daily doses of belladonna, but according to Beauvoir Sartre did not follow the prescription, and his problems continued. "His visual faculties became distorted," she said. "Houses had leering faces, all eyes and jaws," and every clockface that he passed seemed to turn into an owl.[3] Especially persistent was the notion that a lobster was pursuing him; it haunted Sartre for many months. During this difficult period, he spent more time than usual with friends, whose presence "protected him from crabs and similar monsters." To his closest friend, Beauvoir, Sartre confided: "I know what the matter with me is. I'm on the verge of a chronic hallucinatory psychosis." She strongly disagreed, but his anxieties continued until summer. Then he took a long vacation in the countryside of France. After several weeks of solitude, fresh air, and exercise (Sartre liked hiking in the mountains), he abruptly announced that he was "tired of being crazy," Beauvoir reported. "Throughout the trip the lobsters had been trying to trail along behind him, and that evening he finally sent them packing."

Looking back on the episode many years later, Sartre blamed the psychiatrist who administered the mescaline for causing his initial bad reaction: "Since I had been experimenting with Lagache, who's rather saturnine and who said to me, 'What it does to you is terrible,' I ended up having all sorts of unpleasant hallucinations."[4] Concerning his fear of succumbing to a chronic hallucinatory psychosis, Sartre insisted that the drug was *not* primarily responsible. He termed its effect "incidental," as compared to the "profound" cause: a pervasive identity crisis resulting from his passage to adulthood. "It's one thing to be one among many others, as I had been [in school] and in the service. It's quite another to be an individual, such as bourgeois society spawns, burdened with social responsibilities you never asked for, with non-intimate relationships on a social level, isolated completely yet expected to perform certain duties or functions. At that point such a person becomes alienated."

Another significant factor, Sartre added, was his research on the subject of perceptual anomalies for *L'Imaginaire*, his book in progress: "I was forever rummaging around in my own consciousness looking for what I could see, which made my head swim and didn't help matters at all."[5] Beauvour agreed that his "hallucinatory patterns" were the "physical expressions of a deep emotional malaise: Sartre could not resign himself to going on to 'the age of reason,' to full manhood."[6] Neither she nor Sartre, however, were suggesting that his fears had been brought up from the depths of his unconscious to the surface of awareness. Sartre specifically said he was rummaging around in his *consciousness*, not his unconscious, because he denied that the latter existed. A primary tenet of Sartre's philosophical theories is that consciousness only exists when it is conscious *of* something. It is "not entity but a process of attention," explains Hazel Barnes, an expert on Sartre and his philosophy. "Since consciousness is thus only a constant relating, the assuming of a point of view, there is nothing *in* consciousness, certainly no unconscious and no reservoir of determining traits or tendencies."[7] As for the frightening hallucinations Sartre saw in his mescaline session and afterward, Barnes maintains that these were products of his consciousness reflecting on its "residue of experience."[8] Specifically, they were "vivid projections in symbolic form of Sartre's anxious sense of being engulfed in the trappings of the bourgeois world."[9] In other words, the dominant theme of his hallucinations—submarine creatures—represented Sartre's fear of being submerged in the ocean of bourgeois society.

If we accept these explanations as complete, the significance of Sartre's mescaline session is little more than anecdotal. But we should not accept them, for several reasons. One is the fact that until he took the mescaline, Sartre was not troubled with spontaneous hallucinations. Afterward, the particular hallucinations invoked by the drug persisted for several months. Since the physiological effects of mescaline last only for several hours, it is clear that its effect on Sartre's psyche was profound, not "incidental."[10] It is also clear that Sartre's hallucinations were not strictly symbols of his adult identity crisis: a similar confrontation with deep-dwelling monsters is recorded in *Les Mots* (*The Words*), Sartre's 1964 autobiography of his childhood. Sartre tells us that when he was eight years old, he discovered the power of creative writing and found that it "worried" him:

> I would push my little desk against the window. The anguish would come creeping in again.... Then *it* would come, a dizzying, invisible being that fascinated me. In order to be seen, it had to be described. I quickly finished off the adventure

> I was working on, took my characters to an entirely different part of the globe, generally subterranean or underseas, and hastily exposed them to new dangers: as improvised geologists or deep-sea divers, they would pick up the Being's trail, follow it, and suddenly encounter it. What flowed from my pen at the time—an octopus with eyes of flame, a 20-ton crustacean, a giant spider that talked—was myself, a child monster; it was my boredom with life, my fear of death, my dullness and my perversity. I did not recognize myself. No sooner was the foul creature born than it rose up against me, against my brave speleologists. I feared for their lives.[11]

Was this same "Being" incarnated during Sartre's mescaline session? If so, it would seem that he once again failed to recognize himself. We must admit this possibility, and therefore reject the analysis offered by Sartre, Beauvoir, and Barnes. It may be accurate *so far as it goes*; but by denying the unconscious, it does not go far enough.

An alternative analysis proceeds from the theories of Stanislav Grof, M.D., a pioneer researcher in the therapeutic use of LSD and other psychedelic substances. (Since the normal effective doses of LSD and mescaline produce similar physiological and psychological reactions, we can consider the results of LSD research applicable to mescaline as well.)[12] In his seminal book, *Realms of the Human Unconscious: Observations from LSD Research*, Grof describes how his theories developed from his own clinical observation of several thousand psychedelic drug sessions between 1956 and about 1973. To explain how these drugs interact with the mind, he proposed what he refers to as a "useful model" that may or may not reflect the actual structure of the unconscious: "COEX systems" (systems of condensed experience). A COEX system is defined as a "specific constellation of memories consisting of condensed experiences (and related fantasies) from different life periods of the individual. The memories belonging to a particular COEX system have a similar basic theme or contain similar elements and are associated with a strong emotional charge of the same quality."[13] It may be described as a negative or a positive COEX, depending on whether the emotions attached to the memories condensed are pleasant or unpleasant. Psychedelic drugs can serve as a useful therapeutic tool, Grof concluded, because they seem to scan the unconscious like radar and lock onto COEX systems.[14] He first noticed this process when "identical or very similar clusters of visions, emotions, and physical symptoms occurred [to the same individual] in several consecutive LSD sessions. Patients often had the feeling that they were returning again and again to a specific experiential area and each time could get deeper into it. After several sessions, such clusters would then converge into a complex reliving of traumatic memories. When these memories were relived and integrated, the previously recurring phenomena never reappeared in subsequent sessions and were replaced by others."[15]

The most important part of a COEX system is its core—the first experience of a particular kind that was registered in the brain and that laid the foundation, or "memory matrix," for a particular COEX system.[16] Typically, this core is a childhood trauma buried deep in the unconscious. Resolution of the COEX system, therefore, depends on working through its enveloping layers of accumulated memories and reliving the core experience. When a COEX system is engaged in the course of a psychedelic session but not worked through completely, "the subject may remain under the influence of this system after the session in spite of the fact that the effect of the drug has already worn off," resulting in a precarious emotional balance. "This constitutes the mechanism of belated reoccurrences of LSD-like experiences, popularly referred to as 'flashbacks.'"[17]

Applying Grof's theories to Sartre's case, we can speculate that: (1) The mescaline activated a powerful negative COEX system; (2) Sartre failed to achieve a resolution of this COEX before the end of his session; and (3) because the COEX was engaged but remained unresolved, it continued to affect his mood and perceptions for several months. But why was a negative COEX unleashed in the session, rather than a positive one? And what kind of memories did it contain? To discover the answers to these questions, we ourselves must take the role of Sartre's "brave speleologists" and track the original trauma to its lair in his unconscious.

On the day that Sartre was battling with octopuses at Sainte-Anne's Hospital, an intern expressed his amazement at the reaction.[18] The drug's effect on him had been totally different. The intern told Sartre he had "gone romping through flowery meadows, full of exotic houris." Presumably, the mescaline administered to the intern was similar in quality and dosage to that which Sartre was given. Yet Sartre's hallucinations were unpleasant, while the intern's experience was enjoyable. Why?

We have already heard Sartre's explanation: The "saturnine" doctor who injected him with mescaline warned that his reaction might be "terrible," and so it was. We also know the session was conducted in a dimly lit hospital room, where Sartre might well have felt a little out of place. These are not insignificant factors. A number of researchers have observed that psychedelic drugs enhance suggestibility; therefore, "set" (the psychological context, including expectations) and "setting" (the physical environment) can predispose the outcome of a psychedelic session.[19] But probably more influential in Sartre's case was his social situation at the time he took the mescaline. For months, he had been feeling disenchanted with his job as a college philosophy teacher, depressed about his failure to achieve recognition as a writer (a first book had already been rejected and his second was unfinished), and "alienated" by the prospect of becoming an adult among bourgeois adults.[20] As Sartre himself concluded, the mescaline probably made him more sensitive to these emotional problems.

Grof would take it further; his theories suggest that instead of addressing only Sartre's then-current life situation, the drug would have scanned his unconscious in search of a COEX containing a similar charge of emotion.[21] In that case, it is likely that the texture and unpleasant tone of Sartre's mescaline reaction symbolized the theme of this negative COEX—the common emotional charge that united its memories—and not just his "anxious sense of being engulfed in the trappings of the bourgeois world." Since a negative COEX is typically rooted in childhood trauma, its theme must have been a more primitive fear—e.g., of sinking relentlessly down to the level of octopuses, lobsters, and insects. Sartre acknowledged such a level in *The Words*, where he lamented that "things had a horrible underside. When one lost one's reason, one saw it. To die was to carry madness to an extreme and to sink into it."[22] But what *exactly* did he fear? Madness? Death? Or something more general still, like absorption by natural forces?

We require more clues, and we find them in *The Words*. Sartre relates that his father, at the age of 30, died of "intestinal fever" less than two years after Sartre was born in 1905. At about the same time, Sartre contracted enteritis (an intestinal disease) and "sank into a chaotic world full of simple hallucinations." He almost died of this and of "resentment" at the fact that he was weaned prematurely during the illness; his mother's milk dried up when she exhausted herself over worrying about her husband's failing health.[23] When Sartre recovered, he and his mother moved in with her parents, a proper bourgeois couple who lived in the countryside. Sartre's grandfather chose to regard him as a "singular favor of fate, as a gratuitous and always revocable gift."[24] And indeed, Sartre admitted; "[Nobody], beginning with me, knew why the hell I had been born."[25] Nonetheless, he reported that he woke up every morning "dazed with joy at the unheard of luck of having been born into the most united family in the finest country in the world."[26] The family was comfortable financially, and Sartre was brought up, in the absence of brothers or sisters, as a pampered and exalted child. His role in the "family rite" was to charm the adults, so he was free to indulge his desires with little restraint. But the "insipid happiness" of those early years had a "funereal taste," since his freedom was obtained at the cost of his father's death.[27] As time went on, this taste grew stronger. He reported *seeing* death when he was five years old, in the guise of an "old woman, tall and mad, dressed in black." She muttered as she passed: "I'll put that child in my pocket."[28] Sometime after, Sartre recalled: "[We] were visiting Mme. Dupont and her son Gabriel, the composer. I was playing in the garden of the cottage, frightened because I had been told that Gabriel was sick and was going to die. I played horse half-heartedly and capered around the house. Suddenly, I saw a shadowy hole: the cellar had been opened. I do not quite know what manifestation of loneliness and horror blinded me; I turned around and, singing at the top of my voice, ran away."[29] In the following months, his grandmother died of an illness, and the family moved from its pastoral setting to Paris. By the time he was seven years old, Sartre reported:

> I met real Death, the Grim Reaper, everywhere, but it was never there. What was it? A person and a threat. The person was mad. As for the threat, it was this: shadowy mouths could open anywhere, in broad daylight, in the brightest sun, and snap me up... I lived in a state of terror; it was a genuine neurosis. If I seek the reason for it, I find the following: as a spoiled child, a gift of providence, my profound uselessness was all the more manifest to me in that the family rite constantly seemed to me a trumped-up necessity. I felt superfluous; therefore, I had to disappear. I was an insipid blossoming constantly on the point of being nipped in the bud.[30]

Of course, Sartre did not perceive his situation in such complex terms when he was seven years old. His superfluity was non-reflective, something he experienced. He looked in a mirror and discovered himself to be "horribly natural," a "jelly-fish...hitting against the glass of the aquarium."[31] The shock of recognition made his role as a wonderful child seem ludicrous. When other children shunned him as a playmate, Sartre considered it a confirmation: "I had met my true judges, my contemporaries, my peers, and their indifference condemned me. I could not get over discovering myself through them: neither a wonder nor a jelly-fish. Just a little shrimp in whom no one was interested."[32] Rejected and lonely, Sartre took to reading books and writing fiction. He invented fantastic adventures in which he played the role of a powerful hero or tyrant.[33] But sometimes, he recalled: "I would let myself daydream; I would discover, in a state of anguish, ghastly possibilities, a monstrous universe that was only the underside of my omnipotence....Tremulously, always on the point of tearing up the page, I would relate supernatural atrocities....But the imagination was not involved. I did not invent those horrors; I found them, like everything else, in my memory."[34]

What was it Sartre remembered? Something lurking in a "monstrous universe" on the "underside" of his omnipotence, a place that was occupied (according to the passage quoted earlier) by an octopus, a twenty ton crustacean, and a giant spider. If this signifies the presence of the COEX that later emerged in Sartre's mescaline session, then the trauma which comprised its core must have occurred before he started writing fiction at about the age of eight. It even appears to precede his near-fatal bout with enteritis as an infant, when he "sank into a chaotic world full of simple hallucinations" that reminds us again of the COEX. Yet Sartre has acknowledged no earlier trauma. Are we therefore to assume that none occurred? Or if it did, that Sartre forgot it, was never aware of it, or withheld the information?

There is another possibility. According to Grof, many patients who undergo psychedelic therapy and work their way back through a negative COEX eventually seem to relive the original trauma of their birth. Grof's devotion to Freudian principles at first made him doubt that this was possible since reliving birth trauma "lies beyond the realm of psychodynamics as usually understood in traditional psychotherapy."[35] However, the phenomenon so frequently occurred that he was forced to reassess his position, and eventually concluded that negative COEX systems often do appear to be founded on "basic perinatal [memory] matrices" (BPMs) reflecting some phase of the patient's birth trauma.[36] It remains to be determined whether BPMs are symbolic creations of the unconscious or actual memories of the objective event, Grof cautions.[37] But repeatedly, subjects in LSD sessions have exhibited similar symptoms in four basic categories that seem to correspond to four stages of biological birth.[38] These are:

- *BPM I, Primal union with the mother.* Recalling the intrauterine state when the mother and child form one symbiotic unity, this stage often evokes pleasant feelings of serenity and bliss.
- *BPM II, Antagonism with the mother.* Related to the first clinical stage of delivery, when intrauterine contractions squeeze the fetus while the cervix is still closed, this BPM generates feelings of entrapment in a meaningless, claustrophobic world.
- *BPM III, Synergism with the mother.* In the second clinical stage of delivery, the intrauterine contractions continue but the cervix is wide open, and the difficult, agonizing process of propulsion through the birth canal begins. The goal of both mother and child at this point is to terminate their mutually painful condition. Subjects who relive this BPM almost always experience an "atmosphere of titanic struggle."
- *BPM IV, Separation from the mother.* Corresponding to the third clinical stage of delivery, this BPM recalls a sequence of events in which propulsion through the birth canal ends with a crescendo of tension and suffering, followed by sudden relief and relaxation.

Grof emphasizes that these four stages of the birth experience are never relived in chronological sequence during a psychedelic session.[39] Often, because they share some similar emotional content, experiential symptoms of BPM I overlap those of BPM IV, and the symptoms of BPM II tend to cluster with BPM III. It is also unusual for all four BPMs to be relived in the course of a single psychedelic session, and it usually takes more than one session to work through an especially strong BPM.

Although Grof, to my knowledge, has never alluded in print to Sartre's mescaline session, his description of the symptoms and psychodynamics of BPM II makes its relevance to Sartre's case crystal clear.[40] Under the powerful influence of BPM II, subjects perceive a strong link between the agony of birth and the agony of death. They often brood that they were thrown into this world without any choice as to whether, where, when, and to whom they would be born; and that having been born, they have no choice about the fact that they eventually must die. The impermanence of all things in general and human life, in particular, makes everything meaningless for them, resulting in what Grof terms an "existential crisis:"

> For sophisticated individuals, this experience usually results in a fresh understanding and appreciation of existentialist philosophy and the works of such individuals as Martin Heidegger, Soren Kierkegaard, Albert Camus, and Jean-Paul Sartre. Sartre and the other existentialist philosophers seem to be especially tuned in to this experiential complex, without being able to find the only possible solution, which is transcendence. LSD subjects often refer to Sartre's play, *Huit Clos* (*No Exit*), as a brilliant description of the feelings they experienced when they examined their lives and their interpersonal relationships under the influence of the "no exit" stencil of BPM II.[41]

Other symptoms of this matrix are fears of pervading insanity and permanent psychosis.[42] Subjects typically feel as if they have gained "the ultimate insight into the absurdity of the universe and will never be able to return to the merciful self-deception that is a necessary prerequisite for sanity." Also common are feelings of "cosmic engulfment." This experience begins:

> with an overwhelming feeling of anxiety and an awareness of vital threat. The source of danger cannot be clearly identified, and the individual tends to interpret the immediate environment or the whole world in paranoid terms. An intensification of anxiety usually leads to the experience of being sucked into a gigantic whirlpool. A frequent symbolic variant of this phase of engulfment is being swallowed by a terrifying monster—a dragon, whale, tarantula, octopus or crocodile—or of descending into the underworld and encountering its threatening creatures. There is a clear parallel with eschatological visions of the gaping jaws of the gods of death, mouths of Hell, or the descent of heroes into the underworld.[43]

(Similarly, visions and experiences of wild adventures are often reported by those who relive the next sequential matrix, BPM III. An example Grof has mentioned is "encounters of scuba divers with sharks, octopi, and other treacherous sea creatures.")[44] "Agonizing feelings of separation, alienation, metaphysical loneliness, helplessness, hopelessness, inferiority and guilt are standard components of BPM II," Grof observes.[45] It is the matrix that records "all unpleasant life situations in which an overwhelming destructive force imposes itself on the passive and helpless subject," especially those situations that threaten survival.

Armed with the conviction that the negative COEX system engaged during Sartre's mescaline session was founded on a BPM II memory matrix, and the knowledge that this COEX was manifest even in childhood, we can now reconstruct a brief chronology of his emotional life. The first traumatic memories that Sartre himself acknowledged were his brush with death and premature weaning (rejection) at the time that he contracted enteritis in infancy. These memories were filed in a pre-existing memory matrix—BPM II—that reflected the "no exit" stage of Sartre's birth trauma. Later, as a child growing up without siblings or playmates, his loneliness was filed with these memories and helped to build a COEX constellation, as well as his early encounters with death when he visited Gabriel's house, and when his grandmother died. At about the same time, Sartre's family moved from the countryside to Paris (separation). Once in Paris, he found that the other children spurned his company, confirming Sartre's own feeling that he was superfluous.

All these events and frustrations added fuel to the negative COEX, and the strength of its presence in Sartre's unconscious steadily grew. In time, its effect became manifest in Sartre's neurotic fear of death and madness. When he tried to escape his unhappiness by fantasizing wild adventures, he encountered the COEX more directly on the brink of his unconscious; it appeared to him symbolically as submarine and subterranean creatures from the realm beneath awareness. Sartre imagined himself in the role of a powerful hero and faced down the monsters.

At about the age of eight, Sartre concluded that his role as a writer was equally heroic. When he died, Sartre daydreamed, he would surely be remembered and enshrined as a great writer. Thus death became his ally, since the prospect of postmortem fame charged his life with significance;

Sartre lived his own biography.[46] For several years, this sophisticated exercise alleviated Sartre's neurotic fears. But then, he reported: "[My] mandate became my character; my delirium left my head and flowed into my bones, … Previously, I had depicted my life to myself by means of images: it was my death causing my birth, it was my birth driving me toward my death. As soon as I gave up seeing this reciprocity, I *became* it myself; I was strained to the breaking point between those two extremes, being born and dying with each heartbeat."[47]

The tension reached unbearable proportions in February 1935, a few months before Sartre's thirtieth birthday. His plan to rob death of its sting had been stalled by a big complication: he couldn't get published. Instead of becoming a widely-read author, it looked like he might sink into obscurity among the bourgeoisie. In addition, Sartre must have recalled that his father was 30 years old when he died of intestinal fever—a timely reminder that Sartre, who almost died of a similar illness at about the same time, was not immortal. All these factors helped intensify the onus of his darkest fears. And then, with exquisitely bad timing, Sartre took mescaline.

We know in a general way what happened next. The drug invoked the memories contained within Sartre's negative COEX. Their emotional charge overwhelmed his attention, rising up from his unconscious in the form of anxious feelings and unpleasant hallucinations. At the core of this phenomenon were memories of Sartre's own birth, attempting to emerge into consciousness and thus resolve the COEX. (Resolution occurs when the unconscious content is experienced consciously in its original form and full intensity.)[48]

The fact that he continued to suffer anxieties and hallucinations ("flashbacks") for several months would suggest that he failed to relive the COEX memories in their original form, or at least that he did not experience their full intensity. Another possibility is that he *did* relive the memories, perhaps with full intensity, but blocked a resolution by refusing to acknowledge their validity. We should not be confused by Beauvoir's claim that he recovered abruptly one night and "sent the lobsters packing;" this does not necessarily indicate a sudden resolution of the COEX. Sartre more likely came to terms with the particular experience—the mescaline session itself—and continued to resist the emerging unconscious material. That would explain why a lobster returned one year after his supposed recovery.[49] Apparently, he found a way to integrate the session without actually conceding the existence of unconscious memories, by channeling those memories into a "fictional" novel he was writing and assimilating them through the experiences of a protagonist. Grof hints that this is possible when he recommends that subjects should write a detailed account of their psychedelic sessions in order to "greatly facilitate" integration of the experience.[50] Not only diaries and journals, but artistic forms of writing such as poems, plays, and stories can contribute to a deeper understanding of the session.

Sartre's device was his famous first novel, *La Nausée* (*Nausea*), published in 1938. It is structured in the format of a journal kept by Antoine Roquentin, an ostensibly fictional character. But Barnes has pointed out that "[if] we look at *Nausea*, which Sartre had begun writing at the time he took the mescaline, we may discover several reminiscences of that event."[51] Sartre himself, in the mid-1970s, declared: "I wasn't writing *Nausea* during the period I was ill."[52] Also: "What I described in the novel is not something I actually experienced myself."[53] But in *The Words*, he had earlier stated: "At the age of 30, I executed the masterstroke of writing in *Nausea*—quite sincerely, believe me—about the bitter unjustified existence of my fellowmen and of exonerating my own. I *was* Roquentin; I used him to show, without complacency, the texture of my life."[54]

It is certainly true that the context of Roquentin's fictional life parallels that of Sartre's actual life in 1935.[55] He is depicted as 30 years old, an unpublished writer who is working on a scholarly book that represents the "only justification" for his existence.[56] Like Sartre, he is contemptuous of bourgeois society, feels himself getting older, and is starting to fear that his life will be wasted. He is lonely, friendless, and feeling unwanted. The monotony of daily life oppresses him. Early in the novel, Roquentin arranges a meeting with his former lover, Anny, whom he has not seen for several years. He hopes that the meeting will somehow provide him with a *raison d'etre*. But before their reunion takes place, an event unfolds. It starts with what Roquentin calls "precursors of a new overthrow in my life."[57] He is troubled by attacks of nausea (a common complaint during psychedelic sessions) and hallucinatory perceptions. A mirror becomes a "white hole in the wall," a "trap." Gazing into it, Roquentin is enthralled by the sight of his "feverish swelled lips," and of crevices, mole holes, and silky white down covering the "great slopes" of his cheeks. He remembers that when he looked into a mirror as a child, he saw himself as something "at the level of jellyfish."[58] Later, his hand as it rests on a table appears to resemble a crab.[59] At one point, Roquentin runs down the street in a panic, convinced that the houses are watching him with "mournful eyes."[60] Doors especially frighten him; he fears that they will open by themselves. When he gets to the seashore, he imagines that a terrible presence is lurking underneath the waves. "A monster?," he wonders. "A giant carapace, sunk in the mud? A dozen pairs of claws or fins laboring slowly in the slime." The monster rises, but Roquentin runs away before it surfaces.

Finally, he winds up in a park and encounters a chestnut tree.[61] At first, it is only a tree whose roots are sunk into the ground beneath the bench he is sitting on. But "suddenly, suddenly, the veil is torn away," declares Roquentin. "I have understood, I have *seen*." What he has seen, in a transport of "horrible ecstasy," is that the tree is not what people call a tree. It exists in a pure state of Being outside of its name, its characteristics, or the functions by which it is classified. As such, it is essentially superfluous; the tree exists only to be.

In *The Doors of Perception*, Aldous Huxley reported a similar insight that came to him under the influence of mescaline. While gazing at some flowers that seemed to be shining with "their own inner light and all but quivering under the pressure of the significance with which they were charged," he concluded that what they "so intensely signified was nothing more, and nothing less, than what they were—a transience that was yet eternal life, a perpetual perishing that was at the same time pure Being."[62] For Roquentin, however, the existence represented by the chestnut tree has different implications. If a thing has no reason for being except that it is, he reasons, its existence is absurd, not intensely significant. Existence is a "universal burgeoning" of things that exist without reason, so existence itself is absurd.

Sartre then expounds, through Roquentin, on what this portends for society, personal freedom, and consciousness theory. He stresses, however, that such contemplations came *after* the fact. When Roquentin encounters the tree, he apprehends *experientially* "the key to Existence, the key...to [his] own life." In Barnes' opinion, this scene is of "obvious importance in the expression of Sartre's view of the world and of consciousness."[63] She also believes that it "echoes rather specifically the mescaline experience," which implies that a connection can be traced between Sartre's drug-enhanced perceptions and his later philosophical theories. (Beauvoir affirms that Sartre "crystallized the first key concepts of his philosophy" around the time of his mescaline session, though she does not acknowledge a possible connection.)[64] Such a theme is worth exploring, but beyond the immediate scope of this article; readers who want to pursue it are referred to Barnes' *An Existentialist Ethics.*[65] For the present, we are less concerned with Sartre's philosophical theories than we are with his psychology. The question at hand is not, "What philosophical truths are conveyed by Roquentin's contemplation of the chestnut tree?" but, "What did the tree represent, as a personal symbol, for Sartre?" The answer must be that it symbolized the unresolved negative COEX. If *Nausea* was the device by which Sartre integrated deep traumatic memories stirred up in his mescaline session, then the key scene in the book is a logical place to expect him to deal with key psychological issues. There is evidence supporting this conclusion in *The Words*, where Sartre tells us he was "frozen with fear" as a child when he read this account in a newspaper:

> One summer evening, a sick woman, alone on the first floor of a country house, is tossing about in bed. A chestnut tree pushes its branches into the room through the open window. On the ground floor, several persons are sitting and talking. Suddenly someone points to the chestnut tree: "Look at that! Can it be windy?" They are surprised. They go out on the porch. Not a breath of air. Yet the leaves are shaking. At that moment, a cry! The sick woman's husband rushes upstairs and finds his wife sitting up in bed. She points to the tree and falls over dead. The tree is as quiet as ever. What did she see? A lunatic has escaped from the asylum. It must have been he, hidden in the tree, who showed his grinning face....And yet...how is it that no one saw him go up or down? How is it that the dogs didn't bark? ...The writer starts a new paragraph and concludes, casually: "According to the people of the village, it was Death that shook the branches of the chestnut tree." I threw the paper aside, stamped my foot, and cried aloud: "No! No!" My heart was bursting in my chest.[66]

As Roquentin, Sartre again confronts this chestnut tree of Death. It is a difficult, almost heroic mission, one that Roquentin approaches reluctantly.

"I would so like to let myself go, forget myself, sleep," he admits when he enters the park. "But I can't, I'm suffocating; existence penetrates me everywhere, through the eyes, the nose, the mouth..." He sits down on a bench and considers the root of a chestnut tree sunk in the ground. It has no meaning for him, suddenly; he cannot remember its name or its function. "I was sitting, stooping forward, head bowed, alone in front of this black, knotty mass, entirely beastly, which frightened me," Roquentin says. "Then I had this vision." He looks beyond apparent surfaces and finds himself confronting the totality of pure Existence, a world of "soft, monstrous masses, all in disorder—naked, in a frightful, obscene nakedness." The tree root is the focus of this revelation. He regards it with "atrocious joy" but cannot come to terms with its existence. "[It] stayed there, in my eyes, as a lump of food sticks in the windpipe," he says. "I could neither accept nor refuse it."

Then a motion distracts his attention. Roquentin looks up at the treetop and sees that a wind is now stirring its branches. At first, he considers that "movements never quite exist, they are passages, intermediaries between two existences, moments of weakness. I expected to see them come out of nothingness, progressively ripen, blossom," he says. "I was finally going to surprise beings in the process of being born." Three seconds later, all his hopes are "swept aside." Roquentin suddenly perceives that even movements are beings, and either exist or do not at any given moment. If they do, they exist in a "no exit" state of pure Being. Like the tree root, the movements refuse to be born; they are there and already exist at the point where the past meets the future. "I could not attribute the passage of time to the branches groping around like blind men," Roquentin concedes. "The idea of passage was still an invention of man....My eyes only encountered completion."

With mounting distress, he sinks down on the park bench and contemplates beings more closely. Like Sartre as a child, they seem to be nothing but insipid blossomings, totally superfluous. And yet they are everywhere present."[My] ears buzzed with existence [Roquentin complains], my very flesh throbbed and opened, abandoned itself to the universal burgeoning. But why, I thought, why so many existences, since they all look alike? What good are so many duplicates of trees? So many existences missed, obstinately begun again and again—like the awkward efforts of an insect fallen on its back."

"I was one of *those* efforts," Roquentin admits. He tries closing his eyes to escape from the world of Being before it engulfs him. No good; mental images, another form of Being, "immediately leaped up and filled my closed eyes with existence," he reports. There is no place to hide, no retreat from Existence. At last, it is time to stop running from things as they are.

Roquentin shouts out loud. His eyes are open. Something is about to happen.

> Had I dreamed of this enormous presence? [he asks.] It was there, in the garden, toppled down into the trees, all soft, sticky, soiling everything, all thick, a jelly. And I was inside, I with the garden. I was frightened, furious, I thought it was so stupid, so out of place, I hated this ignoble mess. Mounting up, mounting up high as the sky, spilling over, filling up everything with its gelatinous slither, and I could see depths upon depths of it reaching far beyond the limits of the garden. ...I knew it was the World, the naked World suddenly revealing itself, and I choked with rage at this gross, absurd being... I shouted "filth! what rotten filth!" and shook myself to get rid of this sticky filth, but it held fast and there was so much, tons and tons of existence, endless: I stifled at the depths of this immense weariness. And then suddenly the park emptied as through a great hole, the World disappeared as it had come, or else I woke up—in any case, I saw no more of it; nothing was left but the yellow earth around me, out of which dead branches rose upward.

Was this Sartre's account of his birth trauma? If so, it is clear that he experienced more than the "no exit" memories of BPM II. We also detect certain symptoms of BPM III, during which stage of clinical birth "the child can come into contact with various biological material, such as blood, mucus, urine and feces."[67] Finally, what can it mean that the park "emptied [suddenly] as through a great hole," except that Roquentin experienced BPM IV (separation from the mother)? And therefore that Sartre, through his protagonist, completed a full "rite of passage" by reliving his birth from beginning to end? This proposition is supported by the fact that Sartre's mother, like Roquentin's former lover, was named Ann. Thus, when Roquentin leaves the park and has his rendezvous with Anny, she rebuffs his attempts to rekindle their old passion. He must now accept the situation: "There is nothing more which attaches her to me," writes Roquentin. Anny, like Sartre's Ann in real life, had "suddenly emptied herself" of him.[68]

The final question is a pragmatic one: What were the long-term net results of Sartre's mescaline session? We know that it inspired the events in *Nausea*, which launched his career as a writer in 1938. And we have heard that it might have contributed something to Sartre's philosophical theories. When these theories were published in *L'Être et le Néant* (*Being and Nothingness*) in 1943, his reputation was assured. So in one sense, at least indirectly, the mescaline session helped Sartre to achieve the acclaim he was longing for.

In another sense, however, it failed him. As Susan Sontag noticed, Sartre's public fame belied a personal, implacable distress: "Sartre's work is not contemplative, but is moved by a great psychological urgency. His pre-war novel, *Nausea*, really supplies the key to all his work. Here is stated the fundamental problem of the assimilability of the world in its repulsive, slimy, vacuous, or obtrusively substantial *thereness*—the problem that moves all of Sartre's writings. *Being and Nothingness* is an attempt to develop a language to cope with, and record the gestures of, a consciousness tormented by disgust. This disgust, this experience of the superfluity of things and of moral values, is simultaneously a psychological crisis and a metaphysical problem."[69]

According to Grof, the solution to Sartre's crisis was "transcendence."[70] It eluded him. Compelled by his unconscious and its morbid COEX system, he could not transcend the thought of his mortality and blamed it on Existence. He was fixed on the perishing aspect of Being instead of its perpetual rebirth.

Viktor Frankl, the psychiatrist who founded logotherapy (a school of existential psychotherapy), describes in his 1959 book, *Man's Search for Meaning*, how a woman transcended despair by adopting a healthier attitude, ironically, toward the existence of a chestnut tree. He witnessed her death while both were imprisoned in a Nazi concentration camp. Frankl writes,

> There is little to tell and it may sound as if I had invented it; but to me it seems like a poem.
>
> This young woman knew that she would die in the next few days. But when I talked to her she was cheerful in spite of

this knowledge. "I am grateful that fate has hit me so hard," she told me. "In my former life I was spoiled and did not take spiritual accomplishments seriously." Pointing through the window of the hut, she said, "This tree here is the only friend I have in my loneliness." Through that window she could see just one branch of a chestnut tree, and on that branch were two blossoms. "I often talk to this tree," she said to me. I was startled and didn't know quite how to take her words. Was she delirious? Did she have occasional hallucinations? Anxiously, I asked her if the tree replied. "Yes." What did it say to her? She answered, "It said to me, 'I am here—I am here—I am life, eternal life.'"[71]

The Consciousness Research of Stanislav Grof

A Participatory Perspective

Jorge N. Ferrer

To discuss Stanislav Grof's 50 year consciousness research and its profound implications for contemporary psychology, psychiatry, science, philosophy, and society is beyond the scope of this chapter. It should suffice to mention that Grof's research was pivotal for the development of the field of transpersonal psychology,[1] the identification of the perinatal and transpersonal sources of psychopathology and mental health,[2] the understanding of spiritual emergencies,[3] the challenge to the Cartesian-Newtonian scientific paradigm in psychology,[4] and the development of a transpersonal approach to the critical analysis of social, cultural, and global issues,[5] among other contributions.

This chapter focuses on Grof's account of the spiritual experiences and insights that occur during special states of consciousness facilitated by entheogens and holotropic breathwork (i.e., sustained hyperventilation combined with evocative music and bodywork).[6] Specifically, I show that (a) Grof's research provides crucial empirical evidence to potentially resolve one of the most controversial issues in the modern study of mysticism—the question of mediation in spiritual knowledge; (b) Grof interprets his findings as supporting a neo-Advaitin, monistic, esotericist-perspectival version of the perennial philosophy; and (c) a more pluralist participatory vision of human spirituality can harmoniously house Grof's experiential data while avoiding certain shortcomings of perennialism. I conclude by suggesting that a participatory account of Grof's research may not only bring forth richer and more pluralistic spiritual landscapes, but may also have emancipatory potential for spiritual growth and practice.

A Major Debate in the Modern Study of Mysticism

Classical definitions of mysticism explain mystical knowledge in terms of an identification with, or direct experience of, the ultimate ground of Being, which is variously interpreted and described in terms such as God, the Transcendent, the Absolute, the Void, the Noumenal, Ultimate Reality, or, more simply, the Real.[7] These definitions are perennialist insofar as they assume the existence of a single, ready-made ultimate reality that is directly accessed—partially or totally—by mystics of all kinds and traditions. If mystical knowledge is direct and ultimate reality is One, so the reasoning goes, mystical experiences must either be phenomenologically identical, or, if different, corresponding to dimensions, perspectives, expressions, or levels of this singular spiritual ultimate. As Perovich, a contemporary perennialist, put it: "The point [of the perennial philosophers] in insisting on the identity of mystical experiences was, after all, to bolster the claim that the most varied mystics have established contact with 'the one ultimate truth.'"[8]

Perhaps the most influential version of perennialism is the one developed by the traditionalist or esotericist school, whose main representatives are René Guénon, Ananda K. Coomaraswamy, Frithjof Schuon, Huston Smith, and Seyyed H. Nasr.[9] Although with different emphases, these authors claimed that while the exoteric beliefs of the religious traditions are assorted and at times incompatible, their esoteric or mystical core reveals an essential unity that transcends doctrinal pluralism.[10] This is so, traditionalists argued, because mystics of all ages and places transcended the different conceptual schemes provided by their cultures, languages, and doctrines, and consequently accessed a direct, unmediated apprehension of reality (*gnosis*). Most perennialists distinguished between mystical experience, which is universal and timeless, and its interpretation, which is culturally and historically determined.[11] According to this view, the same mystical experience of the nondual Ground of Being would be interpreted as emptiness (*sunyata*) by a Mahayana Buddhist, as Brahman by an Advaita Vedantin, as the union with God by a Christian, or as an objectless absorption (*asamprajnata samadhi*) by a practitioner of Patañjali's yoga. In all cases, the experience is the same, the interpretation different.

This classical perennialist view was strongly challenged by Katz and other contextualist scholars, who argued that all mystical experiences—as with any other human experience—are mediated, shaped, and constituted by the language, culture, doctrinal beliefs, and soteriological expectations of the traditions in which they occur.[12] There are two versions of this thesis of mediation: strong and weak. The *strong thesis of mediation* asserts that all mystical experiences are always entirely mediated—that is, the phenomenology of mysticism can be *fully* explained by resorting to formative variables such as the concepts, doctrines, and expectations that mystics bring to their experiences.[13] In contrast, the *weak thesis of mediation* asserts that most mystical experiences are heavily mediated by contextual variables, but that mystical phenomenology is the product of a complex interaction between formative variables, the creative participation of the mystic, and, in some cases, an encounter with actual ontological realities.[14] In the first half of this chapter, I show that Grof's consciousness research renders the strong thesis of mediation implausible; in the second half, I introduce a participatory vision of human spirituality that embraces a radical version of the weak thesis of mediation.

Regarding Katz's stance, although he has often been misunderstood (and, arguably, unfairly critiqued) as holding the strong version of the mediation thesis, his actual stance seems closer to the weaker version.[15] When asked whether he believed that mystical experiences were entirely made up by mystics' prior beliefs and expectations, he responded: "That's not what I say. I say there is a dialectic between our environment and our experience."[16] However, Katz suggested that it is impossible to transcend cultural context, thereby falling into a mystical version of Popper's *myth of the framework*, which fallaciously holds that human beings are epistemic prisoners of their cultures, religions, and associated conceptual schemes.[17] In addition, Katz stated that the various mystical ultimates are incomplete glimpses of the same absolute reality: "All of us see only one aspect, only one attribute, only a partial vision. And ultimate reality, by its very nature, escapes us because it is ultimate reality, and as human beings, we are partial observers."[18] This statement brings Katz strikingly close to a perspectival perennialism,[19] as well as to a neo-Kantian metaphysical agnosticism regarding the possibility of human knowledge of ultimate reality.[20]

In any event, what contextual and conceptual factors influence, then, is not only the interpretation of mystical states (as most perennialists admit), but also their very phenomenological content: "The experience itself as well as the form in which it is reported is shaped by concepts which the mystic brings to, and which shape, his experience."[21] Therefore, for contextualists, there is not a variously interpreted universal mystical experience, but at least as many distinct types as contemplative traditions.[22] What is more, these types of mysticism do not necessarily correspond to different dimensions or levels of a single spiritual ultimate, but may be independent contemplative goals determined by particular practices, and whose meaning and soteriological power largely depend on their wider religious and metaphysical frameworks.[23] Consequently, as Katz's seminal essay concluded, "'God' can be 'God,' 'Brahman' can be 'Brahman' and *n* can be *n* without any reductionistic attempt to equate the concept of 'God' with that of 'Brahman,' or 'Brahman' with *n*."[24]

Needless to say, such a direct threat to the widely cherished idea of a common spiritual ground for humankind did not go unnoticed or unchallenged. On the contrary, the writings of Katz and his collaborators set the stage for two decades of lively, and often heated, debate between a plethora of perennialist and contextualist-oriented scholars. Although this is not the place to review this debate,[25] I should say here that, as the dialogue between these two camps evolved, their differences have gradually become more of emphasis than of radical disagreement. For example, most perennialist authors recognize today, although against a universalist background, some degree of contextuality in mysticism and a reciprocity between mystical experience and interpretation.[26]

Enter Grof

In the context of the modern study of mysticism, perhaps the most striking and revolutionary finding of Grof's consciousness research is that traditional spiritual experiences, symbolism, and even ultimate principles can allegedly become available during special states of consciousness, such as those facilitated by entheogens, breathwork, or other technologies of consciousness.

What is more, according to Grof, subjects repeatedly report not only having access to but also understanding spiritual insights, esoteric symbols, mythological motifs, and cosmologies belonging to specific religious worlds *even without previous exposure to them.*[27] In Grof's words:

> In non-ordinary states of consciousness, visions of various universal symbols can play a significant role in experiences of individuals who previously had no interest in mysticism or were strongly opposed to anything esoteric. These visions tend to convey instant intuitive understanding of the various levels of meaning of these symbols.
>
> As a result of experiences of this kind, subjects can develop accurate understanding of various complex esoteric teachings. In some instances, persons unfamiliar with the Kabbalah had experiences described in the Zohar and Sepher Yetzirah and obtained surprising insights into Kabbalistic symbols. Others were able to describe the meaning and function of intricate mandalas used in the Tibetan Vajrayana and other tantric systems.[28]

Similarly, Grof explained how Jungian archetypes—in both their primordial essence (e.g., the Great Mother Goddess) and their culturally specific manifestations (e.g., the Virgin Mary, the Greek Hera, the Egyptian Isis)—can be directly experienced during special states of consciousness.[29] According to Grof, the identification with culturally bound archetypal motifs is often independent of cultural background and prior personal learning. For example, Buddhists and Hindus can experience the Western archetypal figure of Christ on the Cross, and Euro-Americans can identify themselves with the Buddha, Shiva, or the Sumerian Goddess Inanna. Commenting on this evidence, Grof wrote: "The encounters with these archetypal figures were very impressive and often brought new and detailed information that was independent of the subject's racial, cultural, and educational background and previous intellectual knowledge of the respective mythologies."[30]

Grof explained this fascinating phenomenon by appealing to the collective unconscious hypothesized by C. G. Jung, which contains "mythological figures and themes from any culture in the entire history of humanity."[31] During special states of consciousness, that is, individuals would draw from the collective unconscious both the archetypal symbols and the necessary hermeneutic keys to decode their meaning. To illustrate the genuinely transpersonal nature of these insights (i.e., that they bring accurate information of cultural and religious systems with which the individuals were personally unfamiliar), Grof described the case of Otto, one of his clients in Prague who suffered from a pathological fear of death.[32] During an LSD session, Otto experienced the vision of a frightening pig-goddess guarding the entrance to the underworld, and then felt compulsively drawn to draft a variety of complex geometrical patterns, as if he was trying to find the correct one for a very specific purpose. The meaning of this imagery remained a mystery until, many years later, Grof discussed the details of the session with the mythologist Joseph Campbell.[33] According to Grof, Campbell immediately identified the imagery of the pig-goddess and the geometrical patterns with a vision of "the Cosmic Mother Night of Death, the Devouring Mother Goddess of the Malekulans in New Guinea."[34] Campbell added that the Malekulans not only believed that a terrifying female deity with distinct pig features "sat at the entrance into the underworld and guarded an intricate labyrinth design," but also practiced the art of labyrinth drawing as a preparation for a successful journey to the afterlife.[35]

An intriguing question raised by these phenomena is why this transpersonal access to cross-cultural spiritual symbolism has not been reported before in the world's religious literature or pictorial history, even by traditions ritually using entheogens, breathing, or other technologies of consciousness modification, such as many shamanic cultures and certain schools of Sufism, Tantra, and Hinduism.[36] In other words, if transpersonal consciousness allows human beings to access transcultural religious symbolism, why cannot scholars find, for example, Buddhist or Christian motifs in Indigenous pictorial art? Or reports of Kabbalistic symbols and experiences in the Tantric or Sufi literature? Many of these traditions were very prone to describe—either pictorially or literarily—their spiritual visions in great detail; had their members encountered the powerful symbolic motifs that Grof's subjects report, it would be reasonable to expect finding some records of them, and to my knowledge scholars have found none.

In the wake of this situation, the modern mind may be tempted to explain away the phenomena reported by Grof in terms of *cryptomnesia* (i.e., the subjects had forgotten their previous exposure to those symbols, and the special state of consciousness simply brings the memories to consciousness).[37] This explanation needs to be ruled out, however, because Grof's subjects reportedly access not only the form of religious and mythological symbols, but also detailed experiential insights into their mystical or esoteric meaning that even ordinary practitioners of those religions do not usually know. Furthermore, although the cryptomnesia hypothesis may account for some of the cases reported by Grof (e.g., self-identification of modern Japanese people with the figure of Christ on the Cross), it would be very difficult to explain in these terms reports of subjects (such as Otto) accessing detailed knowledge of mythological and religious motifs of barely known cultures such as the Malekulans in New Guinea. In these cases, the possibility of previous intellectual exposure to such detailed information is remote enough, I believe, to rule out the cryptomnesia hypothesis as a plausible general explanation of these phenomena.[38]

In some of his lectures, Grof suggested an alternative explanation: contemporary transpersonal access to cross-cultural symbolism may reflect the emergence of a novel evolutionary potential of the human psyche consisting in the capability of accessing transcultural layers of the collective unconscious. The emergence of this potential, he continued, may mirror in the unconscious the greater interconnectedness of human consciousness in our global times (e.g., through media, television, cinema, and especially the World Wide Web). Expanding this account, I conjecture that this transcultural access may have been facilitated by a combination of the following interrelated factors: (a) the lack, after the decline of Christianity, of an unequivocal religious matrix in the modern West that would provide a definite symbolic container for spiritual experiences; (b) the fact that the ritual space for Grof's psychedelic and breathwork sessions is not usually structured according to any specific traditional religious symbolism or soteriological aim (as it was generally the case with the ritual use of entheogens in traditional settings);[39] (c) the "seeking" impetus of the modern spiritual quest in Euro-America, arguably especially strong in individuals who feel drawn to experiment with psychedelics or breathwork (which fits with accounts of modern American spirituality as a "quest culture" composed by a "generation of seekers");[40] and (d) the importation of modern Western values of open inquiry to spiritual pursuits. The combination of these factors may have paved the way for a more open-ended search for, and receptivity to, a larger variety of sacred forms and fostered access to spiritual visions cultivated by different religious traditions.

Whatever explanation results in being more cogent, the fact remains: Grof's subjects were able to directly experience and understand spiritual insights and symbols of different religious traditions without previous knowledge of them.[41] Although Grof's research awaits the more systematic replication necessary to achieve superior scientific status, his data suggest the limitations of the contextualist account of religious diversity and, if appropriately corroborated, constitute an empirical refutation of the strong thesis of mediation (i.e., the cultural-linguistic overdetermination of religious knowledge and experience). Remember that, for some contextualist theorists, religious phenomena are always entirely constructed by doctrinal beliefs, languages, practices, and expectations. As P. Moore put it, "the lack of doctrinal presuppositions might prevent the mystic not only from understanding and describing his mystical states but even from experiencing the fullness of these states in the first place."[42] Whether or not Grof's subjects experience "the fullness" of mystical states and attain a complete understanding of traditional spiritual meanings is an open question. Even if this were not the case, the evidence provided by Grof's case studies is sufficient, I believe, to render the cultural-linguistic strong thesis of mediation questionable on empirical grounds: Grof's subjects report experiences that should *not* take place if the strong thesis of mediation is correct.

Grof's Perennial Philosophy

Grof's *The Cosmic Game* is an exploration of the metaphysical implications of his groundbreaking research into non-ordinary (or *holotropic*, in his terms, which means "moving toward wholeness") states of consciousness. In the introduction, Grof stated that the experiential data he gathered support the basic tenets of the perennial philosophy: "This research...shows that, in its farther reaches, the psyche of each of us is essentially commensurate with all of existence and ultimately identical with the cosmic creative principle itself. This conclusion...is in far-reaching agreement with the image of reality found in the great spiritual and mystical traditions of the world, which the Anglo-American writer and philosopher Aldous Huxley referred to as the 'perennial philosophy.'"[43] He added, "the claims of the various schools of perennial philosophy can now be supported by data from modern consciousness research."[44] I argue below that Grof's experiential data can be consistently explained without appealing to perennialist metaphysics, but first, it may be helpful to identify Grof's specific brand of perennialism.[45]

First, Grof shares with *esotericist* perennialists the belief that true spirituality needs to be sought in an esoteric core purportedly common to all religious traditions. In support of this idea, for example, he wrote: "Genuine religion is universal, all-inclusive, and all-encompassing. It has to transcend specific culture-bound archetypal images and focus on the ultimate source of all forms."[46] Grof's distinction between spirituality and (exoteric) religion is also indicative of this stance: "The spirituality that emerges spontaneously at a certain stage of experiential self-exploration should not be confused with the mainstream religions and their beliefs, doctrines, dogmas, and rituals."[47] Echoing the voice of esotericists such as Schuon or H. Smith,[48] he added: "The really important division in the world of spirituality is not the line that separates the individual mainstream religions from each other, but the one that separates all of them from their mystical branches."[49]

Second, Grof's perennialism is *perspectival* in that he explains the diversity of spiritual ultimates (e.g., personal God, impersonal Brahman, emptiness, the Void, the Tao, pure consciousness) as different ways to experience the same supreme cosmic principle.[50] As I explained elsewhere,[51] perspectival perennialism understands the variety of experiences and accounts of ultimate reality as different perspectives, dimensions, or manifestations of a single supra-ultimate principle or pregiven Ground of Being.[52] In this spirit, Grof wrote, "The ultimate creative principle has been known by many names—Brahman in Hinduism, Dharmakaya in Mahayana Buddhism, the Tao in Taoism, Pneuma in Christian mysticism, Allah in Sufism, and Kether in Kabbalah."[53] Or, as he put it more recently, "The name for this [ultimate] principle could thus be the Tao, Buddha, Shiva (of Kashmir Shaivism), Cosmic Christ, Pleroma, Allah, and many others."[54]

Third, Grof described this esoteric core, supreme cosmic principle, and ultimate source of all religious systems in terms of Absolute Consciousness.[55] This Absolute Consciousness is both the ultimate ground of all that exists (monism) and essentially identical to the individual human soul (nondual). Talking about the nondual relationship between Absolute Consciousness and the individual soul, for example, Grof stated that: "When we reach experiential identification with Absolute Consciousness, we realize that our own being is ultimately commensurate with the entire cosmic network. The recognition of our own divine nature, our identity with the cosmic source, is the most important discovery we can make during the process of deep self-exploration."[56] For Grof, this recognition confirms the truth of the essential message of the Hindu Upanishads: "*Tat twam asi*" or "Thou art that"—that is, the essential unity between the individual soul and the divine.[57] In his own words: "Our deepest identity is with a divine spark in our innermost being (Atman) which is ultimately identical with the supreme universal principle that creates the universe (Brahman). This revelation—the identity of the individual with the divine—is the ultimate secret that lies at the mystical core of all great spiritual traditions."[58]

Fourth, Grof reads some of his experiential data as supporting Sri Aurobindo's notion of involution, according to which the material world is the product of a process of restriction, partition, or self-limitation of Absolute Consciousness.[59] From a state of undifferentiated unity, Grof pointed out, Absolute Consciousness splits and forgets itself to create infinite experiential realities. Although a variety of reasons for this self-forgetting are suggested, Grof usually uses Hindu terminology to explain it and repeatedly suggests that the entire creation is *lila* or a cosmic drama ultimately played out by only one actor: Absolute Consciousness.[60] Talking about the dimensions of the process of creation, for example, Grof wrote: "These are elements that have best been described in ancient Hindu texts which talk about the universe and existence as *lila*, or Divine Play. According to this view, creation is an intricate, infinitely complex cosmic game that the godhead, Brahman, creates himself and within himself."[61]

Fifth, Grof supported Sri Aurobindo's view that evolution is not merely a return to the One, but "the gradual emergence of higher powers of consciousness in the material universe leading to an even greater manifestation of the divine Consciousness Force within its creation."[62] However, Grof also stated that the ultimate goal and zenith of spiritual evolution is self-identification with Absolute Consciousness: "In its farthest reaches, this process [evolution] dissolves all the boundaries and brings about a reunion with Absolute Consciousness."[63] Despite Grof's criticisms of Wilber's work, the endorsement of this particular spiritual goal for all humankind brings

Grof close to Wilber's hierarchical arrangement of spiritual insights and traditions.[64] In this regard, Grof wrote: "Despite the difference in the sources of data, it is not difficult to arrange transpersonal experiences in my classification in such a way that they closely parallel Wilber's description of the levels of spiritual evolution."[65]

Finally, Grof's description of his cosmology included statements suggesting a subtle devaluation of the material world as illusory, imperfect, or even defiled. Summarizing the conclusions of his research, for example, he wrote: "In the light of these insights, the material world of our everyday life, including our own body, is an intricate tissue of misperceptions and misreadings. It is a playful and somewhat arbitrary product of the cosmic creative principle, and infinitely sophisticated 'virtual reality', a divine play created by Absolute Consciousness and the Cosmic Void."[66] A similar tendency can be observed in his description of involution, which, for Grof, implies an increasing loss of contact with the pristine nature of the original unity.[67] To illustrate this process, Grof appealed to Jain cosmology, according to which "the world of creation is an infinitely complex system of *deluded* units of consciousness...trapped in different aspects and stages of the cosmic process. Their pristine nature is contaminated by their entanglement in material reality and, particularly, in biological processes."[68] Finally, Grof repeatedly suggested that his research supports the Hindu view that "the material reality as we perceive it in our everyday life is a product of a fundamental cosmic illusion called *maya*."[69] Or, as he put it elsewhere in the book: "[A]ll boundaries in the material world are illusory and … the entire universe as we know it, in both its spatial and temporal aspects, is a unified web of events in consciousness."[70]

To sum up, Grof interprets his findings as supporting a *neo-Advaitin, esotericist-perspectival version of the perennial philosophy*.[71] The variety of spiritual ultimates is understood as different ways to experience the same universal ground, which can only be found, beyond all archetypal forms, in the esoteric universal heart of all religious traditions. This universal core is described as a monistic Absolute Consciousness, its relationship with human individual consciousness is understood in nondual terms, and the creation of an ultimately illusory material world is explained through the neo-Hindu notion of involution.

A Participatory Account of Grof's Consciousness Research

In this section, I argue that Grof's commitment to a nondual perennialist metaphysics may have been premature. To counter those problems and recast Grof's research onto a more pluralist canvas, I propose that a participatory account of human spirituality can consistently house his experiential data while avoiding the limitations of perennialism.

The many problems of perennialism have been extensively discussed elsewhere and do not need to be repeated here.[72] It should suffice to say that, despite their insistence on the ineffable and unqualifiable nature of a supposedly universal spiritual ultimate, perennialists consistently characterize it as nondual, the One, or Absolute Consciousness. The perennialist Ground of Being, that is, curiously resembles the Neo-Platonic Godhead or the Advaitin Brahman. Besides this reductionist, and often a priori, privileging of a nondual monistic metaphysics, perennialism is contingent upon questionable Cartesian presuppositions (e.g., about the pregiven nature of spiritual ultimate reality), leans toward faulty essentialisms that overlook fundamental spiritual differences among traditions (e.g., positing nonduality or pure consciousness as the ultimate or most fundamental referent for all genuine mysticism),[73] and raises important obstacles for interreligious dialogue and spiritual inquiry (e.g., traditions that do not accept the perennialist vision are regarded as inauthentic, lower, or merely "exoteric").[74] In addition to being textually unwarranted,[75] esotericist universalism has been intersubjectively challenged (refuted?) in the contemporary intermonastic dialogue. Buddhist and Christian monks, for example, acknowledge important differences in both their understandings *and* their experiences of what their respective traditions consider to be ultimate.[76] What is more, even within a single tradition, strong disagreements about the nature of ultimate reality exist among monks, teachers, and contemplative practitioners.[77] In what follows, I outline a participatory understanding of spirituality that can accommodate Grof's experiential data while avoiding these pitfalls.

Briefly, I understand spiritual knowing as a participatory activity.[78] Spiritual knowing is not objective, neutral, or merely cognitive, but rather engages human beings in a participatory, connected, and often passionate activity that can involve the opening of not only the mind, but also of the body, vital energies, the heart, and consciousness. Although particular spiritual events may only involve certain dimensions of human nature, all of them can potentially come into play in the act of participatory knowing—from somatic transfiguration to the awakening of the heart, from erotic communion to visionary cocreation, and from contemplative knowing to moral insight, to mention only a few. In this multidimensional human access to reality I call *participatory knowing*, the role that individual consciousness plays is not one of possession, appropriation, or passive representation of knowledge, but of *communion* and *cocreative participation*. This role is ontologically warranted by the fact that human beings are—whether they know it or not—always participating in the self-disclosure of the mystery of which they are part by virtue of their very existence.

Furthermore, following the groundbreaking work of Varela et al., my understanding of spiritual knowing embraces an enactive paradigm of cognition.[79] Spiritual knowing is not a mental representation of pregiven, independent spiritual objects, but an *enaction*, a "bringing forth" of a world or domain of distinctions cocreated by the different elements involved in the participatory event.[80] Some central elements of spiritual participatory events may include individual intentions and dispositions; the creative power of multidimensional human cognition; cultural, religious, and historical horizons; archetypal energies; encounters with subtle worlds and beings; and the apparently inexhaustible creativity of an undetermined mystery or generative power of life, the cosmos, or reality.

Although I concur with perennialism that most contemplative paths aim at overcoming self-centeredness and the emergence of transconceptual cognition,[81] I maintain that there is a *multiplicity of transconceptual disclosures of reality*. Because of their objectivist and universalist assumptions, perennialists erroneously assume that a transconceptual disclosure of reality must be necessarily "one," and, actually, *the* One metaphysically envisioned and pursued in certain traditional spiritual systems. The mystical evidence, however, strongly suggests that there is a variety of possible spiritual insights and ultimates (e.g., Tao, Brahman, *sunyata*, God, *kaivalyam*) whose transconceptual qualities, although sometimes overlapping, are irreducible and often incompatible (e.g., personal versus impersonal, impermanent versus eternal, dual versus nondual).[82] The typical perennialist move to account for this conflicting evidence is to assume that these qualities correspond to different interpretations, perspectives, dimensions, or levels of a single ultimate reality. However, this move is not only unfounded, but also problematic in its covertly positing a pregiven spiritual ultimate that is then, explicitly or implicitly, situated hierarchically over the rest of spiritual ends. I submit that a more fertile way to approach the plurality of spiritual claims is to hold that the various traditions lead to the enactment—or "bringing forth"—of different subtle worlds, spiritual ultimates, and transconceptual disclosures of reality. Although these spiritual ultimates may apparently share some qualities (e.g., nonduality in *Brahmajñana* and *sunyata*),[83] they constitute independent religious aims whose conflation may prove to be a serious mistake (thus, e.g., the ontological nonduality of individual self and Brahman affirmed by Advaita Vedanta may have little to do with the Mahayana Buddhist nondual insight into *sunyata* as the codependent arising or interpenetration of all phenomena). In other words, the ocean of emancipation has different shores.

To avoid any connotation that such an ocean is a shared end-point for all traditions,[84] in my recent work and lectures, I have preferred to use an arboreal image in which spiritual evolution branches out in many different but potentially intermingled directions (or as an omnicentered rhizome propagating through offshoots and thickenings of its nodes)[85] enacted by both the various traditions and innovative spiritual inquirers.[86] This participatory account should not then be confused with the view that mystics of the various kinds and traditions either reach a singular end-point or access different dimensions or perspectives of a ready-made single ultimate reality. This view merely admits that this pregiven spiritual referent can be approached from different pathways or vantage points. In contrast, the view I am advancing here is that no pregiven ultimate reality exists, and that different spiritual ultimates can be enacted through intentional or spontaneous cocreative participation in an undetermined mystery or generative force of life or reality.[87] Thus, a participatory approach envisions the long-searched-for spiritual unity of humankind, not in any global spiritual megasystem, but in the shared lived experience of communion with the generative dimension of the mystery. In other words, the spiritual unity of humankind may not be found in the heavens (i.e., in mental, visionary, or even mystical visions) but deep down in the earth (i.e., in the embodied connection with a common creative root). As the saying attributed to the thirteenth-century Persian poet and mystic Rumi describes, "Maybe you are searching among the branches for what only appears in the roots." The recognition of such creative root(s) may allow for firmly growing by branching out in countless creative directions without losing a sense of deep communion across differences.

In this context, what Grof's experiential data convincingly show is that, once a particular spiritual shore or branch has been enacted, it becomes potentially accessible—at least to some degree and in special circumstances—to the entire human species. In other words, once enacted, spiritual branches become more easily accessible and, in a way, "given" to some extent for individual consciousness to participate in. In transpersonal cognition, spiritual forms that have been enacted so far are more readily available and tend more naturally to emerge (from *mudras* to visionary landscapes, from liberating insights to ecstatic types of consciousness). The fact that enacted branches become more available, however, does not mean that they are predetermined, limited in number, organized in a transcultural hierarchical fashion, universally sequential in their unfolding, or that no new branches can be enacted through cocreative participation. Like trails cleared in a dense forest, spiritual pathways traveled by others can be more easily crossed, but this does not mean that human beings cannot open new trails and encounter new wonders (and new pitfalls) in the seemingly inexhaustible mystery of being.

Grof's experiential data do not then need to be interpreted as supporting a perspectival account of the perennial philosophy. The various spiritual ultimates accessed during special states of consciousness, rather than being understood as different ways to experience the same Absolute Consciousness—which implicitly establishes a hierarchical ranking of spiritual traditions with monistic and nondual ones such as Advaita Vedanta at the top—can be seen as independently valid enactions of an undetermined mystery. Hierarchical arrangements of spiritual insights or traditions necessarily presuppose the existence of a pregiven spiritual ultimate relative to which such judgments can be made. Whenever objectivist prejudices in spiritual hermeneutics are dropped, however, the very idea of ranking traditions according to a paradigmatic standpoint becomes not only suspect but also fallacious and superfluous.[88]

This more pluralist account of Grof's findings is consistent with the synthesis of the psychedelic evidence offered by Merkur.[89] After indicating that most interpretations of the psychedelic evidence so far have been biased in favor of the idea of a universal mysticism, Merkur emphasized that empirical data have always pointed to a rich diversity of psychedelic spiritual states. Specifically, Merkur distinguished twenty-four types of psychedelic unitive states, suggesting that some of them may be more representative of certain religious traditions than others. What characterizes the psychedelic state, he wrote, is that it "provides access to all."[90] Furthermore, although some of these states can be arranged in terms of increasing complexity, Merkur pointed out that "their development is not unilinear but instead branches outward like a tree of directories and subdirectores on a computer."[91] Thus, Merkur's account of the psychedelic evidence is consistent with the arboreal image of spiritual pluralism presented above.

Conclusion

In this chapter, I situated Grof's consciousness research in the context of the modern study of mysticism. On the one hand, Grof's findings hold the promise to settle one of the most controversial issues disputed by scholars of mysticism for the last few decades—the question of mediation in spiritual knowledge. Specifically, Grof's research provides extensive evidence that individuals can not only access but also understand spiritual experiences, meanings, and symbols belonging to a variety of religious traditions even without previous exposure to them.[92] This empirical finding, if appropriately corroborated, refutes the contextualist strong thesis of mediation, according to which spiritual experience and knowledge are always necessarily mediated by doctrinal beliefs, intentional practices, and soteriological expectations.

On the other hand, Grof suggested that the experiential data gathered during his decades-long consciousness research support the idea of a perennial philosophy—more specifically, a neo-Advaitin, esotericist-perspectival type of perennialism.[93] According to Grof, all religious traditions, at their core, aim at the realization of an Absolute Consciousness which, being identical in essence to human individual consciousness, brings forth an ultimately illusory material world through a process of restriction or involution.[94] The diverse spiritual ultimates espoused by the various religious traditions (e.g., *sunyata*, God, the Tao, *kaivalyam*, Brahman) are simply different ways to name and experience this Absolute Consciousness.

While I argued against this perennialist interpretation, I am not saying that the perennial philosophy cannot find support in Grof's data, or that Grof's research disconfirms perennialist metaphysics. As Grof showed, the psychedelic evidence can be interpreted in ways that are consistent with perennialism (although, I have argued, inconsistent with textual and phenomenological mystical evidence).[95] This should not come as a surprise. In the same way that alternative or even logically incompatible theories can fit the same empirical evidence—as the Duhem-Quine principle of underdetermination of theory by evidence shows[96]—it is very likely that alternative metaphysical systems could fit all possible experiential evidence. What I suggest, in contrast, is that Grof's empirical findings are also consistent with a more pluralist participatory vision of human spirituality that is free from the limitations of perennialist thinking.[97] In other words, given the many problems afflicting perennialism, Grof's appeal to perennialism as the metaphysical framework to organize his experiential data may have been premature.[98] Grof's phenomenological cartography is radically empirical and pluralist in the Jamesian sense, but the perennialist ontology stands in tension with this phenomenological pluralism. Thus I propose that a participatory account of Grof's data may well engender a richer, more fully pluralistic, and for some, more spiritually emancipatory understanding of Grof's revolutionary findings.[99]

Once the dependence on objectivist and essentialist metaphysics is given up, the various spiritual paths can no longer be seen either as purely human constructions (as radical contextualists propose) or as concurrently aimed at a single, pregiven spiritual reality (as perennialists believe). The various spiritual traditions, in contrast, can be better seen as vehicles for the participatory enaction of different subtle worlds, spiritual ultimates, and transconceptual disclosures of reality. In a participatory cosmos, human multidimensional cognition creatively channels and modulates the self-disclosing of the mystery through the bringing forth of subtle worlds and spiritual realities. Spiritual inquiry then becomes a journey beyond any pregiven goal, an endless exploration, and disclosure of the inexhaustible possibilities of an always undetermined mystery. Krishnamurti notwithstanding, spiritual truth is not a pathless land, but an infinitely creative adventure.[100]

Endnotes

NO MORE HORIZONS

This essay is from a lecture (L332) of the same title given at The Cooper Union in New York City on March 1, 1971.

1. Walt Whitman, *Leaves of Grass*, first edition (1855), section 48, ll. 1262-1280, edited with an introduction by Malcolm Cowley (New York: The Viking Press, 1961), pp. 82–83.
2. Bṛhadāranyaka Upaniṣad 1.4.6–10, abridged.
3. The Papyrus of Nebensi, British Museum #9,900, sheets 23, 24. From Ernest Alfred Wallis Budge, *The Book of the Dead: The Chapters of the Coming by Day* (London: Kegan Paul, Trench, Trûbner and Co., 1896), pp. 112-113.
4. Thomas 99:28-30, 95:24-28.
5. Whitman, *Leaves of Grass*, section 52, ll. 1329-1330, p. 86.
6. H. Heras, S.J, "The Problem of Ganapati," *Tamil Culture*, vol. III, no. 2 (Tuticorn, April 1954). For more on Campbell's meeting with Father Heras, cf. Joseph Campbell, *Baksheesh and Brahman: Asian Journals—India* (Novato: New World Library, 2002), p. 168.
7. William Blake, *The Marriage of Heaven and Hell* (Oxford: Oxford University Press, 1975), plate 14.
8. Thomas Merton, "Symbolism: Communication or Communion?" in *New Directions 20* (New York: New Directions, 1968), pp. 11-12.
9. Merton, "Symbolism," pp. 1-2.
10. Merton, "Symbolism, pp. 1, 11.
11. [When this lecture was first published in 1972, Campbell cited Grof's unpublished manuscript, "Agony and Ecstasy in Psychiatric Treatment." This work was never published, but became the basis for five later volumes of Grof's work. Cf. Stanislav Grof, *When the Impossible Happens: Adventures in Non-Ordinary Reality* (Boulder, Colorado: Sounds True, 2006), p. 285.]

12. Aldous Huxley, *The Doors of Perception* (New York: Harper & Row, 1954), p. 54.
13. Grof, "Agony and Ecstasy in Psychiatric Treatment."
14. Matthew 27:46.
15. Ecclesiastes 1:2.
16. Huxley, *The Doors of Perception*, pp. 22–24.
17. Cf. Benjamin Jowett, trans., *The Works of Plato, Vol. IV* (New York: Cosimo Books, 2010), p. 377.
18. Hermes Trismegistus (attributed), *Le Livre des XXIV Philosophes*, Françoise Hudry, ed. (Paris: Jérôme Millon, 1989), p. 152: *Deus est sphaera infinita cuius centrum est ubique, circumferentia nusquam.*

THE PSYCHEDELIC EVIDENCE

This essay was originally published in *Forgotten Truth: The Common Vision of the World's Religions* (HarperSanFrancisco, 1976). The author added the special introduction for this Festschrift in 2001.

1. See Huston Smith, "Wasson's SOMA: A Review Article," *Journal of the American Academy of Religion* 40, no. 4 (1972): 480–499.
2. "LSD, the most powerful psychoactive drug ever known to man, is essentially an unspecific amplifier of mental processes. What we see in LSD sessions is only an exteriorization and magnification of dynamics that underlie human nature and human civilization. Properly used, the drug is a tool for a deeper understanding of the human mind and human nature." Abridged from the writings of Stanislav Grof, cited in endnote 3.
3. His book, the first in a projected five-volume series, is Stanislav Grof, *Realms of the Human Unconscious: Observations from LSD Research* (New York: Viking Press, 1975). His journal articles are: Stanislav Grof, "Beyond Psychoanalysis: I. Implications of LSD Research for Understanding Dimensions of Human Personality," *Darshana International* 10 (1970): 55–73; Stanislav Grof, "LSD Psychotherapy and Human Culture" (Part I), *Journal of the Study of Consciousness* 3 (1970): 100–118; Stanislav Grof, "LSD Psychotherapy and Human Culture" (Part II), *Journal of the Study of Consciousness* 4 (1971): 167–187; Stanislav Grof, "The Use of LSD in Psychotherapy," *Journal of Psychedelic Drugs* 3, no. 1 (1970): 52–62; Stanislav Grof, "Varieties of Transpersonal Experiences: Observations from LSD Psychotherapy," *Journal of Transpersonal Psychology* 4, no. 1 (1972): 45–80; Stanislav Grof, "LSD and the Cosmic Game: Outline of Psychedelic Cosmology and Ontology," *Journal of the Study of Consciousness* 5, no. 2 (1972): 165–193; and Stanislav Grof, "Theoretical and Empirical Basis of Transpersonal Psychology and Psychotherapy: Observations from LSD Research," *Journal of Transpersonal Psychology* 5, no. 1 (1973): 15–53.

4. Grof, "Theoretical and Empirical Basis of Transpersonal Psychology and Psychotherapy." Though his work covers a wide spectrum of psychedelic substances, most of it was with LSD, so we shall limit our references to it.
5. Stanislav Grof, "Theory and Practice of LSD Psychotherapy," unpublished manuscript, p. 68. Instead of being published as a single volume as Grof originally intended, this long, initial report of his study is being reworked for projected issue in five volumes, the first of which, as indicated in endnote 3, appeared in 1975.
6. Grof, "Theory and Practice of LSD Psychotherapy," p. 41.
7. Grof, "Theoretical and Empirical Basis of Transpersonal Psychology and Psychotherapy," p. 21.
8. Grof, "Theoretical and Empirical Basis of Transpersonal Psychology and Psychotherapy," pp. 24-25. On the limited range of the Freudian model, I insert a supporting remark by Gordon Allport, in his latter years the dean of American personality theorists. In his closing years at Harvard, he would invite me to his seminars to register such light as Asian psychology might throw on human nature. One year I organized my remarks around India's "four psychologies," geared respectively to *kama* (pleasure), *artha* (wealth or worldly success), *dharma* (duty), and *moksha* (liberation). Allport's response was: "In the West we have a detailed psychology of pleasure à la Freud's Pleasure Principle. McClelland's 'achievement motivation' has added to this a psychology of success. Respecting duty we have a nickel's worth of Freud's superego, and on the psychology of liberation—nothing."
9. Grof, "Theory and Practice of LSD Psychotherapy," p. 118.
10. Grof, "Theory and Practice of LSD Psychotherapy," p. 118.
11. Grof, "Theoretical and Empirical Basis of Transpersonal Psychology and Psychotherapy," p. 25.
12. Grof, "Theoretical and Empirical Basis of Transpersonal Psychology and Psychotherapy," p. 25.
13. Grof, "Theoretical and Empirical Basis of Transpersonal Psychology and Psychotherapy," p. 25.
14. Grof, "Theory and Practice of LSD Psychotherapy," p. 125.
15. Grof, "Theoretical and Empirical Basis of Transpersonal Psychology and Psychotherapy," p. 25.
16. Grof, "Theory and Practice of LSD Psychotherapy," p. 125.
17. Grof, "Theory and Practice of LSD Psychotherapy," p. 277.
18. Grof, "Theory and Practice of LSD Psychotherapy," p. 279.
19. Grof, "Theory and Practice of LSD Psychotherapy," p. 125.
20. Grof, "Theory and Practice of LSD Psychotherapy," p. 29.
21. Grof, "Theory and Practice of LSD Psychotherapy," p. 29.
22. Cosmologies frequently locate the archetypes and God on separate ontological planes with God as the higher of the two. When this separation is effected, the

principal levels of reality number five instead of the four we are employing. Though the archetypes can be regarded as God's first creations, to keep the paradigm this book presents to simplest possible proportions we are regarding them as his attributes, in the way Plotinus identified Intelligence (*Nous*) with its objects (the Platonic forms) and Augustine saw these forms as God's "divine ideas."

23. Grof, "Theory and Practice of LSD Psychotherapy," p. 267.
24. Grof, "Theory and Practice of LSD Psychotherapy," p. 250.
25. Grof, "Theoretical and Empirical Basis of Transpersonal Psychology and Psychotherapy," p. 17.
26. Grof, "Theory and Practice of LSD Psychotherapy," p. 382. The flyleaf of Rank's book which served as almost the bible for Grof's work in one of its stages carries a quotation from Nietzsche: "The very best...is, not to be born....The next best...is...to die soon."
27. Grof, "Theory and Practice of LSD Psychotherapy," pp. 382-83. [Deletions not indicated.]
28. Grof, "LSD and the Cosmic Game."
29. Grof, "Theoretical and Empirical Basis of Transpersonal Psychology and Psychotherapy," p. 11.

PERCEPTION AND KNOWLEDGE

This essay originally appeared in *Psychedelic Reflections*, edited by Lester Grinspoon, M.D. and James B. Bakalar, J.D. (New York: Human Sciences Press, Inc., 1983).

1. Stanislav Grof, *LSD Psychotherapy* (Pomona, CA: Hunter House, 1980).
2. Frances Vaughan, *Awakening Intuition* (New York: Doubleday, 1979).
3. Carl Jung, *Letters*, edited by Gerhard Adler and Aniela Jaffe (Princeton, NJ: Princeton University Press, 1973).
4. Grof, *LSD Psychotherapy*; Stanislav Grof, *Realms of the Human Unconscious* (London: Souvenir Press, 1993).
5. Roger Walsh and Frances Vaughan, eds., *Beyond Ego: Transpersonal Dimensions in Psychology* (Los Angeles: J.P. Tarcher, 1980); Ken Wilber, *The Spectrum of Consciousness* (Wheaton, IL: Quest, 1977); Ken Wilber, *The Atman Project* (Wheaton, IL: Quest, 1980).

THE PSYCHOLOGY OF BIRTH, THE PRENATAL EPOCH, AND INCARNATION

1. Stanislav Grof and Joan Halifax, *The Human Encounter with Death* (New York: E.P. Dutton, 1977).
2. Stanislav Grof, *Realms of the Human Unconscious* (New York: E. P. Dutton, 1976).

3. Ralph Metzner, "From Harvard to Zihuatanejo," in *Timothy Leary: Outside Looking In*, ed. Robert Forte (Rochester, VT: Inner Traditions International, 1999), pp. 155-196.
4. Grof, *Realms of the Human Unconscious*, pp. 115-121.
5. Grof, *Realms of the Human Unconscious*, pp. 123-134.
6. Gregory Bateson, *Mind and Nature: A Necessary Unity* (New York: E.P. Dutton, 1979).
7. Birth Psychology: The Association for Prenatal and Perinatal Psychology and Health, https://birthpsychology.com
8. R. D. Laing, *The Facts of Life* (New York: Ballantine Books, 1976).
9. Thomas Verny with John Kelly, *The Secret Life of the Unborn Child* (New York: Dell Publishing, 1986); David Chamberlain, *The Mind of Your Newborn Baby* (Berkeley: North Atlantic Books, 1998); David Chamberlain, "Communicating with the Mind of a Prenate: Guidelines for Parents and Birth Professionals," *Journal of Prenatal and Perinatal Psychology and Health* 18, no. 2 (2003): 95-108.
10. Jane English, *Different Doorway: Adventures of a Caesarean Born* (Pt. Reyes, CA: Earth Heart, 1985).
11. David Cheek, *Hypnosis: The Application of Ideo-Motor Techniques* (Boston: Allyn & Bacon, 1993).
12. Winafred Lucas, *Regression Therapy: Handbook for Professionals*, 2 vols. (Crest Park, CA: Deep Forest Press, 1993).
13. Emerson Seminars: Preventing and Healing Prenatal and Birth Trauma in Children, Babies & Adults, https://emersonbirthrx.com.
14. Sheila Fabricant Linn, William Emerson, Dennis Linn, and Matthew Linn, *Remembering Our Home: Healing Hurts & Receiving Gifts from Conception to Birth* (Mahwah, NJ: Paulist Press, 1999).
15. Raymond Castellino, "The Stress Matrix: Implications for Prenatal and Birth Therapy," *Journal of Prenatal and Perinatal Psychology and Health* 15, no. 1 (2000): 31-62.
16. Wilhelm Reich, *Character Analysis* (New York: Farrar, Strauss & Giroux, 1949).
17. Ralph Metzner, *The Unfolding Self: Varieties of Transformative Experience* (Novato, CA: Origin Press, 1998).
18. Linn, Emerson, Linn, and Linn, *Remembering Our Home.*
19. Linn, Emerson, Linn, and Linn, *Remembering Our Home*, p. 37.
20. Metzner, *The Unfolding Self*, chapter 12.
21. Gerald Bongard, *The Near-Birth Experience: A Journey to the Center of the Self* (New York: Marlowe & Co., 2000); Elizabeth Hallett, *Stories of the Unborn Soul: The Mystery and Delight of Pre-birth Communication* (San José, CA: Writer's Club Press, 2002); Sarah Hinze, *Coming from the Light* (New York: Pocket Books, 1994).
22. Lucas, *Regression Therapy*; Michael Newton, *Journey of Souls: Case Studies of Life Between Lives* (St. Paul, MN: Llewellyn Publications, 1994).

23. See Sylvia Browne, *Life on the Other Side: A Psychic's Tour of the After-life* (New York: New American Library, 2000).
24. Caroline Myss, *Sacred Contracts* (New York: Harmony Books, 2002).
25. Browne, *Life on the Other Side.*
26. Linn, Emerson, Linn, and Linn, *Remembering Our Home*, p. 31.
27. Howard Schwartz, *Gabriel's Palace: Jewish Mystical Tales* (New York: Oxford University Press, 1993).
28. Hayim Nahman Bialik and Yohoshua Hana Ravnitzky, *The Book of Legends: Legends from the Talmud and Midrash*," trans. William Braude (New York: Schocken Books, 1992).
29. Daniel C. Matt, trans., *The Zohar: Pritzker Edition* (Palo Alto: Stanford University Press, 2003).
30. Luke 1:19
31. Luke 1:26
32. Schwartz, *Gabriel's Palace*, p. 57.
33. Bialik and Ravnitzky, *The Book of Legends*, p. 575.
34. Matt, *The Zohar*, vol. 2.
35. Schwartz, *Gabriel's Palace*, pp. 57-58.
36. Peter Deadman and Mazin Al-Khafaji, *Manual of Acupuncture* (Hove, East Sussex, England: Journal of Chinese Medicine Publications, 1998), pp. 559-560.
37. Stanislav Grof, *Psychology of the Future: Lessons from Modern Consciousness Research* (Albany, NY: SUNY Press, 2000), p. 32.
38. Stanislav Grof and Christina Grof, *Beyond Death: Gates of Consciousness* (London: Thames & Hudson, 1980); Stanislav Grof, *Books of the Dead: Manuals for Living and Dying* (London: Thames & Hudson, 1994).
39. Schwartz, *Gabriel's Palace*, p. 58.
40. Walter Y. Evans-Wentz, ed., *The Tibetan Book of the Dead*, trans. Kazi Dawa Samdup (Oxford: Oxford University Press, 1960).
41. Timothy Leary, Ralph Metzner, and Richard Alpert, *The Psychedelic Experience: A Manual Based on the Tibetan Book of the Dead* (New Hyde Park, NY: University Books, 1964).
42. Robert A. F. Thurman, trans., *The Tibetan Book of the Dead: The Great Book of Natural Liberation Through Understanding in the Between* (New York: Bantam Books, 1994).
43. Michael Newton, *Destiny of Souls: New Case Studies of Life Between Lives* (St. Paul, MN: Llewellyn Publications, 2000); Browne, *Life on the Other Side.*
44. See Ralph Metzner, "The Buddhist Six-worlds Model of Consciousness and Reality," *Journal of Transpersonal Psychology* 28, no. 2 (1996): 155-166.
45. Evans-Wentz, ed., *The Tibetan Book of the Dead*, p. 169.
46. Evans-Wentz, ed., *The Tibetan Book of the Dead*, p. 175.
47. Evans-Wentz, ed., *The Tibetan Book of the Dead*, p. 176.
48. Evans-Wentz, ed., *The Tibetan Book of the Dead*, p. 183.

49. Evans-Wentz, ed., *The Tibetan Book of the Dead*, p.188.
50. Evans-Wentz, ed., *The Tibetan Book of the Dead*, pp. 190-191.

JOURNEYS BEYOND SPACE AND TIME

1. Fritjof Capra, *The Tao of Physics* (Boulder: Shambhala, 1975).
2. Fritjof Capra, *The Turning Point* (New York: Simon & Schuster, 1982).
3. Fritjof Capra and Pier Luigi Luisi, *The Systems View of Life* (Cambridge, UK: Cambridge University Press, 2014).
4. Stanislav Grof, *Realms of the Human Unconscious* (New York: Dutton, 1976).
5. R. D. Laing, *The Divided Self* (London: Tavistock, 1960).
6. See Fritjof Capra, *The Web of Life* (New York: Anchor/Doubleday, 1996), pp. 112ff.
7. Humberto Maturana and Francisco J. Varela, *The Tree of Knowledge* (Boston: Shambhala, 1987); see also Gregory Bateson, *Mind and Nature* (New York: Dutton, 1979).

PSYCHE AND COSMOS

An earlier version of this chapter was published as the epilogue to Stanislav Grof's *The Way of the Psychonaut: Encyclopedia for Inner Journeys* (Santa Cruz, CA: Multidisciplinary Association for Psychedelic Studies, 2019), 289-320. I am grateful to Stanislav Grof as well as Renn Butler, Max DeArmon, Lilly Falconer, Chad Harris, William Keepin, Sean Kelly, Becca Tarnas, and Yvonne Smith Tarnas for their helpful comments, and to Darrin Drda for his careful work on the illustration on p. 107.

1. C. G. Jung, *Memories, Dreams, Reflections* (1962), (New York: Vintage, 1989), 335.
2. We began our research using Reinhold Ebertin's *Transits* and the transit booklets for individual planets by Frances Sakoian and Louis Acker, followed soon after by the just-published *Planets in Transit* by Robert Hand, Ebertin's *The Combinations of Stellar Influences*, Charles Carter's *Principles of Astrology*, Sakoian and Acker's *Handbook of Astrology*, and several pioneering works by Dane Rudhyar.
3. For the sake of simplicity and brevity, I will include Pluto here as a "planet." The correlations we consistently observed with respect to Pluto did not seem to reflect any tangible difference in archetypal importance compared with correlations involving Neptune, Uranus, Saturn, and the other traditional planets.
4. J. M. Masson, *The Oceanic Feeling: The Origins of Religious Sentiment in Ancient India* (Boston; MA: D. Reidel Publishing, 2012), 33-34.
5. Personal communication, Esalen Institute, March 1974.
6. Besides the *Collected Works of Carl Gustav Jung* (trans. R. F. C. Hull, ed. H. Read, M. Fordham, G. Adler, W. McGuire, Bollingen Series XX [Princeton,

N.J.: Princeton University Press, 1953–79]), especially valuable have been the writings and lectures of James Hillman, notably his magnum opus, *Re-Visioning Psychology* (New York: Harper, 1975), and his remarkable early essay "On Senex Consciousness," first published in *Spring: Annual of Archetypal Psychology and Jungian Thought* (1970), and now available in *Puer and Senex, Uniform Edition of the Writings of James Hillman*, vol. 3 (Thompson, Conn: Spring, 2015).

7. *Cosmos and Psyche: Intimations of a New World View* (New York: Viking, 2006).
8. I have discussed Jung's concept of synchronicity and its relationship to the astrological correlations in greater depth in *Cosmos and Psyche*, pp. 50–79.
9. Plotinus, *Enneads*, II, 3, 7, "Are the Stars Causes?" (c. 268), quoted in Eugenio Garin, *Astrology in the Renaissance*, trans. C. Jackson and J. Allen, rev. C. Robertson (London: Arkana, 1983), p. 117.
10. *The Passion of the Western Mind* (New York: Crown Harmony, 1991; Ballantine, 1993) sets out a narrative history of the Western world view in which the evolution of the archetypal perspective plays a central role, from Plato and the ancient Greeks to Jung and the postmodern. *Cosmos and Psyche* provides a summary overview of the archetypal perspective and the multidimensional and multivalent nature of planetary archetypes before going on to examine the evidence of planetary correlations with the archetypal patterns and cycles of history. Finally, my "Notes on Archetypal Dynamics and Complex Causality," originally written and privately shared with colleagues in 2002, later published in three parts in *Archai: The Journal of Archetypal Cosmology*, Issues 4, 5, and 6 (2012, 2016, 2017), represents a more systematic effort to understand and articulate the unique features of archetypal dynamics observed in planetary correlations and in human experience more generally.
11. While the major aspects between the planets were the most important factors for this research, other factors such as planetary midpoints, minor aspects, progressions, and lunations were often helpful. Of less discernible importance in this context were many factors that are usually focused on in traditional astrology such as the signs, houses, rulerships, and related matters. It was far more important to know that Pluto was transiting in conjunction to the natal Moon than to know whether this happened in Virgo or Libra. The correlations we found of most consequence were thus unrelated to such issues and controversies as the precession of the equinoxes affecting the placement of the zodiacal signs, the two zodiacs (tropical and sidereal), or the multiplicity of potential house systems and rulership systems.

MAPPING THE COURSES OF HEAVENLY BODIES

This article was originally published in the *Journal of Transpersonal Psychology* 23, no. 2 (2000): 103–122.

1. No slight is intended here concerning same-sex unions, as will be clear below. The above statement merely reflects the vast majority of the historical literature.
2. Sex and spirituality seemed more conjoined in ancient times. From what little is known of true indigenous cultures, distinctions between sex and spirituality were not made in the first place. In the established global religions today—Judaism, Christianity, Islam, Hinduism, and Buddhism—esoteric and mainstream conventions may either embrace or reject sex as conducive to spirituality, though it appears that the vast majority of forms deny the role of the body to some extent. Virtually all of them regulate sexual expression, even when it is considered a spiritual path, such as in Judaism and Vajrayana Buddhism. The history of sex and spirituality is beyond the scope of this article. For a brief overview, see Jenny Wade, "The Love that Dares not Speak its Name," in *Transpersonal Knowing: Exploring the Horizon of Consciousness*, eds. Tobin Hart, Peter L. Nelson, and Kaisa Puhakka (Albany, NY: State University of New York Press, 2000), pp. 271-302.
3. Marghanita Laski, *Ecstasy in Secular and Religious Experiences* (Los Angeles: Tarcher, 1961); Abraham H. Maslow, *Motivation and Personality*, revised edition, eds. Robert Frager, James Fadiman, Cynthia McReynolds, and Ruth Cox (New York: Harper & Row, 1987).
4. Harry Maurer, *Sex: An Oral History* (New York: Viking, 1994); Sandra Scantling and Sue Ellin, *Ordinary Women: Extraordinary Sex: Releasing the Passion Within* (New York: Dutton, 1993).
5. Allan L. Smith and Charles T. Tart, "Cosmic Consciousness Experience and Psychedelic Experiences: A First Person Comparison," *Journal of Consciousness Studies* 5, no. 1 (1998): 104.
6. Smith, personal communication, March 1998.
7. Preliminary findings on the transformative powers of these experiences have been reported in Jenny Wade, "Meeting God in the Flesh: Spirituality in Sexual Intimacy," *ReVision* 21, no. 2 (1998): 35-41; Wade, "The Love that Dares not Speak its Name."
8. Stanislav Grof, *Realms of the Human Unconscious: Observations from LSD Research* (New York: Viking, 1975); Stanislav Grof, *The Adventure of Self-Discovery: Dimensions of Consciousness and New Perspectives in Psychotherapy* (Albany, NY: State University of New York Press, 1988).
9. Grof, *Realms of the Human Unconscious*; Stanislav Grof, *Beyond the Brain: Birth, Death and Transcendence in Psychotherapy* (Albany, NY: State University of New York Press, 1985); Grof, *The Adventure of Self-Discovery*; Stanislav Grof, *The Cosmic Game: Explorations of the Frontiers of Human Consciousness* (Albany, NY: State University of New York Press, 1998).
10. Rhea A. White, ed., "Exceptional Human Experience: Studies of the

Unitive/Spontaneous," *Imaginal* 12, no. 2 (1995): 92–134: Grof, *The Adventure of Self-Discovery*, p. 157.

11. Grof, *The Adventure of Self-Discovery*, pp. 42–157.
12. Slightly fuller records have been published previously, and complete ones will be forthcoming. See Wade, "Meeting God in the Flesh;" Wade, "The Love that Dares not Speak its Name."
13. Category titles and descriptions derive, as noted, from Grof, *The Adventure of Self-Discovery*, pp. 42–157, though most are paraphrased, since this version of Grof's represents long prose descriptions of his more concise version in Grof, *Realms of the Human Unconscious*. Grof has expanded his cartography over the years, and continues to do so. Where the 1988 and 1975 versions still coincided closely, paraphrasing as close to his 1975 schema is used owing to its concision.
14. Grof, *Realms of the Human Unconscious.*
15. Mircea Eliade, *Shamanism: Archaic Techniques of Ecstasy*, trans. Willard R. Trask (Princeton, NJ: Princeton University Press, 1964); Holger Kalweit, *Dreamtime and Inner Space: The World of the Shaman*, trans. W. Wunsehe (New York: Summit, 1988); Roger Walsh, *The Spirit of Shamanism* (New York: Tarcher/Perigee, 1990); Fred Alan Wolf, *The Eagle's Quest: A Physicist's Search for Truth in the Heart of the Shamanic World* (New York: Summit, 1991).
16. Incidentally, OOBs in this sample do not include the dissociations of sexually abused individuals; in fact, although the sample includes people who have been abused, they not only reported being present in their bodies during the events, but to have been healed to some degree or another by their embodied yet transcendent sexual encounters.
17. Note: Grof's outline format in terms of hierarchical organization of numbered and lettered categories is not consistent; nevertheless, this paper conforms to his system for consistency and ease of cross-referencing.
18. Grof, *The Adventure of Self-Discovery*, p. 148.
19. Jenny Wade, *Changes of Mind: A Holonomic Theory of the Evolution of Consciousness* (Albany, NY: State University of New York Press, 1996).

FROM 'BAD' RITUAL TO 'GOOD' RITUAL

This article was originally published in an earlier version in *Journal of Prenatal and Perinatal Psychology and Health* 22(2), Winter 2007.

1. Sheila Kitzinger, *Women as Mothers: How They See Themselves in Different Cultures* (New York: Random House, 1978), p. 5.
2. Carol MacCormack, ed., *Ethnography of Fertility and Birth* (London: Academic Press, 1982), p. 10.
3. Robbie Davis-Floyd, "Rituals of the American Hospital," in *Conformity and*

Conflict: Readings in Cultural Anthropology, 8th edition, ed. David McCurdy (Harper/Collins, 1994), p. 325.

4. See Edith Turner, *Experiencing Ritual: A New Interpretation of African Healing* (Philadelphia: University of Pennsylvania Press, 1992).
5. Jack Potter, "Cantonese Shamanism," in *Religion and Ritual in Chinese Society*, ed. Arthur P. Wolf (Stanford: Standford University Press, 1974), pp. 207-231; Lois Paul and Benjamin Paul, "The Maya Midwife as Ritual Specialist: A Guatemalan Case," *American Ethnologist* 2, no. 4 (1975): 707-726; Sheila Kitzinger, "The Social Context of Birth: Some Comparisons between Jamaica and Britain," in *Ethnography of Fertility and Birth*, ed. Carol MacCormack (London: Academic Press, 1982), pp. 181-203; Carol Laderman, *Wives and Midwives: Childbirth and Nutrition in Rural Malaysia* (Berkeley: California University Press, 1983).
6. Gregg Lahood, "An Anthropological Perspective on Near-Death-Like Experiences in Three Men's Pregnancy-Related Spiritual Crises," *Journal of Near-Death Studies* 24, no. 4 (2006): 211-36; Gregg Lahood, "Rumour of Angels and Heavenly Midwives: Anthropology of Transpersonal Events and Childbirth," *Women and Birth* 20, no. 1 (2007): 3-10.
7. Lahood, "An Anthropological Perspective on Near-Death-Like Experiences in Three Men's Pregnancy-Related Spiritual Crises;" Gregg Lahood, "Skulls at the Banquet: Near Birth as Nearing Death," *Journal of Transpersonal Psychology* 38, no. 1 (2006).
8. Stanislav Grof, *The Cosmic Game: Explorations at the Frontiers of Human Consciousness* (Albany, NY: State University of New York, 1998), p. 135.
9. See Robbie Davis-Floyd, *Birth as an American Rite of Passage* (Berkeley: California University Press, 1992); Pamela Klassen, *Blessed Events: Religion and the Home Birth Movement in America* (Princeton and Oxford: Princeton University Press, 2001), p. 104; Susan Sered, "Childbirth as a Religious Experience? Voices from an Israeli Hospital," *Journal for the Feminist Study of Religion* 7, no. 2 (1991): 15.
10. According to philosopher Richard Tarnas, Grof's work is "the most epistemologically significant development in the recent history of depth psychology, and indeed the most important advance in the field since Freud and Jung themselves." Richard Tarnas, *The Passion of the Western Mind: Understanding the Ideas that have Shaped our World View* (New York: Harmony Books, 1991), p. 425.
11. See Lahood, "Rumour of Angels and Heavenly Midwives."
12. See John Heron and Gregg Lahood, "Charismatic Inquiry in Concert: Action Research in the Realm of the Between," in *Handbook of Action Research*, eds. Peter Reason and Hilary Bradbury (London: Sage, 2008), pp. 439-449.
13. Lahood, "An Anthropological Perspective on Near-Death-Like Experiences in Three Men's Pregnancy-Related Spiritual Crises;" Lahood, "Skulls at the Banquet;" Lahood, "Rumour of Angels and Heavenly Midwives."

14. Lahood, "An Anthropological Perspective on Near-Death-Like Experiences in Three Men's Pregnancy-Related Spiritual Crises;" Lahood, "Skulls at the Banquet;" Lahood, "Rumour of Angels and Heavenly Midwives."
15. Lahood, "An Anthropological Perspective on Near-Death-Like Experiences in Three Men's Pregnancy-Related Spiritual Crises."
16. Lahood, "Rumour of Angels and Heavenly Midwives."
17. Various anthropologists have explored non-ordinary states of consciousness, Buddhist meditation techniques, or shamanistic apprenticeships and some have gone on to teach their respective techniques—e.g., Joan Halifax (shamanism and Zen Buddhism), Michael Harner, Felicitas Goodman (neoshamanism), Charles Laughlin (Tibetan Buddhism), Larry Peters (Nepalese shamanism), and Terence McKenna (psychedelic shamanism), to name a few. My own trainings and 'apprenticeships' have been primarily with two Western transpersonal teachers: Stanislav Grof (holotropic breathwork) and John Heron (charismatic co-operative inquiry).
18. Stanislav Grof, *Beyond the Brain: Birth, Death and Transcendence in Psychotherapy* (Albany, NY: State University of New York, 1985).
19. Lahood, "An Anthropological Perspective on Near-Death-Like Experiences in Three Men's Pregnancy-Related Spiritual Crises;" Lahood, "Skulls at the Banquet;" Lahood, "Rumour of Angels and Heavenly Midwives."
20. *Goddess Remembered*, directed by Donna Read (The National Film Board of Canada, 1989), VHS.
21. See Davis-Floyd, *Birth as an American Rite of Passage.*
22. Stanislav Grof, *The Adventure of Self-Discovery* (Albany, NY: State University of New York, 1988); Christopher Bache, *Dark Night, Early Dawn: Steps to a Deep Ecology of Mind* (Albany, NY: State University of New York, 2000); Lahood, "Skulls at the Banquet."
23. Arnold Van Gennep, *The Rites of Passage* (London: Routledge & Keagan Paul, 1960), p. 10.
24. Van Gennep, *The Rites of Passage*, p. 18.
25. Van Gennep, *The Rites of Passage*, pp. 41-64.
26. See Stanislav Grof, "The Implications of Psychedelic Research for Anthropology: Observations from LSD Psychotherapy," in *Symbols and Sentiment: Cross Cultural Studies in Symbolism*, ed. I. M. Lewis (London: Academic Press, 1977), pp. 140-174; Grof, *Beyond the Brain.*
27. Davis-Floyd, *Birth as an American Rite of Passage.*
28. Kitzinger, "The Social Context of Birth," p. 182.
29. Robbie Davis-Floyd and Carol Sargent, eds., *Childbirth and Authoritative Knowledge: Cross-Cultural Perspectives* (University of California Press, 1997), pp. 8-11.
30. Kitzinger, *Women as Mothers*, p. 133. [My emphasis]
31. See Charles Laughlin, "Transpersonal Anthropology: Some Methodological Issues," *Western Canadian Anthropology* 5 (1988): 29-60; Charles Laughlin,

"Psychic Energy & Transpersonal Experience: A Biogenetic Structural Account of the Tibetan Dumo Yoga Practice," *Being Changed by Cross-Cultural Encounters: The Anthropology of Extraordinary Experience*, eds. David Young and Jean-Guy Goulet (Peterborough, Ontario: Broadview Press, 1994), pp. 99-135.
32. Davis-Floyd, *Birth as an American Rite of Passage*, pp. 7-19.
33. Eugene d'Aquili, Charles Laughlin, and John McManus, eds., *The Spectrum of Ritual: A Biogenetic Structural Analysis* (New York: Columbia University Press, 1979); Eugene d'Aquili, "Human Ceremonial Ritual and Modulation of Aggression," *Zygon: Journal of Religion and Science* 20, no. (1) (1985): 21-30.
34. Davis-Floyd, *Birth as an American Rite of Passage*, pp. 11-15.
35. Eugene d'Aquili, Charles Laughlin, and John McManus, eds., *The Spectrum of Ritual*, pp. 173-174.
36. See Laughlin, "Psychic Energy & Transpersonal Experience."
37. Davis-Floyd, *Birth as an American Rite of Passage*, p. 15.
38. Cheryl Beck, "Birth Trauma: In the Eye of the Beholder," *Nursing Research* 53, no. 1 (2004): 28.
39. Penny Simkin and Phyllis Klaus, *When Survivors Give Birth: Understanding and Healing the Effects of Early Sexual Abuse on Childbearing Women* (Seattle, WA: Classic Day Publishing, 2004), p. 92.
40. *DSM-IV: Diagnostic and Statistical Manual of Mental Disorders*, 4th edition (Washington, DC: American Psychiatric Association, 1994).
41. Cheryl Beck, "Birth Trauma."
42. Lahood, "Skulls at the Banquet."
43. Davis-Floyd, *Birth as an American Rite of Passage*, p. 242.
44. Davis-Floyd, *Birth as an American Rite of Passage*, p. 242.
45. Simkin and Klaus, *When Survivors Give Birth*, p. 93.
46. Davis-Floyd, *Birth as an American Rite of Passage*, p. 242.
47. Hayden White, "The Value of Narrative in the Representation of Reality," in *On Narrative*, ed. William John Thomas Mitchell (Chicago: University of Chicago Press, 1980).
48. Davis-Floyd, *Birth as an American Rite of Passage*, p. 242.
49. Douglas Hollan, "Constructivist Models of Mind, Contemporary Psychoanalysis, and the Development of Culture Theory," *American Anthropologist* 102, no. 3 (2000): 539.
50. Hollan, "Constructivist Models of Mind, Contemporary Psychoanalysis, and the Development of Culture Theory," p. 540.
51. Grof, *The Adventure of Self-Discovery*, p. 225.
52. According to Hollan, recent anthropological inquiry into spirit-possession shows that possession idioms are a means "by which otherwise unknowable, suppressed or repressed knowledge...is directly or indirectly expressed." Hollan, "Constructivist Models of Mind, Contemporary Psychoanalysis, and the Development of Culture Theory," p. 539. This is similar to Lévi-Strauss's

argument that the Cuna shaman in their childbirth ritual was expressing 'otherwise inexpressible psychic states.' Claude Levi-Strauss, *Structural Anthropology* (New York: Basic Books, 1963).

53. Richard Katz, "Education for Transcendence: !Kia-healing with the Kalahari !Kung," in *Kalahari Hunters and Gatherers*, eds. Richard B. Lee and Irven DeVore (Cambridge, MA: Harvard University Press, 1976), p. 287.
54. See Tarnas, *The Passion of the Western Mind*, p. 429-31; Neva Walden, "Contributors of Transpersonal Perspectives to Understanding Sexual Abuse," *ReVision* 15, no. 4 (1993): 169.
55. Grof, *The Adventure of Self-Discovery*, p. 227.
56. See Kitzinger, "The Social Context of Birth," p. 181-203; Davis-Floyd, *Birth as an American Rite of Passage*, p. 69; Klassen, *Blessed Events*, p. 181.
57. Davis-Floyd, *Birth as an American Rite of Passage*, p. 69.
58. Davis-Floyd, *Birth as an American Rite of Passage*, p. 69.
59. Brian Bates and Allison Turner, "*Imagery and Symbolism in the Birth Practices of Traditional Cultures*," in The Manner Born: Birth Rites in Cross Cultural Perspective, ed. L. Dundes (Walnut Creek, CA: Altimira Press, 2003), p. 87-97; p. 88.
60. Bates and Turner, "Imagery and Symbolism in the Birth Practices of Traditional Cultures," p. 89.
61. Simkin and Klaus, *When Survivors Give Birth*, p. 71.
62. Simkin and Klaus, *When Survivors Give Birth*, p. 68.
63. Simkin and Klaus, *When Survivors Give Birth*, p. 80.
64. Walden, "Contributors of Transpersonal Perspectives to Understanding Sexual Abuse," p. 173.
65. Walden, "Contributors of Transpersonal Perspectives to Understanding Sexual Abuse," p. 170.
66. Walden, "Contributors of Transpersonal Perspectives to Understanding Sexual Abuse," p. 170. This is not to suggest that sexual abuse can, in any way, be seen as any kind of initiation process. Only that the healing process, like the ritual process, can follow a dialectical pattern based on the pattern of birthing.
67. Lahood, "An Anthropological Perspective on Near-Death-Like Experiences in Three Men's Pregnancy-Related Spiritual Crises;" Lahood, "Skulls at the Banquet;" Lahood, "Rumour of Angels and Heavenly Midwives."
68. Davis-Floyd, "Rituals of the American Hospital," p. 331. [My emphasis]
69. Cheryl Beck, "Birth Trauma," p. 22. [My emphasis]
70. Carol Laderman, "The Ambiguity of Symbols in the Structure of Healing," *Social Science and Medicine* 24, no. 4 (1987): 300.
71. James McClenon, *Wondrous Healing: Shamanism, Human Evolution, and the Origin of Religion* (Dekalb, Illinois: Northern University Press, 2002), p. 53.
72. McClenon, *Wondrous Healing*, pp. 46-57.
73. Grof, *Beyond the Brain*; Lahood, "Skulls at the Banquet."

74. Kitzinger, "The Social Context of Birth," p. 195.
75. Megan Biesele, "An Ideal of Unassisted Birth: Hunting, Healing, and Transformation Among the Kalahari Ju/'hoansi," in *Childbirth and Authoritative Knowledge*, eds. Robbie Davis-Floyd and Carolyn Sargent (Berkeley: University of California Press, 1997), p. 476.
76. Stanislav Grof and Joan Halifax, *The Human Encounter with Death* (London: Souvenir Press, 1977), p. 5.
77. Biesele, "An Ideal of Unassisted Birth."
78. Jeanne Achterberg, "Ritual: The Foundation for Transpersonal Medicine," *Revision* 14, no. 3 (1992): 159.
79. Jurgen Kremer, "The Shadow of Evolutionary Thinking," in *ReVision: A Journal of Consciousness and Transformation* (1996), 19 (1), 41-48.
80. Bruce Kapferer, "Performance and the Structuring of Meaning and Experience," in *The Anthropology of Experience*, eds. Victor Turner and Edward Bruner (Urbana: University of Illinois, 1986), p. 195.
81. Kapferer, "Performance and the Structuring of Meaning and Experience," p. 193.
82. Robert A. McDermott, "Transpersonal Worldviews: Historical and Philosophical Reflections," In *Paths Beyond Ego: The Transpersonal Vision*, eds. Roger Walsh and Frances Vaughn (Jeremy P. Tarcher/Putnam Books, 1993), pp. 206-211.
83. Tarnas, *The Passion of the Western Mind*, p. 419.
84. Robbie Davis-Floyd, *Knowing: A Story of Two Births* (unpublished manuscript, 2002), pp. 10-11.
85. Tarnas, *The Passion of the Western Mind*, p. 437.
86. Kapferer, "Performance and the Structuring of Meaning and Experience," p. 191.
87. Achterberg, "Ritual."
88. See Tarnas, *The Passion of the Western Mind*; Jorge Ferrer, *Revisioning Transpersonal Theory: A Participatory Vision of Human Spirituality* (Albany, NY: State University of New York, 2002).
89. Grof, *The Cosmic Game*; Bache, *Dark Night, Early Dawn*.
90. Grof, "The Implications of Psychedelic Research for Anthropology," p. 167.
91. Bache, *Dark Night, Early Dawn*, p. 9.
92. Edward Whitmont, *Return of the Goddess*. (London: Routledge & Keagan Paul, 1983), pp. 17-18.
93. See Adolphus Elkin, *Aboriginal Men of High Degree* (New York: St. Martin's Press, 1977); Turner, *Experiencing Ritual*.
94. Victor Turner, *The Ritual Process* (New York: Aldine De Gruyter, 1969), pp. 52-53.
95. See Grof, *The Cosmic Game*; Bache, *Dark Night, Early Dawn*.

96. John Heron, *Sacred Science: Person-Centred Inquiry into the Spiritual and the Subtle* (Ross-on-Wye, Herefordshire: PCCS Books, 1998).
97. Heron, *Sacred Science.*
98. Grof, *The Adventure of Self-Discovery*, p. 166.
99. Victor Turner, *Process, Performance and Pilgrimage* (New Delhi: Concept Publishing House, 1979), p. 154.
100. Bache, *Dark Night, Early Dawn*, p. 13.
101. Bache, *Dark Night, Early Dawn*, p. 30; Ferrer, *Revisioning Transpersonal Theory*," p. 149.
102. Richard Katz, "Education for Transcendence," p. 290. [My emphasis]
103. Grof, *The Adventure of Self-Discovery*, p. 207.
104. Davis-Floyd, Birth as an American Rite of Passage, p. 245.

THE USE OF HOLOTROPIC BREATHWORK IN THE INTEGRATED TREATMENT OF MOOD DISORDERS

This article was originally published in *Canadian Journal of Psychotherapy and Counseling* (2011). The names and a few personal details of the persons mentioned in the case vignettes have been adjusted to protect their privacy.

1. Jules Angst and Alex Gamma, "Prevalence of Bipolar Disorders: Traditional and Novel Approaches," *Clinical Approaches in Bipolar Disorders* 1 (2002): 10-14; Hagop S. Akiskala, Marc L. Bourgeois, Jules Angst, Robert Post, Hans-Jürgen Möller, and Robert Hirschfeld, "Re-evaluating the Prevalence of and Diagnostic Composition within the Broad Clinical Spectrum of Bipolar Disorders," *Journal of Affective Disorders* 59, Suppl 1 (2000): S5–S30.
2. Heinz Katschnig, "Are Psychiatrists an Endangered Species? Observations on Internal and External Challenges to the Profession," *World Psychiatry* 9 (2010): 21-28.
3. Irving Kirsch, Brett J. Deacon, Tania B. Huedo-Medina, Alan Scoboria, Thomas J. Moore, and Blair T. Johnson, "Initial Severity and Antidepressant Benefits: A Meta-Analysis of Data Submitted to the Food and Drug Administration," *PLoS Medicine* 5 (2008): 260-268; Erick H. Turner, Annette M. Matthews, Eftihia Linardatos, Robert A. Tell, and Robert Rosenthal, "Selective Publication of Antidepressant Trials and its Influence on Apparent Efficacy," *New England Journal of Medicine* 358 (2008): 252-260.
4. Caroline Mardon, private communication, 2010.
5. Kimberly Goldapple, Zindel Segal, Carol Garson, Mark Lau, Peter Bieling, Sidney Kennedy, and Helen Mayberg, "Modulation of Cortical-Limbic Pathways in Major Depression: Treatment-Specific Effects of Cognitive Behavior Therapy," *Archives of General Psychiatry* 61 (2004): 34-41.
6. Stanislav Grof and Christina Grof, *Holotropic Breathwork: A New Approach to*

Self-Exploration and Therapy (Albany, NY: State University of New York Press, 2010).
7. Grof and Grof, *Holotropic Breathwork.*
8. Paul Grof and Leah Crawford, "Psychiatry and Music: Healing with words and tones," *Psychiatrie* 12, no. 2 (2008): 88-89.
9. Mardon, personal communication, 2010.
10. Grof and Grof, *Holotropic Breathwork.*
11. Grof and Grof, *Holotropic Breathwork.*
12. Paul Grof, *Holotropic Approach to Major Mood Disorders*, Videotape, (Black Productions, Ottawa, ON, 1999).
13. Grof, *Holotropic Approach to Major Mood Disorders.*
14. Kevin C. Martin, Paul Grof, and Arlene Fox, "Which Patients Leave Holotropic Groups Prematurely? A Long-Term Study of a Group of Psychiatric Patients," *The Inner Door* 2 (1996): 1.
15. Grof and Grof, *Holotropic Breathwork.*

BREATHING NEW LIFE INTO SOCIAL TRANSFORMATION

1. William Keepin, Cynthia Brix, and Molly Dwyer, *Divine Duality: The Power of Reconciliation between Women and Men* (Prescott, AZ: Hohm Press, 2007).
2. Keepin, Brix, and Dwyer, *Divine Duality.*
3. Steven Foster and Meredith Little co-founded the School of Lost Borders, a wilderness guide training school.
4. Pride Chigwedere, George Seage III, Sofia Gruskin, Tun-Hou Lee, and M. Essex, "Estimating the Lost Benefits of Antiretroviral Drug Use in South Africa," *Journal of Acquired Immune Deficiency Syndrome* 49, no. 4 (December 2008): 410-415.
5. Stephen Bevan, "African minister ends decade of denial on Aids," The Telegraph, December 11, 2006, http://www.telegraph.co.uk/news/health/news/3346075/African-minister-ends-decade-of-denial-on-Aids.html.
6. Michael Wines, "AIDS Activist Nozizwe Madlala-Routledge Keeps her Convictions but Loses her Job," *New York Times*, September 7, 2007, nytimes.com/2007/09/07/world/africa/07iht-profile.4.7423546.html?pagewanted=all&_r=1.
7. "The Cancer of Patriarchy in the Church, Society and Politics in South Africa" conference in Cape Town, South Africa, March 5-6, 2009.

PERSONAL REFLECTIONS ON THE MYSTERY OF DEATH AND REBIRTH IN LSD THERAPY

1. Stanislav Grof, *LSD Psychotherapy* (Pomona: Hunter House, 1980).
2. Charles Grob, ed., *Hallucinogens* (New York, NY: Jeremy P. Tarcher, 2002);

Stanislav Grof, *Realms of the Human Unconscious* (New York: Dutton, 1976); Stanislav Grof, *LSD Psychotherapy* (Pomona: Hunter House, 1980); Stanislav Grof, *Beyond the Brain: Birth, Death and Transcendence in Psychotherapy* (Albany: State University of New York Press, 1985); Stanislav Grof, *The Cosmic Game: Explorations of the Frontiers of Human Consciousness* (Albany: State University of New York Press, 1998); Stanislav Grof, *The Ultimate Journey* (Ben Lomond, CA: MAPS, 2006); Ralph Metzner, *Sacred Vine of Spirits: Ayahuasca* (Rochester, VT: Park Street Press, 1999); Ralph Metzner, *Sacred Mushroom of Visions* (Rochester, VT: Park Street Press, 2004); Marlene Rios and Oscar Janiger, *LSD, Spirituality, and the Creative Process* (Rochester, VT: Park Street Press, 2003); Rick Strassman, *DMT: The Spirit Molecule* (Rochester, VT: Park Street Press, 2001); Roger Walsh and Charles Grob, eds. *Higher Wisdom* (Albany, NY: State University of New York, 2005).

3. Grof discusses these levels of consciousness in many places starting with *Beyond the Brain* but most fully in his *Psychology of the Future* (Albany: State University of New York Press, 2000), pp. 65-68. I don't think the order in which these levels of consciousness emerged in my sessions reflects a universal or cross-cultural norm. Different spiritual traditions endorse different sequences of initiation, and there is considerable variance of opinion on what constitutes the deeper and deepest levels of reality. (See, for example, Jorge Ferrer, *Revisioning Transpersonal Theory* [Albany: State University of New York Press, 2002]). I am simply marking the stages in how the universe took me in, without suggesting that these are normative for other people in other settings.
4. See Myron Stolaroff's wise and seasoned essay where he discusses the advantages of integrating low doses of LSD (25-50 mcg) into one's contemplative practice. Myron Stolaroff, "Are Psychedelics Useful in the Practice of Buddhism?" *Journal of Humanistic Psychology* 39, no. 1 (1999): 60-80.
5. When I resumed my work after the hiatus, the ocean of suffering resumed *exactly* where it had stopped six years before without missing a beat. This deserves careful attention. In a different period of my life, with fresh expectations and under different astrological transits, this part of my journey resumed exactly where it had stopped, demonstrating, I think, the precision and power of the intelligence guiding these explorations.
6. Christopher Bache, *LSD and the Mind of the Universe* (Rochester, Vermont: Park Street Press, 2019), p. 73.
7. Rupert Sheldrake, *A New Science of Life* (Los Angeles: J.P. Tarcher, 1981); Rupert Sheldrake, *The Presence of the Past* (New York: Vintage, 1988); Rupert Sheldrake, *The Rebirth of Nature* (New York: Bantam, 1991).
8. Christopher Bache, *Dark Night, Early Dawn* (Albany, NY: State University of New York Press, 2000), pp. 86-94.
9. See "Expanding the Narrative—Who Is the Patient?" in Bache, *LSD and the Mind of the Universe*, pp. 135-141. A META-COEX is a collective structure

within the species-mind that gathers the experiences of humanity into giant memory clusters sharing a common emotional theme. Bache, *LSD and the Mind of the Universe*, p. 79. In *Dark Night, Early Dawn*, I called these collective structures Meta-Matrices. Bache, *Dark Night, Early Dawn*, p. 87.

10. Personal communication, October 25, 2006.
11. I will move back and forth between using the terms psychic, subtle, and causal to refer to levels of consciousness and the levels of reality one becomes aware of through these states. Subtle states of consciousness reveal subtle levels of reality, causal states reveal causal levels of reality, and so on. This is the great value of entering these states, that through them we gain access to deeper dimensions of existence.
12. See "Dying and Causal Oneness" in Bache, *LSD and the Mind of the Universe*, pp. 193-195.
13. Bache, *LSD and the Mind of the Universe*, pp. 244-246.
14. I don't know what these impurities consist of exactly. I think of them as the biochemical correlates of our petty judgments, lower sentiments, and self-cherishing emotions. More broadly, they are anything in us that is incompatible with the deeper reality that is emerging in our sessions.
15. Christopher Bache, "Death and Rebirth in LSD Therapy: An Autobiographical Study," *Journal of Transpersonal Research* 7, no. 1 (2015): 89-90; Bache, *LSD and the Mind of the Universe*, pp. 313-316.
16. Sheldrake, *A New Science of Life*; Sheldrake, *The Presence of the Past*; Sheldrake, *The Rebirth of Nature*; Ervin Laszlo, *The Interconnected Universe* (River Edge, NJ: World Scientific, 1995); Ervin Laszlo, *The Connectivity Hypothesis* (Albany: State University of New York Press, 2003); Ervin Laszlo, *Science and the Akashic Field* (Rochester, VT: Inner Traditions, 2004); Ervin Laszlo, *The Self-Actualizing Cosmos* (Rochester, VT: Inner Traditions, 2014).
17. Ego-state psychology has demonstrated that this compartmentalizing of experience is a common feature of our psychological makeup. Many areas of our inner life have this encapsulated, semi-autonomous quality. In drawing this parallel with ego states, however, I want to add an important qualification. Ego-state psychology tends to see ego states as created in reaction to trauma. The shamanic persona, however, is born from a surplus of blessings, from the expanding of our experiential horizons in psychedelic states. On ego-state psychology, see Gordon Emmerson, *Ego State Therapy* (Bethel, CT: Crown House Publishing, 2007); John Rowan, *Subpersonalities* (New York: Routledge, 1990); John Watkins, *Hypnotherapeutic Techniques* (New York: Irvington Publishers, 1987); Thomas Zinser, *Soul-Centered Healing* (Grand Rapids, MI: Union Street Press, 2011).
18. I may be oversimplifying matters by suggesting that there is only one shamanic persona in existence at any given time. Perhaps this is true, but perhaps there are multiple shamanic personas existing within us at one time, each

integrating a specific level of our psychedelic experience (just as there can be multiple ego-states coexisting within the psyche). This diversity would require a more nuanced presentation. We may need to think in terms of clusters of shamanic personas. I don't think this adjustment, however, would change the fundamental recommendation I'm making here concerning the role of the shamanic persona in death and rebirth.

19. Bache, *Dark Night, Early Dawn*, p. 298.
20. Bache, *Dark Night, Early Dawn*, pp. 295-299.
21. Sri Aurobindo, *Savitri*, 3rd ed. (Pondicherry: Sri Aurobindo Ashram, 1970); Satprem, *Sri Aurobindo or The Adventure of Consciousness* (Pondicherry: Sri Aurobindo Ashram, 1993); Eric Weiss, *The Long Trajectory* (Bloomington, IN: iUniverse, 2012).

SEEKERS OF A SECOND BIRTH

I am grateful to Cynthia Morrow and Eric Weiss for many helpful discussions and friendly support throughout the gestation of this paper, to Robert McDermott for his astute reading of the penultimate draft, and to Stan Grof and John Buchanan for inviting me to participate in the seminar for which this paper was originally written.

1. Sean Kelly, *Coming Home: The Birth and Transformation of the Planetary Era* (Great Barrington, MA: Lindisfarne Books, 2010), pp. 118ff.
2. Sean Kelly, *Becoming Gaia: On the Threshold of Planetary Initiation* (Olympia, WA: Integral Imprint, 2021), pp. 121ff.
3. William James, *The Varieties of Religious Experience* (New York: Penguin Books, 1985), p. 231.
4. James, *The Varieties of Religious Experience*, p. 232.
5. James, *The Varieties of Religious Experience*, p. 233.
6. Stanislav Grof, *Beyond the Brain: Birth, Death, and Transcendence in Psychotherapy* (Albany, NY: State University of New York Press, 1985), pp. 1-91.
7. Grof, *Beyond the Brain*, pp. 345-346.
8. Stanislav Grof, *The Adventure of Self-Discovery* (Albany, NY: State University of New York Press, 1988).
9. See Ken Wilber, *Sex, Ecology, Spirituality: The Spirit of Evolution* (Boston: Shambhala, 1995).
10. Ken Wilber, "A More Integral Approach," in *Ken Wilber in Dialogue: Conversations with Leading Transpersonal Thinkers*, eds. Donald Rothberg and Sean Kelly (Wheaton, IL: Quest Books, 1998).
11. James, *The Varieties of Religious Experience*, p. 426.
12. James, *The Varieties of Religious Experience*, p. 426.
13. James, *The Varieties of Religious Experience*, p. 243.

14. Grof, *Beyond the Brain*, p. 137.
15. Stanislav Grof with Hal Zina Bennett, *The Holotropic Mind* (San Francisco: HarperCollins, 1992), pp. 206-207.
16. James, *The Varieties of Religious Experience*, p. 196.
17. James, *The Varieties of Religious Experience*, p. 189.
18. Christina Grof and Stanislav Grof, *The Stormy Search for the Self* (Los Angeles: Jeremy P. Tarcher, 1990).
19. I have argued for the abandonment of the term "prepersonal," as I consider it more consistent and fitting, from an ethical point of view, to grant full personhood to the neonate and infant—which does not mean that I don't recognize, and maintain, the distinction between the pre and the trans in other respects (as in pre-mental egoic and trans-mental egoic, or pre-formal and post- or trans-formal). Sean Kelly, "Revisioning the Mandala of Consciousness," in *Ken Wilber in Dialogue: Conversations with Leading Transpersonal Thinkers*, eds. Donald Rothberg and Sean Kelly (Wheaton, IL: Quest Books, 1998).
20. James, *The Varieties of Religious Experience*, p. 276.
21. James, *The Varieties of Religious Experience*, p. 400.
22. Note the ironic patterning of this realization on the Upanishadic "*tat tvam asi*" (literally, "that, thou art").
23. James, *The Varieties of Religious Experience*, p. 160.
24. James, *The Varieties of Religious Experience*, p. 147.
25. James, *The Varieties of Religious Experience*, p. 148.
26. The setting, one will notice, bears no small resemblance to a session of holotropic breathwork.
27. James, *The Varieties of Religious Experience*, p. 250. For additional examples of passages suggestive of BPM IV, see pages 248, 252, and 255; for BPM III, see pages 265n., 287, and 307.
28. Richard Tarnas, *The Passion of the Western Mind* (New York: Harmony, 1991), p. 429.
29. Sean Kelly, *Individuation and the Absolute: Hegel, Jung, and the Path toward Wholeness* (Mahwah, New Jersey, 1993); Kelly, *Coming Home*.
30. Ken Wilber, *A Brief History of Everything* (Shambhala, 1996).
31. I say ambiguously, since there is always the possibility of the integration of the two in what Wilber calls the "centaur."
32. In terms of Hegel's system, while the metanatal corresponds to the logic, the natal corresponds to the realms of nature and finite spirit, and the transnatal to the realm of absolute Spirit.
33. James, *The Varieties of Religious Experience*, p. 85.
34. James, *The Varieties of Religious Experience*, p. 83.
35. The Mind Cure movement had its origins in nineteenth century "animal magnetism," was a direct influence on Christian Science, and is one of the

sources of the kinds of idealism commonly espoused by many proponents of New Age spirituality.

36. James, *The Varieties of Religious Experience*, p. 102.
37. James, *The Varieties of Religious Experience*, p. 139.
38. James, *The Varieties of Religious Experience*, p. 136.
39. James, *The Varieties of Religious Experience*, p. 139.
40. James, *The Varieties of Religious Experience*, p. 488n.
41. James, *The Varieties of Religious Experience*, p. 135.
42. James, *The Varieties of Religious Experience*, p. 167.
43. Or one could say, following Wilber's distinctions, that the merely once-born, whether healthy-minded or sick-souled, would tend toward predominantly translative and legitimizing (rather than transformative and authentic) forms of religious expression. See Ken Wilber, *A Sociable God* (Boston: Shambhala, 1983).
44. James, *The Varieties of Religious Experience*, p. 187.
45. James, *The Varieties of Religious Experience*, p. 163.
46. This judgment of the healthy-minded worldview is resonant with James's assessment of the truth of his own "anaesthetic revelation" along with his (to my mind, partial mis-) reading of Hegel.
47. James, *The Varieties of Religious Experience*, p. 271.
48. James, *The Varieties of Religious Experience*, p. 272.
49. James, *The Varieties of Religious Experience*, p. 273.
50. James, *The Varieties of Religious Experience*, p. 273.
51. James, *The Varieties of Religious Experience*, p. 273-274.
52. James, *The Varieties of Religious Experience*, p. 285.
53. James, *The Varieties of Religious Experience*, p. 306ff.
54. James, *The Varieties of Religious Experience*, p. 310.
55. Although these virtues, from the point of view of the hard-headed and worldly-wise, are themselves taken to extremes by the saintly types, James considers the saints, in all their excess, to be representative of what Wilber describes as the "advanced mode" or "leading tip" of the collective consciousness. See Ken Wilber, *Up from Eden* (Boulder: Shambhala, 1983). The saints, James says, are "the great torch-bearers..., the tip of the wedge, the clearers of the darkness. Like the single drops which sparkle in the sun as they are flung far ahead of the advancing edge of a wavecrest or a flood, they show the way and are the forerunners. The world is not yet with them, so they often seem in the midst of the world's affairs to be preposterous. Yet they are impregnators of the world, vivifiers and animaters of potentialities of goodness which but for them would lie forever dormant." James, *The Varieties of Religious Experience*, p. 358.
56. James, *The Varieties of Religious Experience*, p. 362.
57. James, *The Varieties of Religious Experience*, p. 365.
58. Grof, *Beyond the Brain*, p. 401.

59. Grof, *Beyond the Brain*, pp. 402-403.
60. James, *The Varieties of Religious Experience*, p. 379.
61. James lists four essential qualities: ineffability, noetic certainty, transiency, and passivity. James, *The Varieties of Religious Experience*, pp. 380-381.
62. James, *The Varieties of Religious Experience*, p. 419.
63. James, *The Varieties of Religious Experience*, p. 416.
64. James, *The Varieties of Religious Experience*, p. 419.
65. James, *The Varieties of Religious Experience*, p. 425.
66. James, *The Varieties of Religious Experience*, p. 425.
67. James, *The Varieties of Religious Experience*, p. 525.
68. Stanislav Grof, *The Cosmic Game* (Albany, NY: State University of New York Press, 1998), p. 10.
69. Grof, *The Cosmic Game*, p. 130.
70. Grof, *The Cosmic Game*, p. 100.
71. Grof, *The Cosmic Game*, p. 110.
72. Grof, *The Cosmic Game*, p. 123; see also pp. 39, 101, and 127.
73. Grof, *The Cosmic Game*, p. 39.
74. Tarnas, *The Passion of the Western Mind*, p. 429. (I have changed this passage to the present tense.)
75. Grof, *Beyond the Brain*, pp. 138-197.

GROF AND WHITEHEAD

1. Stanislav Grof, *Beyond the Brain: Birth, Death and Transcendence in Psychotherapy* (Albany: State University of New York Press, 1985), p. 16.
2. Grof, *Beyond the Brain*, p. 91.
3. Stanislav Grof, *The Cosmic Game: Explorations of the Frontiers of Human Consciousness* (Albany: State University of New York Press, 1998), pp. 2-3. Grof's theories in *The Cosmic Game* are anticipated in a paper he wrote in the early 1970s: Stanislav Grof, "LSD and the Cosmic Game: Outline of Psychedelic Cosmology and Ontology," *Journal for the Study of Consciousness* 5 (1972): 165ff. The original paper is well worth revisiting.
4. Alfred North Whitehead, *Process and Reality: An Essay in Cosmology*, corrected edition, eds. David Ray Griffin and Donald W. Sherburne (New York: Free Press, 1979), p. 3.
5. For more on process philosophy's move towards "domain uniformitarianism," see David Ray Griffin, *Reenchantment without Supernaturalism: A Process Philosophy of Religion* (Ithaca, NY: Cornell University Press, 2001), pp. 18-19.
6. Ken Wilber's theory of *holons,* as presented in *Sex, Ecology, Spirituality,* is perhaps the closest thing to a speculative philosophy to be found in contemporary transpersonal thought. Ken Wilber, *Sex, Ecology, Spirituality: The Spirit of Evolution* (Boston: Shambhala Publications, Inc., 1995). Despite Wilber's frequent

objections to Whitehead's philosophy, Wilber's theories are often surprisingly Whiteheadian. In particular, Wilber's notion of the "holon" bears a striking similarity to Whitehead's conception of the "actual occasion." For a more in-depth discussion of this topic, see John Buchanan, "Whitehead and Wilber: Contrasts in Theory," *The Humanistic Psychologist* 24, no. 2 (1996): 231-256.

7. Grof, *The Cosmic Game*, pp. 11-12.
8. Whitehead, *Process and Reality*, p. 5. Although Whitehead's notion of the "rational" does involve the traditional qualities of coherence and logic (Whitehead, *Process and Reality*, p. 3), the following brief quotations should help to indicate that his vision of a rational philosophy is not inimical to a transpersonal orientation: "Rationalism is an adventure in the clarification of thought, progressive and never final" (Whitehead, *Process and Reality*, p. 9). Furthermore: "If you like to phrase it so, philosophy is mystical. For mysticism is direct insight into depths as yet unspoken. But the purpose of philosophy is to rationalize mysticism: not by explaining it away, but by the introduction of novel verbal characterizations, rationally coordinated," Alfred North Whitehead, *Modes of Thought* (New York: Free Press, 1968), p. 174. I do not think that transpersonal psychology should ever shy away from a maximal effort at conceptual clarity, as long as this process of clarification is not used as an excuse to unjustifiably discount or ignore evidence, or to prematurely close off avenues of thought.
9. For a discussion of three of Whitehead's objections to Einstein's theory of relativity—"(i) the basis for constructing a general theory of relativity; (ii) the underlying uniformity of space-time; and (iii) the basis for measurement," see Victor Lowe, *Alfred North Whitehead: The Man and His Work*, vol. II (Baltimore: John Hopkins University Press, 1985), pp. 123-127.
10. Grof, *The Cosmic Game*, p. 247.
11. Alfred North Whitehead, *Essays in Science and Philosophy* (New York: Rider and Company, 1948), pp. 91-92.
12. Grof, *Beyond the Brain*, p. 22.
13. For Grof's views on science and scientism, see especially chapter one in *Beyond the Brain* and chapter eleven in *The Cosmic Game.*
14. William James, *A Pluralistic Universe*, volume in *The Works of William James*, eds. Frederick Burkhardt and Fredson Bowers (Cambridge: Harvard University Press, 1977); Grof, *Beyond the Brain*, pp. 24-25.
15. Whitehead, *Modes of Thought*, p. 154.
16. Grof, *Beyond the Brain*, pp. 41-48.
17. For a masterful treatment of the mind-body problem from a Whiteheadian perspective, see David Ray Griffin, *Unsnarling the World-Knot: Consciousness, Freedom, and the Mind-Body Problem* (Berkeley: University of California Press, 1998).
18. Grof, *The Cosmic Game*, pp. 236-242.

19. Grof, *Beyond the Brain*, p. 73.
20. Grof, *The Cosmic Game*, p. 3.
21. Grof, *The Cosmic Game*, p. 101.
22. Whitehead, *Process and Reality*, p. 21.
23. Whitehead, *Process and Reality*, p. 21.
24. Whitehead, *Process and Reality*, p. 267.
25. Grof, *The Cosmic Game*, p. 3.
26. Wilber, *Sex, Ecology, Spirituality*, p. 111.
27. Wilber, *Sex, Ecology, Spirituality*, p. 604.
28. Griffin has enumerated three ways in which Whitehead's type of "animism" differs from most premodern forms: (a) "the power of perception and self-movement is not attributed to things such as rocks, lakes and suns. A distinction is made between individuals and aggregates of individuals. Only true individuals are self-moving, perceiving things," (b) "radically different levels of *anima* are assumed. To have experience is not necessarily to have conscious experience, let alone *self-conscious*" experience, and (c) "whereas most premodern animisms (Buddhism was an exception) thought of the basic units of the world as enduring souls, postmodern animism takes the basic units to be momentary experiences." David Ray Griffin, *God and Religion in the Postmodern World: Essays in Postmodern Theology* (Albany: State University of New York Press, 1989), p. 88.
29. See Whitehead, *Process and Reality*, p. 166.
30. Whitehead, *Process and Reality*, p. 116.
31. Whitehead says that the "creativity of the world is the throbbing emotion of the past hurling itself into a new transcendent fact." Alfred North Whitehead, *Adventures of Ideas* (New York: Free Press, 1967), p. 177.
32. Whitehead describes three major modes of perception: (a) "perception in the mode of causal efficacy," which involves a direct feeling connection to another event; (b) "perception in the mode of presentational immediacy," which involves a sheer awareness of the normal sensory display; and (c) "symbolic reference," which is a mixed mode of perception resulting from the integration of the other two pure modes into a meaningful perception of the world. Perception in the mode of presentational immediacy is a primary source of philosophical confusion when assumed to be the only valid source of experiential evidence.
33. Whitehead, *Process and Reality*, p. 308; Alfred North Whitehead, *Science and the Modern World* (New York: Free Press, 1967), p. 150.
34. Hartshorne goes even further in his evaluation of the conceptual power of Whitehead's "prehension": "Causality, substance, memory, perception, temporal succession, modality, are all but modulations of one principle of creative synthetic experiencing, feeding entirely upon its own prior products. This I regard as the most powerful metaphysical generalization ever accomplished."

Charles Hartshorne, *Creative Synthesis and Philosophic Method* (Lanham, MD.: University Press of America, 1983), p. 107. For Griffin's edifying discussion of the significance of Hartshorne's claim, see David Ray Griffin, John B. Cobb, Jr., Marcus P. Ford, Pete A. Y. Gunter, and Peter Ochs, *Founders of Constructive Postmodern Philosophy: Peirce, James, Bergson, Whitehead, and Hartshorne* (Albany: State University of New York Press, 1993), 208-210.

35. Please do not be disconcerted by my phrase, "past experience"; in most cases, the past I am speaking of is only somewhere between a mere micro- to milli-seconds from the present. Whitehead's reason for saying that we feel only *past* experience is that he believes something must be *fully actual* in order for it to be felt by another experience. Since what is actual must have *already* attained its moment of reality, new moments of experience can only draw information from such past, complete events.
36. Griffin, *Reenchantment without Supernaturalism*, pp. 58-59.
37. See Whitehead, *Process and Reality*, pp. 242-243.
38. Whitehead, *Process and Reality*, p. 162.
39. Grof, *The Cosmic Game*, p. 267.
40. Whitehead, *Science and the Modern World*, p. 185.
41. Grof, *Beyond the Brain*, p. 64.
42. Stanislav Grof, *The Adventure of Self-Discovery: Dimensions of Consciousness and New Perspectives in Psychotherapy and Inner Exploration* (Albany: State University of New York Press, 1988), p. 102.
43. Alfred North Whitehead, *Function of Reason* (Boston: Beacon Press, 1958), pp. 77-78.
44. Stanislav Grof, *Realms of the Human Unconscious: Observations from LSD Research* (Albany: State University of New York Press, 1976), p. 157.
45. In a recent personal communication, Grof clarifies that while transpersonal experiences definitely provide access to a subjective counterpart to the objects of our world, it is difficult to determine from these experiences the exact quality or degree of awareness that is inherent to the actualities in themselves, that is, whether they share the *same kind* of consciousness that the person feels within the transpersonal state of identification. In addition, Grof notes that verifiable precognitive or clairvoyant transpersonal experiences are very rare.
46. It should be added that process philosophy's position against the attribution of consciousness to the realm of plant life is not a dogmatic one, but rather is based on Whitehead's cosmological views combined with the best available empirical evidence. Process thought could easily entertain other ideas on this question if convincing evidence arises in support of this possibility.
47. Grof, *The Cosmic Game*, pp. 43-47.
48. Griffin, *Reenchantment without Supernaturalism*, p. 91.
49. Henry Corbin, *Creative Imagination in the Sufism of Ibn 'Arabi*, trans. Ralph Manheim (Princeton: Princeton University Press, 1981), p. 20.

SUBTLE CONNECTIONS

This article was first published as "Cosmic Connectivity: Toward a Scientific Foundation for Transpersonal Consciousness," *International Journal of Transpersonal Studies* 23, no. 1 (2004): 21-31.

1. K. Ramakrishna Rao and John Palmer, "The Anomaly Called Psi: Recent Research and Criticism," *Behavioral and Brain Sciences* 10, no. 4 (1987): 539-551.
2. Russel Targ and Harold Puthoff, *Mind-reach: Scientists Look at Psychic Abilities* (New York: Delacorte Press, 1977).
3. Harold Puthoff and Russel Targ, "A Perceptual Channel for Information Transfer over Kilometer Distances: Historical Perspective and Recent Research," *Proceedings of the IEEE* 64 (1976): 329-354; Russell Targ and Keith Harary, *The Mind Race* (New York: Villard Books, 1984); Russel Targ and Harold Puthoff, "Information Transmission under Conditions of Sensory Shielding," *Nature* 251 (1974): 602-607.
4. Montague Ullman and Stanley Krippner, *Dream Studies and Telepathy: An Experimental Approach* (New York: Parapsychology Foundation, 1970); Michael A. Persinger and Stanley Krippner, "Dream ESP Experiments and Geomagnetic Activity," *Journal of the American Society for Psychical Research* 83 (1989): 101-116.
5. Jacobo Grinberg-Zylverbaum, Manuel Delaflor, M. E. Sanchez Arellano, Miguelangel A. Guevara, and M. Perez, "Human Communication and the Electrophysiological Activity of the Brain," *Subtle Energies* 3, no. 3 (1993), 25-43.
6. R. Olistiche, *Report* (Milan, Italy: Cyber, 1992).
7. Daniel J. Benor, *Healing Research*, Vol. 1 (London: Helix Editions, 1993); William Braud and Marilyn Schlitz, "Psychokinetic Influence on Electrodermal Activity," *Journal of Parapsychology* 47 (1983): 95-119; Larry Dossey, *Recovering the Soul: A Scientific and Spiritual Search* (New York: Bantam, 1989); Larry Dossey, *Healing Words: The Power of Prayer and the Practice of Medicine* (San Francisco, CA: Harper. 1993); Charles Honorton, Rick Berger, Mario P. Varvoglis, Marta Quant, P. Derr, E. I. Schechter, and D. C. Ferrari, "Psi Communication in the Ganzfeld: Experiments with an Automated Testing System and a Comparison with a Meta-Analysis of Earlier Studies," *Journal of Parapsychology* 54 (1990): 99-139; Robert Rosenthal, "Combining Results of Independent Studies," *Psychological Bulletin* 85 (1978): 185-193; Mario Varvoglis, "Goal-directed and Observer-dependent PK: An Evaluation of the Conformance-behavior Model and the Observation Theories," *Journal of the American Society for Psychical Research* 80 (1986): 137-162.
8. Braud and Schlitz, "Psychokinetic Influence on Electrodermal Activity."
9. Stanislav Grof, *The Adventure of Self-Discovery* (Albany, NY: SUNY Press, 1988).
10. William James, *The Varieties of Religious Experience* (New York: Modern Library, 1929), p. 378.

11. Stanislav Grof, "Healing and Heuristic Potential of Non-ordinary States of Consciousness: Observations from Modern Consciousness Research," unpublished manuscript (1996).
12. Grof, *The Adventure of Self-Discovery*; Stanislav Grof, *The Holotropic Mind* (San Francisco, CA: Harper, 1993).
13. Grof, *The Adventure of Self-Discovery*; Stanislav Grof, *The Holotropic Mind* (San Francisco, CA: Harper, 1993); Grof, "Healing and Heuristic Potential of Non-ordinary States of Consciousness."
14. Carl Gustav Jung, *The Collected Works of C. G. Jung*, Vol. 8 (Princeton: Princeton University Press, 1958), para. 417.
15. Jung, *The Collected Works of C. G. Jung*, Vol. 8, para. 420.
16. Jung, *The Collected Works of C. G. Jung*, Vol. 8, para. 417-418.
17. Carl Gustav Jung, *The Collected Works of C. G. Jung*, Vol. 14 (Princeton: Princeton University Press, 1958), para. 767.
18. Charles R. Card, "The Emergence of Archetypes in Present-day Science and its Significance for a Contemporary Philosophy of Nature," in *Philosophia Naturalis: Beiträge zu einer zeitgemässen Naturphilosophie*, eds. Thomas Artzt, Maria Hippius-Grafin Durkheim, and Roland Albert Dollinger (Wurzburg, Germany: Konigshausen and Neumann, 1996).
19. László Gazdag, "Combining of the Gravitational and Electromagnetic Fields," *Speculations in Science and Technology* 16, no. 1 (1993): 20-25; László Gazdag, *A Relativitás Elméleten Túl* (*Beyond Relativity Theory*) (Budapest, Hungary: Szenci Molnar Tarsasag, 1995).
20. Bernhard Haisch, Alfonso Rueda, and Harold E. Puthoff, "Inertia as a Zero-point-field Lorentz Force," *Physical Review* 49, no. 2 (1994): 679-694.
21. Anatolij Akimov, "Heuristic Discussion of the Problem of Finding Long-Range Interactions, EGS-Concepts" (Moscow: Center of Intersectoral Science, Engineering, and Venture, Non-Conventional Technologies [CISE VENT], Preprint No. 74, 1991); Gennady I. Shipov, "A Theory of Physical Vacuum," unpublished manuscript (1995).
22. Center of Intersectoral Science, Engineering, and Venture, Non-Conventional Technologies (CISE VENT), *Consciousness and Physical World* (Moscow: CISE VENT, 1995).
23. Peter Petrovich Gariaev, K. V. Grigor'ev, Anatoly Vasiliev, and Vladimir Poponin, "Investigation of the Fluctuation Dynamics of DNA Solutions by Laser Correlation Spectroscopy," *Bulletin of the Lebedev Physics Institute* (1999): 11-12, 23-30; V. P. Poponin, "Modeling of NLE Dynamics in One Dimensional Anharmonic FPU-lattice," Physics Letters A, 1994.
24. T. S. Elliot, "The Wasteland," in *The Complete Poems and Plays: 1909-1950* (New York: Harcourt Brace & Co., 1952), p. 38.

GROF'S PERINATAL MATRICES AND THE SĀṂKHYA *GUṆAS*

I am grateful to Darrin Drda for the illustrations of *puruṣa* and *prakṛti*, and to James Ryan and Christiane Weissbach for their comments on an earlier draft of this chapter.

1. Buddhism rejects the metaphysical essentialism implicit in *triguṇa*, although in the *Kālacakra-tantra*, understood to be a syncretistic text, connections are made between the twelve signs of the zodiac and the twelve members of dependent origination, with each of the twelve associated with one of *tamas*, *rajas*, and *sattva*. Alex Wayman, "Buddhist Dependent Origination and the *Sāṃkhya Guṇas*," *Ethnos* 27, no. 1-4 (1962): 14-22. The *Buddhacarita* contains multiple Sāṃkhyan terms, but the three *guṇas* here seem to relate to three states of being (*bhāvas*); Kent argues that the development of the concept of *triguṇa* parallels the development of Buddhist notions of the roots of good (*kuśalamūla*) and evil (*akuśalamūla*). Stephen A. Kent, "Early Sāṃkhya in the Buddhacarita," *Philosophy East and West* 32, no. 3 (July 1982): 259-78.
2. *Puruṣa* might also be translated as "self" or "spirit," but "pure consciousness" is most consistent with the metaphysics Sāṃkhya expounds. See, for example, Gerald James Larson, Ram Shankar Bhattacharya, and Karl H. Potter, *Encyclopedia of Indian Philosophies, Volume 4: Sāṃkhya: A Dualist Tradition in Indian Philosophy* (Delhi: Motilal Banarsidass, 1987), p. 49.
3. *Sāṃkhya Kārikā* 20. Mikel Burley, *Classical Samkhya and Yoga* (Oxford: Routledge, 2012). I will be basing my reading of the *guṇas* on the *Sāṃkhya Kārikā*, the earliest known scripture of the Sāṃkhya school and considered its representative text. Many scriptures of this school have been lost: for historical accounts, see Gerald James Larson, *Classical Sāṃkhya: An Interpretation of its History and Meaning* (Delhi: Motilal Banarsidass, 2011), pp. 15-74; Lallanji Gopal, *Retrieving Sāṃkhya History: An Ascent from Dawn to Meridian* (New Delhi: D K Printworld, 2000).
4. The analogy of the transparent crystal, as consciousness, illusorily appearing to be colored is found in multiple Eastern texts, including the *Sāṃkhya Sūtras* 2.35. James R. Ballantyne, *The Sāṃkhya Aphorisms of Kapila* (London: Trübner & Co, 1885). While this illustration provides an intuitive understanding of the interplay of the *guṇas* and *puruṣa* in both Sāṃkhya and in the reading of Grof suggested, the role of the intellect (*buddhi*) in *presenting puruṣa* with the *guṇas*' play adds a layer of complexity here: "These specifications of the *guṇas*, distinct from one another, present the whole [world] to *buddhi*, illuminating it like a lamp for the sake of *puruṣa*." Sāṃkhya Kārikā 36. Burley, *Classical Samkhya and Yoga*.
5. *Sāṃkhya Kārikā* 21. Burley, *Classical Samkhya and Yoga*.
6. It is the norm in English translations to translate *prakṛti* along the lines of

"primordial materiality" (see, for example, Larson, Bhattacharya, and Potter, *Encyclopedia of Indian Philosophies*, p. 49), but such a primordial materiality would have to be bifurcated into gross matter and some form of subtle matter constituting the psyche; no such bifurcation is stated in the *Sāṃkhya Kārikā*. Burley argues: "The fundamental dualism for which Sāṃkhya [is] well known is often characterized as one between subject and object... [but] while this object may be taken to represent a world external to the mind, the objectual representation itself appears within consciousness, and is to that extent mental rather than mind-independent. Nowhere in the Sāṃkhya-Yoga schema do we find any explicit references to a world of objects whose reality transcends their being objects for a subject." Burley, *Classical Samkhya and Yoga*, p. 125. For a similar argument, see Jajneswar Ghosh, "Introduction," in *The Sāṃkhya-sūtras of Pañcaśikha and the Saṃkhyatattvāloka*, ed. Swami Hariharānanda (Delhi: Motilal Banarsidass, 1977), p. 7.

7. Burley, *Classical Samkhya and Yoga*, p. x.
8. *Atharva Veda Samhita* 10.8.43a. R. L. Kashyap, trans., *Atharva Veda, Kāṇdas 8, 9 & 10* (Bengaluru, Karnataka: Sri Aurobindo Kapali Sastry Institute of Vedic Culture, 2011). Kashyap's translation with the exception that he translates *guṇas* as "strands."
9. *Chāndogya Upaniṣad*, 6.4.1-4. Swami Nikhilananda, ed. and trans., *The Upanishads: A New Translation*, vol. 4 (New York: Ramakrishna-Vivekananda Center, 2008).
10. *Chāndogya Upaniṣad* 6.4.5. Nikhilananda, *The Upanishads*, vol. 4.
11. *Chāndogya Upaniṣad* 6.4.6-7. Nikhilananda, *The Upanishads*, vol. 4.
12. *Śvetāśvatara Upaniṣad* 1.3a. Swami Nikhilananda, ed. and trans., *The Upanishads: A New Translation*, vol. 2 (New York: Ramakrishna-Vivekananda Center, 2004).
13. See Swami Nikhilananda, ed. and trans., *The Upanishads: A New Translation*, vol. 2 (New York: Ramakrishna-Vivekananda Center, 2004), pp. 108-109 for discussion of this. *Ajā*, from the root √*jan*, to be born, can translate as the name of *prakṛti*, as "what is unborn," as "female," or as "she-goat." The feminine noun *prakṛti* can also refer to a woman, mother, or goddess.
14. *Śvetāśvatara Upaniṣad* 4.5a. Nikhilananda, *The Upanishads*, vol. 2.
15. *Maitrāyaṇīya Upaniṣad* 5.2. Robert E. Hume, ed. and trans., *The Thirteen Principal Upanishads* (Oxford: Oxford University Press, 1921). Hume's translation with the exception that he translates *tamas*, *rajas*, and *sattva* as "Darkness," "Passion," and "Purity," respectively.
16. *Bhagavad Gītā* 14.4-5. Swami Chinmayananda, *Shrimad Bhagavad Gītā* (Mumbai: Central Chinmaya Mission Trust, 2018). Chinmayananda's translation with the exception that he translates *guṇas* as "qualities" and *tamas*, *rajas*, and *sattva* as "inertia," "passion," and "purity," respectively.
17. See in relation to this the discussion of Kālī in the Śaivite Trika tradition

later in this chapter. The word *rajas* derives from the root √*raj*, one of whose translations is to redden. The stages of alchemy, when considered to be three rather than four, are blackening, whitening, and reddening, and the three play roles not only in specific tantric traditions but as the three forms in which tantra itself can be categorized. A seminal study on color terms concluded that all cultures have terms for dark/black and white/bright; if a known culture then has a third, it is invariably red. Brent Berlin and Paul Kay, *Basic Color Terms: Their Universality and Evolution* (Stanford: Center for the Study of Language and Information, 1999), p. 104. And in the *Bhagavad Gītā*, discussed later this chapter, on the bloody field of Kurukṣetra stand Krishna (the word *kṛṣṇa* translatable as black) and Arjuna (the word *arjuna* translatable as white).

18. The Yoga school, understood to be closely affiliated to Sāṃkhya, understands *triguṇa* in a similar manner, although the sheer volume of terminology in the *Yoga Sūtras* does mean that the *guṇas* do not occupy the centrality in the text that they do in the *Sāṃkhya Kārikā*. In Advaita Vedānta *tamas* is associated with the concealing power (*avaraṇa*) of *maya*, *rajas* with its projecting power (*vikṣepa*), and *sattva* with its revealing power; Kashmir Śaivism relates *sattva, rajas,* and *tamas* to pleasure, pain, and ignorance (*moha*), respectively, and *triguṇa* does not play a central role in its metaphysics. See, for example, *Tantrāloka* 9.220b-222. Satya Prakash Singh and Swami Maheshvarananda, trans., *Tantrāloka: Chapters 8-13, Volume 3* (Varanasi, India: Indian Mind, 2016). Neither of these latter conceptions is consistent with the specific formulation I am suggesting in this chapter, as the essential building blocks of a metaphysics of experience, and I have accordingly referenced specifically Sāṃkhya rather than Hindu *guṇas* in the title of this chapter. See, however, Judit Törzsök, "Tantric Goddesses and their Supernatural Powers in the Trika of Kashmiṛ," *Rivista Degli Studi Orientali* 73, fasc. 1/4 (1999): 134-136 and later in this chapter for the apparent connection between the *guṇas* and the three manifestations of Kāli in the Śaivite Trika tradition, in which *triguṇa* appears to undergo a personification in the form of the goddess. Sanderson argues in a similar vein: "[T]he three goddesses were contemplated as the three fundamental constituent powers of creation which was consciousness only." Alexis Sanderson, "Śaivism and the Tantric Traditions," in *The World's Religions*, eds. Peter Clarke, Friedhelm Hardy, Leslie Houlden, and Stewart Sutherland (London: Routledge and Kegan Paul, 1988), p. 674. Such personification is common, with *tamas, rajas,* and *sattva* frequently identified with Kāli or Durgā, Lakṣmī, and Sarasvatī, respectively.

19. *Sāṃkhya Kārikā* 12-13. Burley, *Classical Samkhya and Yoga*.

20. Verses 12-13 of the *Sāṃkhya Kārikā* are the only verses to provide an explicit description of the *guṇas'* characteristics, but elsewhere the *Kārikā* states: "The upper realm is pervaded by luminosity (*sattva*), and the base is pervaded by opacity (*tamas*); the middle is pervaded by activity (*rajas*)...." *Sāṃkhya Kārikā*

54. Burley, *Classical Samkhya and Yoga*. There are more than evocations here of Vedic cosmology's three-fold division of the world into upper, middle, and lower regions. S. G. Weerasinghe, *The Sāṅkhya Philosophy: A Critical Evaluation of its Origins and Development* (Delhi: Sri Satguru Publications, 1993), p. 143. The *guṇas* defined in relation to this cosmology challenges their employment as the base constituents of a metaphysics of experience. We might note, however, that verse 54 is not providing definitions of the *guṇas* themselves. It is rather the characteristics of the three realms that are being described here, in a manner that perhaps evokes the *Gītā's* division of multiple phenomena into their tamasic, rajasic, and sattvic forms.

21. Possibly as a result of this problem, Burley posits the *guṇas* as *capacities* under which experience itself is possible: *rajas*, for example, is then the capacity of mutability. Burley, *Classical Samkhya and Yoga*, pp. 101-107. But this would mean performing a marked interpolation into the *Sāṃkhya Kārikā*.
22. It is true that if the *gunas* are understood in a materialist sense, with *tamas*, *rajas*, and *sattva* as matter in a state of inertia, motion, and perhaps order, respectively (see, for a similar example, Larson, Bhattacharya, and Potter, *Encyclopedia of Indian Philosophies*, p. 66), this would be both pivotal and comprehensive in such a materialist metaphysic, since all matter must logically be in one of these states. As argued, however, the mutual dependence of *puruṣa* and *prakṛti* renders materialist conceptions of *prakṛti* highly problematic.
23. Stanislav Grof, *LSD: Doorway to the Numinous* (Rochester, VT: Park Street Press, 2009), pp. 107-108.
24. Grof, *LSD: Doorway to the Numinous*, p. 118. [Emphasis added.]
25. *Grof, LSD: Doorway to the Numinous*, p. 127.
26. Grof, *LSD: Doorway to the Numinous*, p. 134. [Emphasis added.]
27. Grof, *LSD: Doorway to the Numinous*, pp. 142-3. [Emphasis added.]
28. The *mokṣa* of the non-dual schools is more compatible with Grof's BPM I than is Sāṃkhya's liberated *prakṛti*, which remains separate from *puruṣa* following its liberation. Grof's description of the universe as "an infinite web of adventures in consciousness" would perhaps need reframing from a Sāṃkhyan perspective as an infinite web of *puruṣa's* adventures in *prakṛti*. Stanislav Grof, *Beyond the Brain: Birth, Death, and Transcendence in Psychology* (Albany, NY: State University of New York Press, 1985), p. 50.
29. Stanislav Grof, "The Adventure of Self-Discovery with Stanislav Grof, M.D," interview with Dr. Jeffrey Mishlove, 1998, http://www.williamjames.com/transcripts/grof.htm.
30. Stanislav Grof, *Psychology of the Future* (Albany, NY: State University of New York Press, 2000), p. 29.
31. Grof, *Psychology of the Future*, p. 32.
32. This is an important difference between Grof's model and *mokṣa*, which is an inviolable state of unity and bliss. What Grof describes as negative BPM

I experiences would be classified in terms of the *guṇas* under *tamas*, but Grof's model is helpful in distinguishing hellish experiences where the horror is *constituted in* the absence of boundaries between subject and object (for example, in parasitical attack) from hellish experiences where an external object is inflicting pain on the subject (as in the torturer and the tortured). More generally Grof's BPM I, while accommodating the state of absolute unity, also accommodates specificity—the specific baby in the specific womb—where *mokṣa* does not.

33. Grof, *Beyond the Brain*, p. 102. The natural objection here is that the achievement of *mokṣa*, the result of lifetimes of spiritual toil, is hardly compatible with the achievement of a fetus, and the experience of absolute wisdom surely cannot be the experience of absolute ignorance. This is an example of what Ken Wilber would call the pre-trans fallacy. Ken Wilber, *Sex, Ecology, Spirituality: The Spirit of Evolution* (Boston, MA: Shambala, 2000), pp. 209-212. The following passage is from an account of a BPM I experience in a Grofian session: "I experienced feelings of basic identity and oneness with the universe; it was the Tao, the Beyond that is Within, the Tat tvam asti (Thou art That) of the Upanishads. I lost my sense of individuality; my ego dissolved; and I became all of existence. Sometimes this experience was intangible and contentless, sometimes it was accompanied by many beautiful visions—archetypal images of Paradise, the ultimate cornucopia, golden age, or virginal nature." Grof, *LSD: Doorway to the Numinous*, p. 116. In its intangible and contentless form, this is evocative of a transcendental state such as *mokṣa*. Yet the first sentence above contains a clause prior to it: "During what seemed to be episodes of reliving of positive memories of fetal existence." Grof, *LSD: Doorway to the Numinous*, p. 116. The dialectic here is hence a paradoxical one, but it might be noted that what wisdom and ignorance can share is an unsullied joy in the creative process.
34. Grof, *Beyond the Brain*, p. 105.
35. Grof, *Beyond the Brain*, p. 111.
36. Grof, *Beyond the Brain*, p. 105.
37. Grof, *Beyond the Brain*, p. 116.
38. Grof, *Beyond the Brain*, p. 105.
39. Grof, *Beyond the Brain*, pp. 122-123.
40. Grof, *Beyond the Brain*, p. 105.
41. Sri Aurobindo, Essays on the Gita (Twin Lakes, WI: Lotus Press, 2003), p. 53. [Emphasis added.] It is true that the Hindu tradition does not generally speak of tamasic states as possessing the level of horror of Grof's second perinatal matrix. When the Hindu tradition speaks of a person in the grip of tamas it will talk of powerlessness and inertia, but this does not immediately evoke the trapped and terrified quality of Grof's BPM II. Sri Aurobindo's description of the tamasic person's "dull laziness, slowness, procrastination, looseness, want

of vigour or of sincerity" (Aurobindo, *Essays on the Gita*, p. 502), for example, appears to be of a person in the grip only of the more superficial aspects of BPM II. Yet it is perhaps noteworthy that Sri Aurobindo here only describes how the tamasic person appears to others, rather than describing the actual subjective state: what it feels like, rather than looks like, to be in a tamasic place. Someone whose psyche is heavily under the influence of tamas can certainly, to an external observer, appear derisorily ineffectual. But this ineffectuality is a symptom: a symptom of being psychically stuck in the original terrible experience of a phase of birth when all of us were ineffectual.

42. Aurobindo, *Essays on the Gita*, p. 502. [Emphasis added.]
43. Aurobindo, *Essays on the Gita*, pp. 428-429. [Emphasis added.]
44. In the *Devī Māhātmya* Śiva throws himself under Kālī's feet in an attempt to stop her killing spree.
45. Stanislav Grof, *The Adventure of Self-Discovery* (Albany, NY: State University of New York Press, 1988), pp. 27-35. The *guṇas* recontextualized as specific positions taken in relation to the maternal body, and Grof's own seminal encounter with Kālī, means that his work deserves to be entered into dialogue with the Kālī-worshipping Kālīkula tradition prevalent in North India. The need in Grofian work to internalize the experience of BPM III as much as possible may find a counterpoint in the rituals of the related Kaula school: see Alexis Sanderson, *Meaning in Tantric Ritual* (New Delhi: Tantra Foundation, 2006), pp. 82-83.
46. *Siddhayogeśvarīmata* 6.27a. Törzsök, "Tantric Goddesses and their Supernatural Powers in the Trika of Kashmiṛ."
47. Alexis Sanderson, "Śaivism and the Tantric Traditions," p. 674.
48. Parāparā is placed between Parā and Aparā both physically in ritual and in terms of her characteristics; Törzsök notes also the place that *rajas* (associated with her color red) occupies in the *triguṇa* hierarchy. Törzsök, "Tantric Goddesses and their Supernatural Powers in the Trika of Kashmiṛ," p. 134. Understood in the context of Grof's matrices, Parāparā would lie between Parā and Aparā in a temporal rather than a qualitative sense.
49. *Siddhayogeśvarīmata* 6.21-22. Sanderson, "The Visualization of the Deities of the Trika."
50. Stanislav Grof. *LSD Psychotherapy*. (San Jose, CA: Multidisciplinary Association for Psychedelic Studies, 2008), p. 80b.
51. Alexis Sanderson, "Śaivism and the Tantric Traditions," p. 674.
52. Grof, *Beyond the Brain*, p. 214.
53. *bhagaṃ bhagavatī jñeyā dakṣiṇā triguṇeśvarī* [my translation]. *Niruttara Tantra* 2.4. R. Śukla, ed., *Niruttara Tantram* (Sanskrit. Prayāg: Kalyāna Mandira Prakāśana, 1979).
54. In terms of assessing this contribution, the fact that Grof, without any prior knowledge of the *guṇas*, derived their qualities unassisted is certainly worthy

of note. It should be stressed that there is no suggestion that Hindu scripture stands in any need of such an endorsement and no suggestion that Grofian work is a substitute for yogic or meditative praxis. The point of connection posited here is clearly one that deserves to be published and discussed; it does not, in itself, demonstrate anything that is beyond reasonable doubt other than that Grof's findings here independently derive facets of truths already known.

55. It should be noted that in Sāṃkhya there is not considered to be such a thing as the pure *guṇa*: experience is always a combination of the three. Yet considering the analogy of the color palette, the permutation of maximum red and minimum green and blue is nevertheless a specific permutation of red, green and blue. In a similar vein, we could argue that the extremes of *tamas* endured in Grof's second perinatal matrix, for example, still remains one specific permutation of the *guṇas*: it is just that permutation where there is maximum *tamas* and minimum *rajas* and *sattva*.
56. *Bhagavad Gītā* 14.10. Chinmayananda, *Srimad Bhagavad Gītā.*
57. The Sāṃkhya school discussed in this chapter is tangential to the Sāṃkhya referred to in the *Gītā*, the latter to a great extent being representative of knowledge as opposed to action rather than a reference to a specific school. The characteristics of the *guṇas* as expounded in the *Gītā* are those of the *Sāṃkhya Kārikā*, but its non-dualist metaphysics is not that of the dualist Sāṃkhya school.
58. See, for example, Aurobindo, *Essays on the Gita*, p. 12.
59. The multiple forms of address in this and the below quotations should be understood as from Arjuna to Krishna and vice-versa.
60. *Bhagavad Gītā* 1.32b-33. [Emphasis added.] Chinmayananda, *Srimad Bhagavad Gītā.*
61. *Bhagavad Gītā* 2.2-3. Chinmayananda, *Srimad Bhagavad Gītā.* Many readers have a sense that Arjuna's motivations in not engaging in battle seem to have more laudable roots than Krishna here accords them. It is interesting that Arjuna's state of mind (see *Bhagavad Gītā* 1.29-46) seems to evoke the way that the Saturn archetype in contemporary Western astrology can manifest: "as depression but also as discipline, as gravity in the sense of heaviness and weight but also as gravity in the sense of seriousness and dignity." Richard Tarnas, *Cosmos and Psyche: Intimations of a New World View* (New York: Plume, 2007), p. 87. The dialogue might therefore lend itself to a contemporary reading of Arjuna crying, "I am beset by the gravitas of Saturn, whether for good or ill I do not know," while Krishna responds, "No: this is *tamas*; there is nothing good about this state at all." This reading shows how the different perspectives upon the perinatal matrices suggested here and those held by Western archetypal astrology, with (for example) the attributes of Grof's BPM II noted to be the more challenging attributes of Saturn, can mutually nourish and

enrich one another. Stanislav Grof, "Holotropic Research and Archetypal Astrology," *Archai: The Journal of Archetypal Cosmology* 1 no. 1 (2009): 76.

62. *Bhagavad Gītā* 14.7. Chinmayananda, *Srimad Bhagavad Gītā.*
63. *Bhagavad Gītā* 2.47. Chinmayananda, *Srimad Bhagavad Gītā.*
64. *Bhagavad Gītā* 3.36-37. Chinmayananda, *Srimad Bhagavad Gītā.* Chinmayananda's translation with the exception that he translates *rajas guṇa* as "active."
65. *Bhagavad Gītā* 2.37b-8. Chinmayananda, *Srimad Bhagavad Gītā.*
66. *Bhagavad Gītā* 3.39-41. Chinmayananda, *Srimad Bhagavad Gītā.*
67. Sri Aurobindo, *The Doctrine of Passive Resistance* (Calcutta: Arya Publishing House, 1949), p. 30.
68. *Bhagavad Gītā* 2.37-8. Chinmayananda, *Shrimad Bhagavad Gītā.*
69. The British authorities "came in the early revolutionary years [of the early twentieth century] to regard the *Gītā* as an anarchist manifesto, and to regard anyone possessing more than one copy as in all probability a revolutionary ringleader." Eric Sharpe, "Avatara and Shakti: Traditional Symbols in the Hindu Renaissance," Scripta Instituti Donneriani Aboensis 7 (January 1975): 60. See also Lawrence Dundas, *The Heart of Āryāvarta* (Chester, NY: Anza Publishing, 2005), pp. 97-108.
70. For the most famous such reading, see Mohandas K. Gandhi, *Discourses on the Gita,* trans. V. G. Desai (Ahmedabad: Navajivan Publishing House, 1960).
71. The Kurus were the family to which both sides in the war belonged. Kurukṣetra is referred to in the first verse of the *Gītā* as *dharma-kṣetra*, the field of *dharma*, perhaps here best translatable as "justice." If the field of the Kurus is, in this narrative, specifically the field of *dharma* this also supports an allegorical reading.
72. *Bhagavad Gītā* 13.2a. Chinmayananda, *Srimad Bhagavad Gītā.*
73. *Bhagavad Gītā* 3.1. Chinmayananda, *Srimad Bhagavad Gītā.*
74. Grof calls these threads of experience "systems of condensed experience" (COEX systems): the COEX systems and the perinatal matrices form the cornerstone of his model. See Grof, *LSD: Doorway to the Numinous*, chapter 3.
75. *Bhagavad Gītā* 3.36. Chinmayananda, *Srimad Bhagavad Gītā.*
76. *Bhagavad Gītā* 3.37-41. Chinmayananda, *Srimad Bhagavad Gītā.*
77. *Bhagavad Gītā* 2.37, 2.45, and 2.47. Chinmayananda, *Srimad Bhagavad Gītā.*
78. George Orwell, *1984* (New York: New American Library, 1977), 245. Orwell seems to have written these words in 1947, the year that India's battle for independence was finally realized. It is not to romanticize the effect a scripture can have upon our actions to presume an effect the sheer power of Krishna's injunction to fight has had upon those fighting in its name. One could understand that those so empowered would find it hard to believe these words were most fundamentally applicable to the experience of birth.

The reading of the *Gītā* I am suggesting may go a certain way to resolving in what sense, and when, *rajas* is to be engaged with. But for as long as there can be the need to engage in an external battle outside of the birth process, the reading does not fully do so; India's fight for independence is one of the strongest examples of the need for such a battle, outside allegory, and outside the original experience of the *guṇas* at birth.

All that those of us advocating for the importance of Grof's teachings can do here is to highlight how, in thousands of hours of clinical work, the *desire* to engage in war could be convincingly shown to be the result of the experience of *rajas* at birth, a desire which dramatically reduced when the birth *rajas* was fully experienced and worked through. But this does not address the problem that the decision *not* to fight can result in more *tamas* and more *rajas*; and the painful dilemmas of how best to secure the *sattva* of ourselves and others perhaps highlight a truth the scriptures patiently repeat: even our *sattva* can be a form of attachment to the results of our actions; even our *sattva* can be a source of paradox and pain. Grof's contribution means that *rajas* can be viewed with a more discerning gaze, but his work does not close off the possibility that there might still be times when its embrace, severed from an attachment to its fruits, may lead to a less rajasic future.

79. Other praxes by which one does so may be subtler than Grof's. The concept in the Yoga school of *pratiprasava*, a return to an original state (*Yoga Sūtras* 2.10 and 4.34) has been described by Chapple as encapsulating the entire Yoga praxis and telos. Christopher Chapple, "Activity, Cessation, and a Return to Origins in the Yoga Sūtra," in Christopher Chapple and Ana Maderey eds., *Thinking With the Yoga Sūtra of Patañjali* (Lanham, MA: Lexington Books, 2019), p. 195. Grof's model effectively argues for a more explicitly temporal and biographical interpretation of what is entailed in this process.
80. This passage is worth comparing to Sri Aurobindo's exegesis on *sattva* quoted above.
81. Richard Tarnas, *The Passion of the Western Mind: Understanding the Ideas that Shaped Our World-View* (New York: Ballantine Books, 1991), pp. 426-427.
82. *Sāṃkhya Kārikā* 62-63. Burley, *Classical Samkhya and Yoga.*
83. The extent to which this form of seeing evokes *Sāṃkhya Kārikā* 66: "'I have seen her,' says the spectating one; 'I have been seen,' says the other, desisting; although the two remain in conjunction, there is no initiation of [further] emergence" is worth considering. Burley, *Classical Samkhya and Yoga.*
84. Eknath Easwaran, ed. and trans., *The Bhagavad Gītā* (Tomales, CA: Nilgiri Press, 2017), p. 47.

SARTRE'S RITE OF PASSAGE

This article was first published in *Journal of Transpersonal Psychology* 14, no. 2 (1982), 105–123.

1. Except where noted otherwise, all biographical references to Sartre's mescaline session and its aftermath are from Simone de Beauvoir, *The Prime of Life* (New York: Harper & Row, 1976), pp. 168–178; and Francis Jeanson, *Sartre dans sa Vie* (Paris: Editions du Seuil, 1974), pp. 75–76 (translated for this article by Dawn Scheffner).
2. Jean-Paul Sartre, *Sartre by Himself* (New York: Urizen Books, 1978), p. 38.
3. Simone de Beauvoir, Letter to the author dated October 6, 1982: "*C'est bien de belladonna qu'il sagit et nous n'avons pas compris Ie scns de cette ordonance que Sartre n'a d'ailleurs pas sui vi.*"
4. Sartre, *Sartre by Himself*, pp. 37–38.
5. Sartre, *Sartre by Himself*, p. 38.
6. Beauvoir, *The Prime of Life*, p. 170.
7. Hazel E. Barnes, *An Existentialist Ethics* (Chicago: The University of Chicago Press, 1978), p. 13.
8. Barnes, *An Existentialist Ethics*, p. 253.
9. Barnes, *An Existentialist Ethics*, p. 231.
10. Lester Grinspoon and James B. Bakalar, *Psychedelic Drugs Reconsidered* (New York: Basic Books, 1979), p. 21.
11. Jean-Paul Sartre, *The Words* (New York: George Braziller, 1964), p. 152.
12. Grinspoon and Bakalar, *Psychedelic Drugs Reconsidered*, p. 21; Stanislav Grof, *LSD Psychotherapy* (Pomona, CA: Hunter House, 1980), p. 262.
13. Stanislav Grof, *Realms of the Human Unconscious: Observations from LSD Research* (New York: Dutton, 1976), pp. 46–49.
14. Grof, *Realms of the Human Unconscious*, p. 216.
15. Grof, *Realms of the Human Unconscious*, p. 20.
16. Grof, *Realms of the Human Unconscious*, pp. 70–73.
17. Grof, *Realms of the Human Unconscious*, p. 93.
18. Beauvoir, *The Prime of Life*, p. 169.
19. Grinspoon and Bakalar, *Psychedelic Drugs Reconsidered*, p. 106.
20. Beauvoir, *The Prime of Life*, pp. 170–171.
21. Grof, *Realms of the Human Unconscious*, p. 215.
22. Sartre, *The Words*, p. 96.
23. Sartre, *The Words*, pp. 15–16.
24. Sartre, *The Words*, p. 23.
25. Sartre, *The Words*, p. 87.
26. Sartre, *The Words*, p. 34.
27. Sartre, *The Words*, p. 29.
28. Sartre, *The Words*, p. 94.

29. Sartre, *The Words*, pp. 94-95.
30. Sartre, *The Words*, p. 96.
31. Sartre, *The Words*, pp. 109-110.
32. Sartre, *The Words*, p. 134.
33. Sartre, *The Words*, pp. 146-48.
34. Sartre, *The Words*, pp. 148-49.
35. Grof, *Realms of the Human Unconscious*, p. 51.
36. Grof, *Realms of the Human Unconscious*, pp. 100-101.
37. Grof, *Realms of the Human Unconscious*, p. 222.
38. Grof, *Realms of the Human Unconscious*, pp. 104-149.
39. Grof, *Realms of the Human Unconscious*, 149-150.
40. Grof, *Realms of the Human Unconscious*, pp. 115-123.
41. Grof, *Realms of the Human Unconscious*, p. 118.
42. Grof, *Realms of the Human Unconscious*, p. 119.
43. Stanislav Grof and Christina Grof, *Beyond Death: The Gates of Consciousness* (New York: Thames & Hudson, 1980), p. 26.
44. Grof, *Realms of the Human Unconscious*, p.129.
45. Grof, *Realms of the Human Unconscious*, p. 118.
46. Sartre, *The Words*, pp. 169-171.
47. Sartre, *The Words*, p. 230.
48. Grof, *LSD Psychotherapy*, p. 157.
49. Beauvoir, *The Prime of Life*, p. 219.
50. Grof, *LSD Psychotherapy*, p. 159.
51. Barnes, *An Existentialist Ethics*, p. 231.
52. Sartre, *Sartre by Himself*, p. 38.
53. Sartre, *Sartre by Himself*, p. 41.
54. Sartre, *The Words*, p. 251.
55. Jean-Paul Sartre, *Nausea* (New York: New Directions, 1964).
56. Sartre, *Nausea*, p. 70.
57. Sartre, *Nausea*, p. 5.
58. Sartre, *Nausea*, pp. 16-17.
59. Sartre, *Nausea*, pp. 98-99.
60. Sartre, *Nausea*, pp. 78-79.
61. Sartre, *Nausea*, pp. 126-135.
62. Aldous Huxley, *The Doors of Perception* (New York: Harper, 1954), pp. 17-18.
63. Barnes, *An Existentialist Ethics*, pp. 232-233.
64. Beauvoir, The *Prime of Life*, p. 168.
65. Barnes, *An Existentialist Ethics*, pp. 226-256.
66. Sartre, *The Words*, pp. 150-151.
67. Grof, *Realms of the Human Unconscious*, p. 124.
68. Sartre, *Nausea*, pp. 135-154.

69. Susan Sontag, *Against Interpretation* (New York: Dell, 1981), pp. 97-98.
70. Grof, *Realms of the Human Unconscious*, p. 118.
71. Viktor Frankl, *Man's Search for Meaning* (New York: Pocket Books, 1985), p. 90.

THE CONSCIOUSNESS RESEARCH OF STANISLAV GROF

The author first presented a longer version of this chapter in 2000 at the Esalen Institute's "Grof Transpersonal Conference" (Big Sur, California). He presented an updated version at the 2014 conference, "Expanding and Reenchanting the Psyche: The Pioneering Thought of Stanislav Grof," at California Institute of Integral Studies (CIIS), San Francisco, California.

1. Stanislav Grof, *Beyond the Brain: Birth, Death, and Transcendence in Psychotherapy* (Albany, NY: SUNY Press, 1985); Stanislav Grof, *Psychology of the Future: Lessons from Modern Consciousness Research* (Albany, NY: SUNY Press, 2000); Stanislav Grof, *Healing our Deepest Wounds: The Holotropic Paradigm Shift* (Newcastle, WA: Stream of Consciousness, 2012).
2. Stanislav Grof, *Realms of the Human Unconscious: Observations from LSD Research* (New York: Dutton, 1975); Grof, *Beyond the Brain*.
3. Stanislav Grof, *The Adventure of Self-Discovery: Dimensions of Consciousness and New Perspectives in Psychotherapy and Inner Exploration* (Albany, NY: SUNY Press, 1988); Stanislav Grof and Christina Grof, eds., *Spiritual Emergency: When Personal Transformation Becomes a Crisis* (Los Angeles: J. P. Tarcher, 1989); Christina Grof and Stan Grof, *The Stormy Search for the Self* (Los Angeles: J. P. Tarcher, 1990).
4. Grof, *Beyond the Brain*; Grof, *Psychology of the Future*; Grof, *Healing our Deepest Wounds*; Stanislav Grof, "Revision and Re-Enchantment of Psychology: Lessons from Half a Century of Consciousness Research," in *The Wiley-Blackwell Handbook of Transpersonal Psychology*, eds. Harris L. Friedman and Glenn Hartelius (Malden, MA: John Wiley and Sons, 2013), pp. 91-120.
5. Grof, *Healing our Deepest Wounds*; Stanislav Grof, "Modern Consciousness Research and Human Survival," in *Human Survival and Consciousness Evolution*, ed. Stanislav Grof (Albany, NY: SUNY Press, 1988), pp. 57-78.
6. See Grof, *Healing our Deepest Wounds*; Stanislav Grof and Christina Grof, *Holotropic Breathwork: A New Approach to Self- Exploration and Therapy* (Albany, NY: State University of New York Press, 2010).
7. See John Hick, *An Interpretation of Religion: Human Responses to the Transcendent* (New Haven: Yale University Press, 1989); Aldous Huxley, *The Perennial Philosophy* (New York: Harper and Row, 1945); Frithjof Schuon, *The Transcendent Unity of Religions* (Wheaton, IL: Quest Books, 1984); Evelyn Underhill, *Mysticism* (New York, NY: Meridian, 1955).
8. Anthony N. Perovich, Jr., "Mysticism and the Philosophy of Science," *The Journal of Religion* 65 (1985): 75.

9. Jean Borella, "René Guénon and the Traditionalist School," in *Modern Esoteric Spirituality,* vol. 21 of *World Spirituality: An Encyclopedic History of the Religious Quest*, eds. Antoine Faivre and Jacob Needleman (New York: Crossroad, 1995), pp. 330-358; Martin Lings and Clinton Minnaar, eds. *The Underlying Religion: An Introduction to the Perennial Philosophy* (Bloomington, IN: World Wisdom, 2007); Harry Oldmeadow, *Traditionalism: Religion in the Light of the Perennial Philosophy* (Colombo, Sri Lanka: Sri Lanka Institute of Traditional Studies, 2000); William W. Quinn, Jr., *The Only Tradition* (Albany, NY: SUNY Press, 1997). For two accounts of this movement, see the partisan study by Oldmeadow and the far superior work by Sedgwick. Harry Oldmeadow, *Journeys East: 20th Century Western Encounter with Eastern Religious Traditions* (Bloomington, IN: World Wisdom, 2004); Mark J. Sedgwick, *Against the Modern World: Traditionalism and the Secret Intellectual History of the Twentieth Century* (New York, NY: Oxford University Press, 2004).
10. Schuon, *The Transcendent Unity of Religions*; Quinn, *The Only Tradition*; Huston Smith, *Forgotten Truth: The Primordial Tradition* (New York, NY: Harper and Row, 1976).
11. See Ninian Smart, "Interpretation and Mystical Experience," in *Understanding Mysticism*, ed. Richard Woods (Garden City, NY: Doubleday, 1980), pp. 78-91.
12. Robert M. Gimello, "Mysticism in its Contexts," in *Mysticism and Religious Traditions*, ed. Steven T. Katz (New York, NY: Oxford University Press 1983), pp. 61-88; Steven T. Katz, ed. *Mysticism and Philosophical Analysis* (New York: Oxford University Press, 1978); Steven T. Katz, ed. *Mysticism and Religious Traditions* (New York, NY: Oxford University Press, 1983).
13. See Gimello, "Mysticism in its Contexts."
14. See Michael Stoeber, *Theo-monistic Mysticism: A Hindu-Christian Comparison* (New York, NY: St. Martin's Press, 1994).
15. Katz, ed. *Mysticism and Philosophical Analysis*; Katz, ed. *Mysticism and Religious Traditions.*
16. Quoted in John Horgan, *Rational Mysticism: Spirituality Meets Science in the Search for Enlightenment* (New York, NY: Mariner Books, 2004), p. 46.
17. Karl Popper, *The Myth of the Framework: In Defense of Science and Rationality*, ed. M. A. Notturno (New York, NY: Routledge, 1994); Jorge N. Ferrer, *Revisioning Transpersonal Theory: A Participatory Vision of Human Spirituality* (Albany, NY: SUNY Press, 2002); Jorge N. Ferrer, *Participation and the Mystery: Transpersonal Essays in Psychology, Education, and Religion* (Albany, NY: SUNY Press, 2017).
18. Quoted in Horgan, *Rational Mysticism*, p. 47.
19. See Ferrer, *Revisioning Transpersonal Theory.*
20. Both this perspectival perennialism and metaphysical agnosticism are overcome by the participatory approach's adoption of the enactive paradigm:

enactive spiritual cognition is an antidote for perspectival perennialism, and affirmation of the nonduality of the mystery and its enactions counteracts metaphysical agnosticism. For an updated summary of Katz's contextualism, see Steven T. Katz, "Diversity and the Study of Mysticism," in *New Approaches to the Study of Religion: Textual, Comparative, Sociological, and Cognitive Approaches*, vol. 2, eds. Peter Antes, Armin W. Geertz, and Randi R. Warne (Berlin, Germany: Walter de Gruyter, 2004), pp. 189-210. For a revealing discussion of Katz's metaphysical agnosticism as central to his weak version of mediation, see Torben Hammersholt, "Steven T. Katz's Philosophy of Mysticism Revisited," *Journal of the American Academy of Religion* 81, no. 2 (June 2013): 467–490. See Ferrer's *Revisioning Transpersonal Theory* for a more detailed analysis of the Cartesian and neo-Kantian roots of perennialism and contextualism, respectively. Essentially, I argued that whereas perennialism subscribes to the myth of the given (Wilfrid Sellars, *Science, Perception, and Reality* [New York, NY: Humanities Press, 1963]) and contextualism to the myth of the framework (Popper, *The Myth of the Framework*), Davidson's dismantling of the dualism of framework and reality (or scheme-content) renders both myths unintelligible and paves the way for more participatory considerations of spiritual knowing. Donald Davidson, *Inquiries into Truth and Interpretation* (New York, NY: Oxford University Press, 1984). See also Jorge N. Ferrer and Jacob H. Sherman, "Introduction: The Participatory Turn in Spirituality, Mysticism, and Religious Studies," in *The Participatory Turn: Spirituality, Mysticism, Religious Studies*, eds. Jorge N. Ferrer and Jacob H. Sherman (Albany, NY: State University of New York Press, 2008), pp. 1-78.

21. Steven T. Katz, "Language, Epistemology, and Mysticism," in *Mysticism and Philosophical Analysis*, ed. Steven T. Katz (New York, NY: Oxford University Press, 1978), p. 26.
22. Philip C. Almond, *Mystical Experience and Religious Doctrine: An Investigation of the Study of Mysticism in World Religions* (New York: Mouton Publishers, 1982); John Y. Fenton, "Mystical Experience as a Bridge for Cross-Cultural Philosophy of Religion: A Critique," in *Religious Pluralism and Truth: Essays on Cross-Cultural Philosophy of Religion*, ed. Thomas Dean (Albany, NY: State University of New York Press, 1995), pp. 189-204; Jess B. Hollenback, *Mysticism: Experience, Response, and Empowerment* (University Park, PA: Pennsylvania State University Press, 1996).
23. S. Mark Heim, *Salvations: Truth and Difference in Religion* (Maryknoll, NY: Orbis, 1995); Stephen Kaplan, *Different Paths, Different Summits: A Model for Religious Pluralism* (Lanham, MD: Rowman and Littlefield, 2002).
24. Katz, "Language, Epistemology, and Mysticism," p. 66.
25. See Jorge N. Ferrer, "The Perennial Philosophy Revisited," *The Journal of Transpersonal Psychology* 32, no. 1 (2000): 7-30; Ferrer, *Revisioning Transpersonal Theory*.

26. Robert K. C. Forman, ed., *The Problem of Pure Consciousness: Mysticism and Philosophy* (New York, NY: Oxford University Press, 1990); Robert K. C. Forman, ed., *The Innate Capacity: Mysticism, Psychology, and Philosophy* (New York, NY: Oxford University Press, 1998); Robert K. C. Forman, *Mysticism, Mind, Consciousness (Albany, NY:* State University of New York Press, 1999).
27. Grof, *Beyond the Brain*; Grof, *The Adventure of Self-Discovery*; Stanislav Grof, *The Cosmic Game: Explorations of the Frontiers of Human Consciousness* (Albany, NY: SUNY Press, 1998).
28. Grof, *The Adventure of Self-Discovery*, p. 139.
29. Grof, *The Cosmic Game.*
30. Grof, *The Cosmic Game*, pp. 23-24.
31. Grof, *The Cosmic Game*, pp. 18-19. Note that transcultural similarities in myths that cannot be explained by historical or cultural diffusion are thought by many to derive, not from any Jungian "collective unconscious" or related notions of human "psychic unity," but rather from our "shared humanity" or "shared life experience" (e.g., birth, body, sexual desire, procreation, parenting, pain, loss, death). See Wendy Doniger, *The Implied Spider: Politics and Theology in Myth* (New York, NY: Columbia University Press, 1998), p. 61. Some of Grof's reported cases, however, arguably challenge this mainstream account and, in my view, should be carefully studied by modern comparative mythologists. Grof, *The Cosmic Game*; Stanislav Grof, *When the Impossible Happens: Adventures in Non-Ordinary Realities* (Louisville, CO: Sounds True, 2006).
32. Grof, *The Cosmic Game*; Grof, *When the Impossible Happens.*
33. For Campbell's changing views on Jung's approach to mythology, see Ritske Rensma, *The Innateness of Myths: A New Interpretation of Joseph Campbell's Reception of C. G. Jung* (New York, NY: Continuum, 2009).
34. Grof, *The Cosmic Game*, p. 20.
35. Grof, *The Cosmic Game*, p. 20.
36. Grof, *The Adventure of Self-Discovery*; Robert Jesse, "Entheogens: A Brief History of Their Spiritual Use," *Tricycle* 6, no. 1 (1996): 60-64; Ralph Metzner, *Green Psychology: Transforming Our Relationship to the Earth* (Rochester, VM: Park Street Press, 1999).
37. See Grof, *When the Impossible Happens.*
38. Although a few of Grof's reported cases dramatically challenge both the strong thesis of mediation in mysticism studies and the "shared humanity" model in comparative mythology, many scholars may legitimately request additional evidence before accepting alternative transpersonal explanations. In other words, a systematic research program is imperative to turn these anecdotal cases into thoroughly documented evidence that can be presented

to the scientific community. Grof, *The Cosmic Game*; Grof, *When the Impossible Happens*.

39. Grof, *Realms of the Human Unconscious*; Grof, *The Adventure of Self-Discovery*.
40. See Wade Clark Roof, *A Generation of Seekers: The Spiritual Lives of the Baby Boom Generation* (San Francisco: HarperSanFrancisco, 1993).
41. Grof, *The Adventure of Self-Discovery*; Grof, *The Cosmic Game*.
42. Peter Moore, "Mystical Experience, Mystical Doctrine, Mystical Technique," in *Mysticism and Philosophical Analysis*, ed. Steven T. Katz (New York, NY: Oxford University Press, 1978), p. 112.
43. Grof, *The Cosmic Game*, p. 3. Grof's proposal closely follows Huxley's belief that his psychedelic self-experimentation provided corroboration for both a transcultural mystical experience and attendant perennialist metaphysics. Aldous Huxley, *The Doors of Perception and Heaven and Hell* (New York: Harper and Row, 1954); Michael Horowitz and Cynthia Palmer, eds. *Moksha: Aldous Huxley's Classical Writings on Psychedelics and the Visionary Experience* (Rochester, VT: Park Street Press, 1999). In addition to Zaehner's classical Christian critique of Huxley's psychedelic perennialism, see Halbfass and Wezler for two critical accounts of Huxley's views on Indian religion shaped by his psychedelic experiences. Robert Charles Zaehner, *Mysticism Sacred and Profane: An Inquiry into Some Varieties of Preternatural Experience* (New York, NY: Oxford University Press, 1980); Wilhelm Halbfass, "Mescaline and Indian Philosophy: Aldous Huxley and the Mythology of Experience," in *Aldous Huxley between East and West*, ed. C. C. Barfoot (New York, NY: Rodopi, 2001), pp. 221-236; Wezler, A. (2001). "'Psychedelic' Drugs as a Means to Mystical Experience: Aldous Huxley Versus Indian Reality," in *Aldous Huxley between East and West*, ed. C. C. Barfoot (New York, NY: Rodopi, 2001), pp. 191-220. The latest version of this idea comes from Langlitz, who suggested that contemporary psychedelic research "point[s] to a new form of perennialism that reconciles biology and spirituality"—a *neurobiologia perennis* or mystic materialism based on the universality of the human brain, DNA, or neurochemical makeup. Nicolas Langlitz, *Neuropsychedelia: The Revival of Hallucinogenic Research Since the Decade of the Brain* (Berkeley, CA: The University of California Press, 2013), p. 260.
44. Grof, *The Cosmic Game*, p. 4.
45. For a typology of perennialisms, see Ferrer, *Revisioning Transpersonal Theory*.
46. Grof, *The Cosmic Game*, p. 24.
47. Grof, *The Adventure of Self-Discovery*, p. 269.
48. Schuon, *The Transcendent Unity of Religions*; Smith, *Forgotten Truth*.
49. Grof, *The Adventure of Self-Discovery*, p. 270.
50. Grof, *The Cosmic Game*.
51. Ferrer, *Revisioning Transpersonal Theory*.
52. Abramson claimed that the participatory postulation of both a diversity of

spiritual ultimates and a mystery (out of which those ultimates are enacted through human cocreative participation) is equivalent to perspectival perennialism. John Abramson, "The Emperor's New Clothes: Ferrer isn't Wearing Any—Participatory is Perennial. A Reply to Hartelius," *Transpersonal Psychology Review* 17, no. 1 (2015): 38-48. In my view, three things differentiate the participatory approach from perspectival perennialism. First, the participatory approach rejects the myth of the given. (Note that even when traditionalist scholars speak about an ineffable or transconceptual spiritual Ultimate, they immediately—and arguably contradictorily—qualify it, stating that it is nondual or that Advaita Vedanta offers, through its notion of *nirguna Brahman*, the best articulation of the perennial wisdom, and so forth). Second, the participatory approach adopts an enactive paradigm of cognition, according to which the various spiritual ultimates are not perspectives of a single spiritual Ultimate, but rather enactions bringing forth ontological realities. Third, and perhaps most crucially, by overcoming of the dualism of the mystery and its enactions, the participatory approach avoids the traditionalist (and neo-Kantian) duality between religions' relative absolutes and the Absolute supposedly existing behind them. See Ferrer, *Participation and the Mystery.* For a thorough rebuttal of Abramson's view, see Glenn Hartelius, "A Startling New Role for Wilber's Integral Model: Or, How I Learned to Stop Worrying and Love Perennialism. (A Response to Abramson)," *Transpersonal Psychology Review* 17, no. 1 (2015a): 25-37; Glenn Hartelius, "Participatory Thought has No Emperor and No Absolute—A Further Response to Abramson," *Transpersonal Psychology Review* 17, no. 1 (2015b): 49-53. See also Merlo for some interesting reflections on how a hypothetical nondogmatic, nonhierarchical, and noninclusivist perennialism may be consistent with the participatory approach. V. Merlo, "Sobre el Giro Participativo [About the Participatory Turn]," *Journal of Transpersonal Research* 3 (2011): 55-58.

53. Stanislav Grof, *The Holotropic Mind: The Three Levels of Human Consciousness and How They Shape Our Lives* (New York: HarperCollins, 1993), p. 164.
54. Grof, "Revision and Re-Enchantment of Psychology," p. 93.
55. Grof, *The Cosmic Game.*
56. Grof, *The Cosmic Game*, p. 38.
57. Applying an entheogenic, psycho-historical lens to Grof's allegiance to a Hindu cosmology, it is noteworthy that his two reportedly most transformational experiences (including his first ego-death) took place through entheogenic encounters with the Hindu God Shiva. Grof, *The Cosmic Game.* In Grof's own words: "I considered Shiva to be the most important personal archetype because the two most powerful and meaningful experiences I have ever had in my psychedelic sessions involved this Indian deity" (Grof, *When the Impossible Happens*, p. 30).
58. Grof, "Revision and Re-Enchantment of Psychology," p. 92.

59. Grof, *The Cosmic Game*; Sri. Aurobindo, *The Life Divine*, 6th ed. (Pondicherry, India: Sri Aurobindo Ashram, 2001).
60. Grof, *Psychology of the Future*; Grof, "Revision and Re-Enchantment of Psychology;" Grof, *The Cosmic Game.*
61. Grof, *Psychology of the Future*, p. 278.
62. Grof, *The Cosmic Game*, p. 79; Aurobindo, *The Life Divine.*
63. Grof, *The Cosmic Game*, p. 79.
64. Grof, "Revision and Re-Enchantment of Psychology;" Ferrer, *Participation and the Mystery*; Stanislav Grof, "Ken Wilber's Spectrum Psychology: Observations from Clinical Consciousness Research," in *Ken Wilber in Dialogue: Conversations with Leading Transpersonal Thinkers*, eds. Donald Rothberg and Sean Kelly (Wheaton, IL: Theosophical Publishing House, 1996), pp. 85-116; Ken Wilber, *Sex, Ecology and Spirituality: The Spirit of Evolution* (Boston, MA: Shambhala, 1995); Ken Wilber, *The Atman Project: A Transpersonal View of Human Development*, 2nd ed. (Wheaton, IL: Quest Books, 1996).
65. Grof, "Revision and Re-Enchantment of Psychology," p. 101.
66. Grof, *The Cosmic Game*, p. 39.
67. Grof, *The Cosmic Game*, p. 50.
68. Grof, *The Cosmic Game*, p. 52.
69. Grof, *The Cosmic Game*, p. 66.
70. Grof, *The Cosmic Game*, p. 85.
71. Grof, "Revision and Re-Enchantment of Psychology;" Grof, *The Cosmic Game.*
72. See Ferrer, "The Perennial Philosophy Revisited;" Ferrer, *Revisioning Transpersonal Theory*; Ferrer, *Participation and the Mystery*; Paul J. Griffiths, *An Apology for Apologetics: A Study in the Logic of Interreligious Dialogue* (New York: Orbis Books, 1991); Olav Hammer, "Same Message from Everywhere: The Sources of Modern Revelation," *New Age Religion and Globalization*, ed. Mikael Rothstein (Aarhus, Denmark: Aarhus University Press, 2001), pp. 42-57; Wouter J. Hanegraaff, *New Age Religion and Western Culture: Esotericism in the Mirror of Secular Thought* (Albany, NY: SUNY Press, 1998); Katz, ed. *Mysticism and Philosophical Analysis*; Sallie B. King, "The *Philosophia Perennis* and the Religions of the World," in *The Philosophy of Seyyed Hassein Nasr*, vol. XXVIII of *The Library of Living Philosophers*, eds. Lewis E. Hahn, Randell E. Auxier and Lucian W. Stone Chicago: Open Court, 2001), pp. 203-20.
73. The most nuanced defense of a qualified version of the "common core" theory is due to Forman, who, together with his collaborators, presented compelling evidence for the occurrence of a "pure consciousness event" in many, although by no means all, contemplative traditions. Forman, ed., *The Problem of Pure Consciousness*; Forman, ed., *The Innate Capacity*; Forman, *Mysticism, Mind, Consciousness.* However, the ontological and spiritual status of such an event is valued quite differently across traditions; for example, many theistic traditions and mystics considered a dual I/Thou relationship with God

spiritually superior to monistic nondual states such as pure consciousness. See Martin Buber, *Between Man and Man* (London: Routledge, 2006); William Harmless, *Mystics* (Oxford: Oxford University Press, 2008); Bernard McGinn, *The Foundations of Mysticism*, vol. 1 of *The Presence of God: A History of Western Christian Mysticism* (New York: Crossroad, 1991).

74. Ferrer, *Revisioning Transpersonal Theory*.
75. Hollenback, *Mysticism*; Robert C. Neville, ed. *Ultimate Realities. A Volume in the Comparative Religious Ideas Project* (Albany, NY: State University of New York Press, 2001).
76. Donald W. Mitchell and James Wiseman, eds., The Gethsemani Encounter: A Dialogue on the Spiritual Life by Buddhist and Christian Monastics (New York: Continuum Intl Pub Group, 1997); Susan Walker, ed., *Speaking of Silence: Christians and Buddhists on the Contemplative Way* (New York, NY: Paulist Press, 1987).
77. Consider, for example, the Buddhist tradition. For discussions highlighting the diversity of Buddhist views about the nature of ultimate reality, see Jeffrey Hopkins, "Ultimate Reality in Tibetan Buddhism," *Buddhist-Christian Studies* 8 (1988): 111-129; Francis H. Cook, "Just This: Buddhist Ultimate Reality," *Buddhist-Christian Studies* 9 (1989): 127-142; Hans Küng, "Response to Francis Cook: Is It Just This? Different Paradigms of Ultimate Reality in Buddhism," *Buddhist-Christian Studies* 9 (1989): 143-156. See also W. L. King for an analysis of differences between Mahayana and Theravada accounts of *nirvana*, Chen for distinctions among Buddhist enactions of emptiness (*sunyata*), and Komarovski for an account of Tibetan Buddhism's conflicting views of ultimate reality and its realization. Winston L. King, "Zen as a Vipassana-type Discipline," in *Asian Religions: History of Religions: 1974 Proceedings*, ed. H. B. Partin (Tallahassee, FL: American Academy of Religion, 1974), pp. 62-79; C. M. Chen, *The Subtle Discrimination Between the Practices of Sunyata in Hinayana, Mahayana and Vajrayana*, Chenian booklet series, trans. Jivaka (Kalinpang, India: Chen, 1972); Yaroslav Komarovski, *Tibetan Buddhism and Mystical Experience* (New York, NY: Oxford University Press, 2015). For a general introduction to some of the controversies among Buddhist schools regarding the substantiality of ultimate reality, see Mangala R. Chinchore, *Anattā/Anātmatā: An Analysis of Buddhist Anti-substantialist Crusade* (Delhi, India: Sri Satguru, 1995).
78. Ferrer, *Revisioning Transpersonal Theory*; Jorge N. Ferrer, "Transpersonal Knowledge: A Participatory Approach to Transpersonal Phenomena," in *Transpersonal Knowing: Exploring the Farther Reaches of Consciousness*, eds. Tobin Hart, Peter L. Nelson, and Kaisa Puhakka (Albany, NY: State University of New York Press, 2000b), pp. 213-252; Jorge N. Ferrer, "Spiritual Knowing as Participatory Enaction: An Answer to the Question of Religious Pluralism," in *The Participatory Turn: Spirituality, Mysticism, Religious Studies*, eds. Jorge N.

Ferrer and Jacob H. Sherman (Albany, NY: State University of New York Press, 2008), pp. 135-169.

79. Francisco J. Varela, Evan Thompson, and Eleanor Rosch, *The Embodied Mind: Cognitive Science and Human Experience* (Cambridge, MA: The MIT Press, 1991). See also Evan Thompson, *Mind in Life: Biology, Phenomenology, and the Sciences of the Mind* (Cambridge, MA: Harvard University Press, 2007).

80. Varela et al. understand *enaction* as an embodied action that brings forth a domain of distinctions as the result of the mutual specification of organism and its natural environment, limiting thereby the scope of their proposal to the perceptual cognition of the sensoriomotor world. Varela, Thompson, and Rosch, *The Embodied Mind.* Participatory formulations adapt and extend the enactive paradigm to account for the emergence of ontologically rich spiritual realms (i.e., subtle worlds or domains of distinctions) cocreated by human multidimensional cognition and the generative force of life or the mystery. Ferrer, *Revisioning Transpersonal Theory*; *Participation and the Mystery*; Ferrer, "Spiritual Knowing as Participatory Enaction;" Ferrer and Sherman, "Introduction;" Sean M. Kelly, "Participation, Complexity, and the Study of Religion," in *The Participatory Turn: Spirituality, Mysticism, Religious Studies*, eds. Jorge N. Ferrer and Jacob H. Sherman (Albany, NY: State University of New York Press, 2008), pp. 113-133; Lee Irwin, "Esoteric Paradigms and Participatory Spirituality in the Teachings of Mikhaël Aïvanhov," in *The Participatory Turn: Spirituality, Mysticism, Religious Studies*, eds. Jorge N. Ferrer and Jacob H. Sherman (Albany, NY: State University of New York Press, 2008), pp. 197-224. For further refinements of the enactive paradigm, see Evan Thompson, "The Mindful Body: Embodiment and Cognitive Science," in *The Incorporated Self: Interdisciplinary Perspectives of Embodiment*, ed. Michael O'Donovan-Anderson (Lanham, MD: Rowman and Littlefield, 1996), pp. 127-144; Thompson, *Mind in Life.* See also Frisina for a lucid account of both classic and contemporary, Asian and Western perspectives of the emerging nonrepresentational paradigm of cognition. Warren G. Frisina, *The Unity of Knowledge and Action: Toward a Nonrepresentational Theory of Knowledge* (Albany, NY: State University of New York Press, 2002).

81. I am merely suggesting that liberation from self-centeredness is an ideal (or aspiration) shared by virtually all contemplative traditions. Whether such an ideal is ever actualized in practice is an empirical question to be explored through both the historical study of religious communities and biographies, as well as research into the effectiveness of contemporary religious practices in fostering such a transformation.

82. Ferrer, *Revisioning Transpersonal Theory*; Heim, *Salvations*; Hollenback, *Mysticism*; Kaplan, *Different Paths, Different Summits.*

83. See David Loy, *Nonduality: A Study in Comparative Philosophy* (New Haven, CT: Yale University Press, 1988).

84. Jules Evans, *What's Wrong with the Perennial Philosophy?* Philosophy for Life (October 2, 2014), https://www.philosophyforlife.org/blog/exploring-the-multiverse-of-spiritual-pluralism
85. Gilles Deleuze and Félix Guattari, *A Thousand Plateaus: Capitalism and Schizophrenia*, trans. Brian Massumi (Minneapolis: University of Minnesota Press, 1987).
86. See Ferrer, *Participation and the Mystery*; Jorge N. Ferrer, "A Participatory Vision of the Mystical Unity of Religions," International Transpersonal Conference 2017: Beyond Materialism, Prague, Czech Republic (September 2017), https://slideslive.com/38903388/a-participatory-vision-of-the-mystical-unity-of-religions
87. What does happen after death? If one accepts the feasibility of some sort of postmortem survival, would not such an experience validate a single religious after-death scenario (e.g., Christian Heaven) and refute the others (e.g., Buddhist Pure Lands or merging with the Absolute), thus providing ultimate verification of the superiority of one religion's cosmology over the rest? Not necessarily. Although I have not in my work discussed this crucial question, I concur with Barnard's important participatory reflections, which deserve to be quoted at length:

    It seems quite likely to me that how we envision postmortem existence, the variety of religious beliefs (both conscious and pre-conscious) that we have about what will happen after death (e.g., reincarnation, sensual heavens, ghostly shadow worlds, merger with God, etc.), as well as our own conscious and subconscious beliefs about what we deserve and how compassionate or not we believe the "unseen world" to be, will strongly shape the quality and form of our postmortem experience. I do not think that there is just some utterly objective after death realm of experience "out there" or "up there" waiting for us. Instead, I am convinced that our level of imagination (or lack of it), the audacity of our hopes or the tenacity of our fears, the complex layering of our beliefs (especially our subconscious beliefs and assumptions) will help to create a unique after-death quality of experience for each of us.

    I want to be clear, however. I do not think that our postmortem experience will be utterly plastic, that it will be simply subjective. Instead, similar to our current level of experience, I think that our postmortem existence will be a subjective *and* objective *co-created* reality, or more accurately, that it will be a reality in which the subject/object dichotomy itself will be transcended even more completely and obviously than is currently possible on this level of reality. I imagine that after death we will "arrive" in a dimension of reality that has a degree of "otherness," that "pushes back" in a way that is similar to our current level of physical existence; however, this dimension of experience will be powerfully shaped and formed by the ongoing creative continuity of our memories, hopes, fears, beliefs—on both conscious and subconscious levels. What this co-created, participatory notion of after-death experience implies is that our after-death experience will most likely be extremely variable—a variability that the plurality of different cultural beliefs about the afterlife both reflects, and more subtly, helps to shape. (G. William Barnard, *Living Consciousness: The Metaphysical Vision of Henri Bergson* [Albany, NY: State University of New York Press, 2011], p. 261.)

To appreciate the diversity of postmortem scenarios in one single tradition (Buddhism), see Carl B. Becker, *Breaking the Circle: Death and the Afterlife in Buddhism* (Carbondale, IL: Southern Illinois University Press, 1993). For a fascinating visual tour of different visions of paradise across traditions, see Alessandro Scafi, *Maps of Paradise* (Chicago, IL: The University of Chicago Press, 2013). See also Zuleski for an account of how historical, cultural, and social variables shape (enact?) the experiences of visionary journeys to the afterlife. Carol Zuleski, *Otherworld Journeys: Accounts of Near-Death Experience in Medieval and Modern Times* (New York, NY: Oxford University Press, 1987).

88. I am not saying that spiritual insights are incommensurable, but merely that it may be seriously misguided to compare them according to any preestablished spiritual hierarchy. Although "higher" and "lower" spiritual insights may exist both within and between religious traditions, I suggest that these qualitative distinctions need to be elucidated through spiritual inquiry, inter-religious dialogue, and the assessment of their emancipatory power of self, relationships, and world, and not determined from the perspective of overarching metaphysical schemes telling, in an a priori and ultimately doctrinal manner, which insights and traditions are superior or inferior. As I see it, the validity of spiritual insights does not rest in their accurate matching with any pregiven content, but in the quality of selfless, integrated, and eco-socio-political awareness disclosed and expressed in perception, thinking, and action. In this way, both pernicious relativisms and religious anarchies are effectively undermined. See Ferrer, *Revisioning Transpersonal Theory*; Ferrer, *Participation and the Mystery*; Ferrer, "Spiritual Knowing as Participatory Enaction."
89. David Merkur, *The Ecstatic Imagination: Psychedelic Experiences and the Psychoanalysis of Self-Actualization* (Albany, NY: SUNY Press, 1998).
90. Merkur, *The Ecstatic Imagination*, p. 155.
91. Merkur, *The Ecstatic Imagination*, p. 98. Merkur's psychoanalytic account of psychedelic experiences is ultimately reductionistic. For Merkur, all psychedelic experiences are pseudohallucinations explicable in terms of intense intrapsychic fantasying. In his own words: "psychedelic unions are all experiences of imagination" (Merkur, *The Ecstatic Imagination*, p. 156). What

Merkur does not perhaps realize is that many mystics sharply distinguished the noetic faculty known as the active Imagination from "imagination" or merely mental fantasizing. According to these mystics, the function of active Imagination is to raise sensual/perceptual experience to an imaginal level in that isthmus between physical and spiritual realms that Corbin called *mundus imaginalis*. Henry Corbin, "*Mundus Imaginalis:* Or the Imaginary and the Imaginal," in *Swedenborg and Esoteric Islam*, ed. Henry Corbin, trans. Leonard Fox (West Chester, PA: Swedenborg Foundation, 1995), pp. 1-33. Henry Corbin, "A Theory of Visionary Knowledge," in *The Voyage and the Messenger: Iran and Philosophy*, trans. Joseph H. Rowe (Berkeley, CA: North Atlantic Books, 1998), pp. 117-134. On the epistemic status of imagination in Islamic mysticism, see Chittick (1994b). For a contemporary, nonreductive account of human imagination and its centrality in promoting human transformation, see Francisco J. Varela and Natalie Depraz, "Imagining: Embodiment, Phenomenology, and Transformation," in *Buddhism and Science: Breaking New Ground*, ed. B. Alan Wallace (New York, NY: Columbia University Press, 2003), pp. 195-230.

92. Grof, *Beyond the Brain*; Grof, *The Adventure of Self-Discovery*; Grof, *The Cosmic Game.*
93. Grof, "Revision and Re-Enchantment of Psychology;" Grof, *The Cosmic Game.*
94. Grof, *The Cosmic Game.*
95. Grof, *The Cosmic Game.*
96. Pierre Duhem, *The Aim and Structure of Physical Theory* (Princeton, NJ: Princeton University Press, 1953); Willard Van Orman Quine, *From a Logical Point of View* (New York: Harper and Row, 1953).
97. Grof, *Realms of the Human Unconscious*; Grof, *The Adventure of Self-Discovery*; Grof, *The Cosmic Game.* Note that a participatory framework can be also seen as more coherent with the open-ended approach of Grof's Holotropic Breathwork practice (Grof, *The Adventure of Self-Discovery*; Grof and Grof, *Holotropic Breathwork*), whose deep trust in the organic unfolding of each person's psychospiritual trajectory is fully consistent with the participatory emphasis on enacted spiritual paths, cosmological hybridization, and spiritual individuation. In this light, one might argue that a non-essential tension exists between Grof's theory (neo-Advaita cosmology) and praxis (Holotropic Breathwork).
98. Grof, "Revision and Re-Enchantment of Psychology;" Grof, *The Cosmic Game.*
99. For a more general participatory account of psychedelic consciousness, see Matthew Segall, "Participatory Psychedelia: Transpersonal Theory, Religious Studies and Chemically-Altered (Alchemical) Consciousness," *Journal of Transpersonal Research* 5, no. 2 (2013): 86-94.
100. Jiddu Krishnamurti, *Total Freedom: The Essential Krishnamurti* (New York, NY: HarperSanFrancisco, 1996).

Contributor Biographies

Editors

Sean Kelly is professor of Philosophy, Cosmology, and Consciousness at the California Institute of Integral Studies (CIIS). He is the author of *Becoming Gaia: On the Threshold of Planetary Initiation*; *Coming Home: The Birth and Transformation of the Planetary Era*; and co-editor of *The Variety of Integral Ecologies: Nature, Culture, and Knowledge in the Planetary Era*; and *Ken Wilber in Dialogue: Conversations with Leading Transpersonal Thinkers*. Along with his academic work, Sean teaches t'ai chi and is a facilitator of the group process *Work that Reconnects* developed by Joanna Macy.

Richard Tarnas is the founding director of the graduate program in Philosophy, Cosmology, and Consciousness at the California Institute of Integral Studies, where for the past three decades he has taught courses in the history of ideas, depth psychology, archetypal studies, and the evolution of religion. He has also frequently taught at Pacifica Graduate Institute, and was formerly the director of programs at Esalen Institute. He is the author of *The Passion of the Western Mind* and *Cosmos and Psyche: Intimations of a New World View*. He is a past president of the International Transpersonal Association and served on the Board of Governors for the C. G. Jung Institute of San Francisco.

Contributors

Christopher M. Bache is professor emeritus in the department of philosophy and religious studies at Youngstown State University, adjunct faculty at the California Institute of Integral Studies, Emeritus Fellow at the Institute of Noetic Sciences, and on the Advisory Board of Grof Legacy Training. An award-winning teacher, he has written four books including *LSD and the Mind of the Universe*, the story of his twenty-year journey with LSD.

John H. Buchanan received his master's degree in humanistic/transpersonal psychology from West Georgia College, and his doctorate from the Graduate Institute of the Liberal Arts at Emory University. He has been trained and certified as a Holotropic Breathwork practitioner by Stanislav and Christina Grof. A book based upon his continuing interests in process philosophy and transpersonal psychology is being published in 2022. Buchanan has written a number of chapters on these same interests, and was a contributing co-editor for *Rethinking Consciousness: Extraordinary Challenges for Contemporary Science.* Dr. Buchanan also serves as president of the Helios Foundation.

Joseph Campbell was one of the foremost scholars of world mythology in the twentieth century. For thirty-seven years he was a professor of literature at Sarah Lawrence College where he specialized in comparative literature and comparative mythology. As well as being a prolific public lecturer and editor, he was the author of, among other works, *The Hero with a Thousand Faces,* the four-volume *Masks of God, The Mythic Image, The Inner Reaches of Outer Space,* and the *Historical Atlas of World Mythology.*

Fritjof Capra, physicist and systems theorist, is the author of several international bestsellers, including *The Tao of Physics, The Turning Point,* and *The Web of Life.* He is coauthor, with Pier Luigi Luisi, of the multidisciplinary textbook, *The Systems View of Life.*

Rick Doblin is executive director of the Multidisciplinary Association for Psychedelic Studies (MAPS) which he founded in 1986. He received his doctorate in Public Policy from Harvard's Kennedy School of Government where he wrote his dissertation on the regulation of the medical uses of psychedelics and marijuana. He studied with Stanislav Grof and was among the first to be certified as a holotropic breathwork practitioner.

Jorge N. Ferrer is a clinical psychologist, author, and educator, best known for his participatory approach to transpersonal psychology, religious studies, contemplative education, and intimate relationships. He was a professor of psychology for more than twenty years at California Institute of Integral Studies, where he also served as chair of the Department of East-West Psychology. His books include *Revisioning Transpersonal Theory: A Participatory Vision of Human Spirituality*, *Participation and the Mystery: Transpersonal Essays in Psychology, Education, and Religion*, and *Love and Freedom: Transcending Monogamy and Polyamory.*

Arlene Fox completed her studies in Eastern religion at the University of British Columbia, and worked as a Research Associate, Mood Disorders Centre of Ottawa. Now retired, she is exploring socialism through Scandinavian Studies at the University of Columbia.

Charles S. Grob is Professor of Psychiatry at the UCLA School of Medicine. He has conducted approved clinical research with a variety of psychedelic drugs and is the editor of numerous articles and books on the topic, including the recently published *Handbook of Medical Hallucinogens.*

Paul Grof is a Canadian research psychiatrist and clinician involved in transpersonal activities for six decades. He utilized Holotropic Breathwork and psychedelics with clients suffering from severe mood disorders, meditators, and psychiatric residents. Presently a Professor of Psychiatry at the University of Toronto, he is also Director of Mood Disorders Center of Ottawa.

William Keepin is a mathematical physicist with a research background in quantum physics, sustainable energy to abate global warming, and transpersonal psychology. He is a Board member of the Grof Legacy Project, and co-founder of Gender Equity and Reconciliation International which has conducted 275 trainings in twelve countries, and Satyana Institute which has organized five international conferences on inter-spirituality. His books include *Divine Duality: The Power of Reconciliation Between Women and Men*, and *Belonging to God: Spirituality, Science, and a Universal Path of Divine Love.*

Gregg Lahood is a transpersonal anthropologist and integrative psychotherapist. He has published numerous papers dealing with childbirth shamanism and relational spirituality. He currently lives in Byron Bay, Australia where he is involved in a long-term participatory action research group.

Ervin Laszlo is Founder and President of the Club of Budapest, and Founder and Director of the Laszlo Institute of New Paradigm Research. He has received Honorary degrees from the United States, Canada, Finland, and Hungary, and was awarded the Japan Peace prize (the Goi Award) in 2001, and the Luxembourg Peace Prize in 2017. He is the author of more than one hundred books and several hundred articles and studies. His most recent book is *The Wisdom Principles.*

Ralph Metzner was a recognized pioneer in psychological, philosophical, and cross-cultural studies of consciousness and its transformations. He obtained his PhD in Clinical Psychology at Harvard University, and while at Harvard he collaborated with Timothy Leary and Richard Alpert in studies of psychedelics in the 1960s, co-authoring with them *The Psychedelic Experience.* He was a psychotherapist in private practice in the San Francisco Bay Area and a professor for thirty years at the California Institute of Integral Studies. Author of over one hundred scientific papers and scholarly essays, he published some twenty-four books on psychology, psychedelics, and European mythology.

Michael Mithoefer, with his wife Annie, has been conducting Clinical Trials of MDMA-assisted therapy for PTSD for over twenty years, sponsored by The Multidisciplinary Association for Psychedelic Studies (MAPS). The inspiration for this research and the approach to facilitating the therapeutic process in these studies stems from their experience in the Grof Training. Michael is now a Senior Medical Director at MAPS Public Benefit Corporation. He has been board certified in Psychiatry, Emergency Medicine, and Internal Medicine, and is a Fellow of the American Psychiatric Association, and Affiliate Assistant Professor Department of Psychiatry and Behavioral Sciences Medical University of South Carolina.

Thomas Purton holds an M.A. in Psychoanalysis and received his PhD in Philosophy and Religion from the California Institute of Integral Studies. His research interests include Sanskrit computational linguistics, Kleinian and Lacanian psychoanalysis, and the Kashmir Śaivite tradition.

Thomas J. Riedlinger is a licensed professional counselor and inpatient group therapist. He compiled and edited *The Sacred Mushroom Seeker: Essays for R. Gordon Wasson,* and has written numerous articles and essays on psychedelic substances for academic journals, anthologies, and magazines. He and his wife Beverly Jenden-Riedlinger live in Olympia, Washington.

Huston Smith was one of the most distinguished and influential scholars of world religions for many decades. A professor of religious studies at Washington University, Massachussetts Institute of Technology, Syracuse University, and the University of California Berkeley, he was the author of many books, including *The World's Religions, Beyond the Postmodern Mind,* and *Cleansing the Doors of Perception.*

Frances Vaughan was a psychotherapist, teacher, pioneer in transpersonal psychology, and student of multiple contemplative practices. She was also president of two national psychology associations, and a faculty member at both the Institute for Transpersonal Psychology and the University of California Medical School at Irvine. She published over one hundred articles and nine books, including *Shadows of the Sacred, The Inward Arc,* and *Awakening Intuition.*

Jenny Wade is a professor at the California Institute of Integral Studies, researcher, organization and leadership development consultant, and developmental psychologist who specializes in the structuring of normal and naturally-occurring states of consciousness and their application. She is the author of numerous popular, trade, and academic articles and the books *Transcendent Sex: When Lovemaking Opens the Veil* and *Changes of Mind: A Holotronomic Theory of the Evolution of Consciousness.*

Bibliography

Books by Stanislav Grof

Realms of the Human Unconscious: Observations from LSD Research (New York: Viking, 1975; E. P. Dutton, 1976). Republished as *LSD: Doorway to the Numinous* (Rochester, Vermont: Inner Traditions, 2009).

LSD Psychotherapy (Pomona, California, Hunter House, 1980; Santa Cruz, California: Multidisciplinary Association for Psychedelic Studies, 2001).

Beyond Death: Gates of Consciousness (with Christina Grof), (London: Thames & Hudson, 1980).

Ancient Wisdom and Modern Science (Albany: State University New York Press, 1984).

Beyond the Brain: Birth, Death, and Transcendence in Psychotherapy (Albany: State University of New York Press, 1985).

Human Survival and Consciousness Evolution (Albany: State University of New York Press, 1988).

Spiritual Emergency: When Personal Transformation Becomes a Crisis (ed. with Christina Grof) (New York: Penguin, 1989).

The Stormy Search for the Self: A Guide to Personal Growth Through Transformational Crises (with Christina Grof) (New York: Penguin, 1991)

Books of the Dead: Manuals for Living and Dying (London: Thames and Hudson, 1994).

The Cosmic Game: Explorations of the Frontiers of Human Consciousness (Albany: State University of New York Press, 1998).

Psychology of the Future: Lessons from Modern Consciousness Research (Albany: State University of New York Press, 2000).

The Consciousness Revolution: A Transatlantic Dialogue (with Ervin Laszlo and Peter Russell) (London and Las Vegas: Elf Rock, 2003).

When the Impossible Happens: Adventures in Non-Ordinary Realities (Louisville, Colorado: Sounds True, 2006).

The Ultimate Journey: Consciousness and the Mystery of Death (Santa Cruz, California, MAPS Publications, 2006).

Holotropic Breathwork: A New Approach to Self-Exploration and Therapy (Albany: State University of New York Press, 2011).

Healing Our Deepest Wounds: The Holotropic Paradigm Shift (Newcastle, Washington: Stream of Experience, 2012).

Modern Consciousness Research and the Understanding of Art (Santa Cruz, California: MAPS Publications, 2006).

The Way of the Psychonaut, Volumes I and II (Santa Cruz, California: MAPS Publications, 2019).

The above books have been translated into German, French, Italian, Spanish, Portuguese, Dutch, Swedish, Danish, Russian, Romanian, Czech, Polish, Bulgarian, Hungarian, Latvian, Greek, Turkish, Korean, Japanese, and Chinese.

PSYCHE UNBOUND